FUNDAMENTALS OF SPACE MEDICINE

Figure by Philippe Tauzin.

THE SPACE TECHNOLOGY LIBRARY

Published jointly by Microcosm Press and Springer

An Introduction to Mission Design for Geostationary Satellites, J.J. Pocha
Space Mission Analysis and Design, 1st edition, James R. Wertz and Wiley J. Larson
Space Mission Analysis and Design, 2nd edition, Wiley J. Larson and James R. Wertz
Space Mission Analysis and Design, 3rd edition, James R. Wertz and Wiley J. Larson
Space Mission Analysis and Design Workbook, Wiley J. Larson and James R. Wertz
Handbook of Geostationary Orbits, E.M. Soop
Spacecraft Structures and Mechanisms, From Concept to Launch, Thomas P. Sarafin
Spaceflight Life Support and Biospherics, Peter Eckart
Reducing Space Mission Cost, James R. Wertz and Wiley J. Larson
The Logic of Microspace, Rick Fleeter
Space Marketing: A European Perspective, Walter A.R. Peeters
Fundamentals of Astrodynamics and Applications, David A. Vallado
Influence of Phychological Factors on Product Development, Eginaldo S. Kamata
Essential Spaceflight Dynamics and Magnetospherics, Boris V. Rauschenbakh, Michael Yu.
Ochinnikov and Susan McKenna-Lawlor
Space Psychology and Psychiatry, Nick Kanas and Dietrich Manzey

---

The Space Technology Library Editorial Board

Fundamentals of Space Medicine

by

Gilles Clément

Centre National de la Recherche Scientifique (CNRS),
Toulouse, France

Published jointly by
Microcosm Press
El Segundo, California

A.C.I.P. Catalogue record for this book is available from the Library of Congress.

Additional meterial to this book can be downloaded from http://extras.springer.com.
ISBN 1-4020-3246-3 (PB)
ISBN 1-4020-1598-4 (HB)
ISBN 1-4020-3434-2 (e-book)

Springer,
P.O. Box 17, 3300 AA Dordrecht, The Netherlands.

Sold and distributed in North, Central and South America
by Microcosm
401 Coral Circle, El Segundo, CA 90245-4622 U.S.A.
and Springer,
101 Philip Drive, Norwell, MA 02061, U.S.A.

In all other countries, sold and distributed
by Springer,
P.O. Box 322, 3300 AH Dordrecht, The Netherlands.

Illustrations (Figures 1-02, 1-16, 1-18, 1-19, 1-22, 2-05, 2-06, 2-07, 2-11,
2-20, 2-22, 2-23, 3-06, 3-07, 3-08, 3-18, 3-21, 4-04, 4-05, 4-06, 4-08,
4-19, 4-20, 5-04, 5-06, 5-07, 5-08, 7-21):

Philippe Tauzin
Service Commun Multimedia
Université Paul Sabatier, Toulouse III,
CHU Rangueil, Avenue Jean Poulhès
F-31043 Toulouse Cedex, France

*Cover photo of Earth from Space: View of Africa and the Indian Ocean taken in Dec. 1972,
by Apollo 17, the last of the Apollo missions to explore the Moon, Photo courtesy of NASA.*

Cover design by Jeanine Newcomb and Joy Sakaguchi.

Printed on acid-free paper

Printed in the Netherlands.

To my parents,
To my brother and sister,
To Josiane, Guillaume, and Jean-Loup

Contents

Preface by Dr. Didier Schmitt   xv

Preface by Dr. Douglas Hamilton   xvii

Foreword   xix

Acknowledgments   xxiii

CHAPTER 1: INTRODUCTION TO SPACE LIFE SCIENCES   1
1   Space Life Sciences: What Is It?   1
    1.1   Objectives   1
    1.2   The Space Environment   4
       1.2.1   Microgravity   4
       1.2.2   Other Factors of the Space Environment   6
    1.3   Justification for Human Spaceflight   8
       1.3.1   Humans versus Robots   8
       1.3.2   Space Science   9
    1.4   Where We Are   11
2   The Legacy of Space Life Sciences Research   13
    2.1   Major Space Life Sciences Events   13
       2.1.1   The Pioneers   13
       2.1.2   Spaceflights with Animals   15
       2.1.3   Humans in Space   17
       2.1.4   Space Life Sciences Investigations   18
       2.1.5   Today's Access to Space   22
    2.2   Surviving the Odyssey   25
    2.3   Life Support Systems   28
3   Challenges Facing Humans in Space   30
    3.1   Astronauts' Health Maintenance   30
       3.1.1   Preflight   30
       3.1.2   In-flight   31
       3.1.3   Postflight   33
    3.2   Environmental Health during Space Missions   33
    3.3   Human Mars Mission   35
    3.4   Countermeasures   37
    3.5   Artificial Gravity   40
    3.6   A New Science is Born   42
4   References   45

CHAPTER 2: SPACE BIOLOGY   47
1.   What is Life?   47
        1.1   Life on Earth   47
        1.2   Life on Mars   51
2   Gravitational Biology   53
        2.1   Questions   53
        2.2   Results of Space Experiments   56
                2.2.1   Suspended Cultures   57
                2.2.2   Attached Cells   58
                2.2.3   Human Blood Cells   60
        2.3   Bioprocessing in Space   61
3   Development Biology   64
        3.1   Questions   64
        3.2   Results of Space Experiments   68
                3.2.1   Invertebrates   68
                3.2.2   Lower Vertebrates   70
                3.2.3   Mammals   72
4   Plant Biology   75
        4.1   Questions   75
        4.2   Results of Space Experiments   75
                4.2.1   Graviception   75
                4.2.2   Development of Plants   76
5   Radiation Biology   79
        5.1   Ionized Radiation in Space   79
        5.2   Biological Effects of Radiation   80
6   ISS Facilities for Space Biology   83
7   References   87

CHAPTER 3: THE NEURO-SENSORY SYSTEM IN SPACE   91
1   The Problem: Space Motion Sickness   92
2   Vestibular Function   94
        2.1   The Vestibular System   95
                2.1.1   The Vestibular End Organs   95
                2.1.2   Linear Acceleration and Gravity   96
                2.1.3   Changes in the Vestibular Receptors
                          during Spaceflight   100
        2.2   The Other Senses   102
                2.2.1   Vision   102
                2.2.2   Hearing   105
                2.2.3   Smell and Taste   106
                2.2.4   Proprioception   107
3   Effects of Spaceflight on Posture and Movement   109

	3.1	Rest Posture	109
	3.2	Vestibulo-Spinal Reflexes	110
	3.3	Locomotion	113
	3.4	Body Movement	116
	3.5	Eye Movement	118
4		Effects of Spaceflight on Spatial Orientation	123
	4.1	Visual Orientation	123
	4.2	Cognition	124
	4.2.1	Navigation	124
	4.2.2	Mental Rotation	126
	4.2.3	Mental Representation	128
5		What Do We Know?	130
	5.1	Space Motion Sickness (SMS) Experience	130
	5.2	Theories for Space Motion Sickness	131
	5.3	Countermeasures	133
6		References	136

CHAPTER 4: THE CARDIO-VASCULAR SYSTEM IN SPACE 139

1		The Problem: Postflight Orthostatic Intolerance	139
2		Physiology of the Cardio-Vascular System	143
	2.1	Basics .	143
	2.2	Control Mechanisms	145
	2.2.1	Control of Blood Pressure	145
	2.2.2	Baroreceptor Reflexes	147
	2.2.3	Fluid Volume Regulation	148
3		Effects of Spaceflight on the Cardio-Vascular System	149
	3.1	Launch Position	150
	3.2	Early On-Orbit	151
	3.2.1	Fluid Shift	151
	3.2.2	Blood Pressure	153
	3.3	Later On-Orbit	154
	3.3.1	Fluid Shift	154
	3.3.2	Maximal Exercise Capability	155
	3.3.3	Extra Vehicular Activity	156
	3.3.4	Heart Rhythm	157
	3.4	Postflight	158
4		What Do We Know?	159
	4.1	Orthostatic Intolerance	159
	4.2	Pulmonary Function	162
	4.3	Bed Rest	163
5		Countermeasures	164
	5.1	In-flight	165

	5.1.1	Exercise	165
	5.1.2	Lower Body Negative Pressure	167
	5.1.3	Monitoring	168
5.2		End of Mission	169
6		References	170

CHAPTER 5: THE MUSCULO-SKELETAL SYSTEM
            IN SPACE   173
1   The Problem: Muscle Atrophy and Bone Loss   173
       1.1   Muscle Atrophy   173
       1.2   Bone Loss   175
2   Muscle and Bone Physiology   177
       2.1   Muscle Physiology   177
       2.2   Bone Physiology   180
3   Effects of Spaceflight on Muscle   183
       3.1   Decrease in Body Mass   183
       3.2   Decrease in Muscle Volume and Strength   184
       3.3   Changes in Muscle Structure   186
4   Effects of Spaceflight on Bone   188
       4.1   Human Studies   188
       4.2   Animal Studies   190
5   What Do We Know?   192
       5.1   Muscle Atrophy   192
       5.2   Bone Demineralization   193
6   Countermeasures   196
       6.1   Muscle   196
       6.2   Bone   198
              6.2.1   Exercise   198
              6.2.2   Mechanical Countermeasures   198
              6.2.3   Nutritional Countermeasures   199
              6.2.4   Pharmacological Countermeasures   200
       6.3   Aging and Space   200
7   References   202

CHAPTER 6: PSYCHO-SOCIOLOGICAL ISSUES OF
            SPACEFLIGHT   205
1   The Problem: Reaction to Stress   205
       1.1   Analogs   208
       1.2   Space Simulators   210
       1.3   Actual Space Missions   212
       1.4   Rules   213

2   Individual Selection   214
        2.1   Select-Out Criteria   214
        2.2   Select-In Criteria   216
        2.3   Psychological Profile of Astronauts and Cosmonauts 218
3   Crew Selection   221
        3.1   Sociological Issues   221
                3.1.1   Confinement and Personal Space   221
                3.1.2   Mixed Gender Issues   224
                3.1.3   Multi-Cultural Issues   225
        3.2   Selection Issues   225
                3.2.1   Compatibility   228
                3.2.2   Crew Composition   229
                3.2.3   External Factors   230
4   Assessment of Crew Behavior and Performance   232
5   Psychological Training and Support   233
        5.1   Training   234
        5.2   Support   235
                5.2.1   Soviet/Russian Experience   235
                5.2.2   ISS Psychological Support   236
                5.2.3   Unsolved Issues   239
6   References   241

CHAPTER 7: OPERATIONAL SPACE MEDICINE   245
1   What Is It?   245
        1.1   Objectives   247
        1.2   Risk Assessment   248
2   Astronaut Selection and Training   249
        2.1   Crew Position   249
        2.2   Physical Requirements for Astronaut Selection   251
        2.3   Astronaut Training   255
3   Prevention: Health Hazards in Space   259
        3.1   Medical Events during Spaceflight   261
        3.2   Medical Aspects of Extra-Vehicular Activity   264
        3.3   Medical Problems of Radiation in Space   267
                3.3.1   Space Radiation Environment   268
                3.3.2   Spacecraft Radiation Environment   269
                3.3.3   Medical Effects of Radiation Exposure   271
                3.3.4   Exposure Limits   272
                3.3.5   Radiation Countermeasures   275
                3.3.6   Strategies in Radiation Shielding   277
                3.3.7   Conclusion on Radiation Issues   278
        3.4   Conclusion on Space Health Hazards   279

4   Treatment: Space Medical Facilities   279
         4.1   Crew Health Care System (CHeCS)   281
                  4.1.1   The Health Maintenance System   282
                  4.1.2   The Environmental Health System   285
                  4.1.3   Typical ISS On-Orbit Medical Assessment   286
         4.2   Telemedicine   286
         4.3   Emergency and Rescue   287
                  4.3.1   Crew Medical Officer (CMO)   288
                  4.3.2   Surgery in Space   289
                  4.3.3   Evacuation   291
5   Future Challenges   292
         5.1   Human Needs for Long-Duration Missions   293
                  5.1.1   Environment and Hygiene   293
                  5.1.2   Human Needs   295
                  5.1.3   Nutrition Requirements   296
         5.2   Controlled Ecological Life Support System   299
                  5.2.1   Life Support System Fundamentals   299
                  5.2.2   Engineering Solutions for LSS   301
                  5.2.3   CELSS for Long-Duration and
                           Exploratory Missions   303
         5.3   Terraforming   305
         5.4   Conclusion   308
6   References   308

CHAPTER 8: SPACE LIFE SCIENCES INVESTIGATOR'S GUIDE   313
1   Resources and Constraints of Space Life Sciences Missions   313
         1.1   Opportunities for Space Life Sciences Experiments   314
                  1.1.1   International Space Station   314
                  1.1.2   SpaceHab   315
                  1.1.3   Shuttle *Small Payload* Flight Experiments 317
                  1.1.4   Biosatellites (Photon)   318
                  1.1.5   Ground-Based Investigations   319
         1.2   Constraints   320
                  1.2.1   Ethical Considerations   321
                  1.2.2   Other Considerations   321
                  1.2.3   Space Shuttle Constraints   322
                  1.2.4   International Space Station Constraints   324
                  1.2.5   Soyuz "Taxi" Flights Constraints   325
                  1.2.6   Constraints of Pre- and Post-Mission Studies 326
2   How to "Fly" an Experiment   327
         2.1   Flight Experiment Selection   327
         2.2   Experiment Design   329

	2.3	Hardware Selection	330
	2.4	Feasibility	332
	2.5	Experiment Integration	334
		2.5.1 Key Documentation	334
		2.5.2 Reviews	335
	2.6	Crew Science Training	336
	2.7	In-Flight Science Operations	337
		2.7.1 Organization	337
		2.7.2 Communications	340
		2.7.3 Re-Planning	341
	2.8	Data Analysis	343
3		References and Documentation	344
Index			345

PREFACE

Didier Schmitt, M. D., Ph. D.
Head, Space Life Sciences
European Space Agency

Even before the actual beginnings of spaceflight people were interested in the question of what happens to human beings when they enter the extreme and fascinating environment of space. Decades of experimentation with biosatellites and more than 240 crewed spaceflights with about 450 astronauts from various countries have led to a solid knowledge base on many effects that the space environment has on living organisms. In many cases, many of which are mentioned in this book, the results obtained have significantly changed our concepts of physiological mechanisms. While advance in knowledge, as always, leads to further questions and yet some basic mechanisms are still unknown, the field of space life sciences has reached a maturity which allows to add on top of basic research the aspect of applying "space" knowledge for the benefit of the citizen on the ground. In Europe, where traditionally the majority of (physiological) space life sciences research is carried out by researchers who are also involved in "normal" physiological or clinical research on the ground, this application and transfer of "space knowledge" is a constant characteristic that defines our work.

At the same time, as in any field that matures, complexity is rising and specialists develop new experiments in order to make further progress. However, the need for "the big picture" regarding the various physiological systems and their interactions and interdependencies is growing at a similar rate. This book is remarkable in providing exactly this vast overview. In spanning an arc from cell biology via plants, and human physiology up to psychology the author collects a vast amount of information that will serve as fountain of knowledge and motivation to students in the life sciences with an interest in space and extreme environments. Also, the very practical advices on the preparation of space experiments, which builds on many years of first hand experience by the author, will prove invaluable to future space life science experimenters.

With the International Space Station providing for continuous international presence of human beings in space since October 2000, we have a unique tool at hands to perform further research. When thinking about the future of human spaceflight, one of the most exciting challenges is posed by a planetary mission that will carry human beings to the surface of our neighbor planet, Mars. While naturally the knowledge and understanding that we have accumulated so far will provide a solid basis for the preparation of a mission to Mars, such a mission will be very different in nature compared with the experience in low Earth orbit.

Factors like the interplanetary radiation environment or the psychological effects of prolonged isolation and confinement will necessitate very different measures than what is currently used to counter spaceflight effects. Thus, space life sciences research continues to expand and has gained an additional third layer, which is the targeted preparation of mankind's future in space.

Didier Schmitt, M. D., Ph. D.
Head, Space Life Sciences
European Space Agency

28 May 2003

PREFACE

Douglas R. Hamilton, M. D., Ph. D.
Flight Surgeon

I am writing this preface on February 2nd, 2003 at 2am, because I can't sleep. My 7-year old daughter, Keltie, is sleeping beside me on the sofa and won't go to bed. She understands that today something very bad has happened to Iain Clark's mom, Laurel, and senses that I am very distraught. I have to tell her that "Miss Laurel" has gone away and that her schoolmate, Iain, will be without a mother. Just a few days ago, I was in the NASA Mission Control Center sitting at the SURGEON console with Laurel's husband, Jon, who also is a flight surgeon. We were supporting the Space Shuttle flight STS-107, laughing and joking around while we watched Laurel on a television monitor. She was living an experience that Jon and I could only dream about. I remember our biomedical engineers on our support console joking over the intercom that Jon should allow the Flight Activities Officer to remove the private family videoconference with Laurel from the timeline since he already had watched her for 12 hours during the previous shift. Thank goodness, Jon and Iain had that special private time with Laurel.

I was in the Mission Control Room helping execute the emergency procedures after the breakup of STS-107 earlier today. The Shuttle flight and payload controllers were working the issues like we had trained so many times before. I want you to know how proud I am of those remarkable men and women who stood at their consoles and remained calm and professional throughout the whole day. They knew when to run and when to walk. It was indeed an impressive sight.

Having been an attending physician in a hospital, I have felt the emotional hardship of escorting a patient's family out of the hospital after the death of a loved one, …only to turn around, walk back into the hospital and onto the ward, face your medical team with a calm demeanor, and with a steady hand, reach for the next patient's chart off the rack in the nursing unit. Rick, Willie, Mike, KC, Dave, Ilan, and Laurel will be sorely missed. The real measure of the space community will be how we pick ourselves up after being knocked down so hard.

If anyone knows how to take science from the bench and perform it in space, it is Gilles. He has been performing research with crewmembers in space and on the ground for more than 20 years. He knew Jon and Laurel because of their participation in science experiments and also because Jon is a board certified neurologist who, along with Gilles, has a passion for the neurological aspects of space travel. I know that Gilles did not just lose science today but also very close friends.

I first met Gilles in 1992 when I was attending the International Space University (ISU) summer session in Kitakyushu, Japan. He taught us all about the neurological aspects of space travel. Since then, he and I have lectured together for the Space Life Sciences department at summer sessions from 1997 until the present (Vienna, Houston, Cleveland, Ratchasima, Valparaiso, Bremen, and Pomona). During this time, I have been able to familiarize Gilles with the operational aspects of space medicine and he acquainted me with the challenges facing scientific investigators such as himself. Together, we designed a series of space life sciences lectures which are intended for a multi-cultural audience who is not educated in life sciences. The magic of International Space University is that it uses this pedagogical approach for all aspects of space. To communicate the complex issues associated with space to such a diverse audience requires the natural teaching ability of educators like Gilles and many of the other department lecturers at ISU.

Right after this tragic event, I thought a document that describes how we medically support a crew like STS-107 and what they were trying to scientifically accomplish needs to be written. Serendipitously, Gilles emailed me his initial manuscript for this book just a few weeks ago and asked me to write a preface. Hence I find myself writing this preface at 2am in the morning, after one of the saddest days of my life. Laurel and Jon Clark are both flight surgeons and pilots. They represent the "pointy end of the spear" when it comes to medical support in space. Laurel as a physician astronaut and Jon as a console flight surgeon. The medical issues they and others need to deal with and the scientific questions behind these complex problems are clearly represented in Gilles marvelous book. I have read many books and articles about physiology and space life sciences, but none of them successfully make the connection between space life sciences and medical operations. When you read this book, you will understand what the space medical community is all about and the real challenges that face the flight surgeon and life science investigator.

Gilles, thank you for writing this book and allowing me to be part of it.

Now, I must put my daughter back into bed...and say goodbye to some close friends.

Douglas R. Hamilton
M. D., Ph. D., M. Sc. E. Eng
P. E., P. Eng., ABIM, FRCPC
Flight Surgeon / Electrical Engineer

FOREWORD

When preparing my courses for the International Space University (ISU), where I teach in the Department of Space Life Sciences since 1989, I realized it was difficult to get the information I needed regarding the effects of spaceflight on humans. Looking for this information into the original papers published in scientific journals, finding the essential, and making clear presentations for students (with often limited background in biology) requires a lot of time and effort. Most reviews on this topic are either addressed to specialists (Churchill 1997, Colin 1990, DeHart 1985, DeHart and Davis 2002, Moore et al. 1996) or to laymen (Lujan and White 1994, Nicogossian and Parker 1982, Oser and Battrick 1989, Stine 1997). Other reviews are hard to get or most of their content is outdated (Clavin and Gazenko 1975, Johnston and Dietlein 1977, Link 1965, Nicogossian 1977, Nicogossian et al. 1993, Parker and Jones 1975). Lastly, most of these reviews are lacking the basic principles of human physiology and a structured presentation directly useful for educational lectures.

Thus, I decided to write "Fundamentals of Space Medicine" as a textbook for professors and undergraduate or graduate students. Why this title? *Space Medicine* and *Space Physiology* are often viewed as two aspects of space life sciences, with the former being more operational, and the latter being more investigational. *Space Medicine* tries to solve medical problems encountered during space missions. These problems include some adaptive changes to the environment (microgravity, radiation, temperature, and pressure) and also some non-pathologic changes that become mal-adaptive on return to Earth (e.g., bone loss). *Space Physiology* tries to characterize body responses to space, especially microgravity. It provides the necessary knowledge, hence the "fundamentals", required for an efficient space medicine.

As a neurophysiologist participating in space research since 1982, with experiments manifested on Salyut-7, Mir, and 31 Space Shuttle flights, I know what it takes to collect data during relatively simple space experiments, and then try to make sense of the sparse, often contradictory, results in a scientific paper. Many articles in space life sciences include a discussion which goes far beyond the results actually obtained. The interpretations proposed by one author may some day prove incorrect as new data are collected. In this textbook, I have tried to compile these scientific facts, and apologize to the authors if all their interpretations are not included.

Now that the International Space Station (ISS) is being assembled, the opportunities for conducting biomedical research in orbit are reduced because of the limited crew time. Until the laboratories in the ISS are fully operational, this is a good time to review what we learnt from previous studies. This book reflects *what we do know* in space life sciences at the beginning of the 21st century. It also points to the missing data, i.e., *what we don't know* and *what we should know* before committing to a larger access for humans (i.e., space tourists, by contrast with the current, professional astronauts) in space and for longer duration exploratory missions.

The structure of this book is such that it reviews step by step the changes in the major body functions during spaceflight, from the cellular level to the behavioral and cognitive levels:

Chapter 1 starts with an introduction to the environmental challenges that spaceflight poses to the human body, and continues with a short history of space life sciences research.

Chapter 2 reviews the effects of microgravity and radiation at the cellular level on bacteria, animals, plants, and humans, including the issues of reproduction and development.

The following chapters each review the effects of spaceflight on the major human body functions: Chapter 3: Neuro-sensory function (the brain in space); Chapter 4: Cardio-vascular function (the heart in space); Chapter 5: Musculo-skeletal function (the muscle and bone in space); Chapter 6: Psycho-sociological issues (the mind in space).

However, every system or process must ultimately be viewed in the context of the entire body. The consequences of the fore mentioned changes at a function level on the health and well being of the astronauts are therefore described in the Chapter 7: Operational Space Medicine.

Chapter 8 concludes this review with some tips from the author of this book on how to proceed for proposing and planning a space experiment which utilizes humans as test subjects, given the available resources and constraints of space missions.

Each of these chapters corresponds to one core lecture of the ISU Space Life Sciences Department. These lectures were developed with the help of many people from all over the world in a collegial and collaborative environment. In particular, the areas related to the medical effects of spaceflight are significantly due to the contribution of Doug Hamilton.

Some space-related changes and interpretations for these changes are sometimes described in the text in greater details than what is required for a plenary lecture. For this reason, a PowerPoint version of each of these lectures with the key concepts presented in bullet-form is included on the CD-ROM bound to this book. These PowerPoint presentations also include related colorful illustrations and video clips.

References:

Churchill SE (ed) (1997) *Fundamentals of Space Life Sciences*, Volumes 1 and 2. Malabar, FL: Krieger

Clavin M, Gazenko OG (eds) (1975) *Foundations of Space Biology and Medicine*, NASA SP-374. Washington, DC: National Aeronautics and Space Administration Scientific and Technical Information Office

Colin J (ed) (1990) *Médecine Aérospatiale*, Paris, France: Expansion Scientifique Française

DeHart RL (ed) (1985) *Fundamentals of Aerospace Medicine*, Philadelphia, PA: Lea and Febiger

DeHart RL, Davis JR (eds) (2002) *Fundamentals of Aerospace Medicine*, Third edition, Philadelphia, PA: Lippincott Williams & Wilkins

Johnston RS, Dietlein LF (eds) (1977) *Biomedical Results from Skylab*, NASA SP-377. Washington, DC: National Aeronautics and Space Administration Scientific and Technical Information Office

Link MM (ed) (1965) *Space Medicine in Project Mercury*, NASA SP-4003. Washington, DC: National Aeronautics and Space Administration Scientific and Technical Information Office

Lujan BF, White RJ (1994) *Human Physiology in Space*. Teacher's Manual. A Curriculum Supplement for Secondary Schools. Houston, TX: Universities Space Research Association

Moore D, Bie P, Oser H (eds) (1996) *Biological and Medical Research in Space*, Berlin, Germany: Springer

Nicogossian AE (ed) (1977) *The Apollo-Soyuz Test Project. Medical Report*, NASA SP-411. Washington, DC: National Aeronautics and Space Administration Scientific and Technical Information Office

Nicogossian AE, Parker JF (1982) *Space Physiology and Medicine*, NASA SP-447. Washington, DC: National Aeronautics and Space Administration Scientific and Technical Information Branch

Nicogossian AE, Huntoon CL, Pool SL (eds) (1993) *Space Physiology and Medicine*, 3rd edition. Philadelphia, PA: Lea and Febiger

Oser H, Battrick B (eds) (1989) *Life Sciences Research in Space*, ESA SP-1105. Noordwijk, The Netherlands: European Space Agency Publication Division

Parker JF, Jones WL (eds) (1975) *Biomedical Results from Apollo*, NASA SP-368. Washington, DC: National Aeronautics and Space Administration Scientific and Technical Information Office

Stine HG (1997) *Living in Space*. New York, NY: M. Evans & Co

ACKNOWLEDGMENTS

[Sitting atop the Saturn V rocket] I am everlasting thankful that I have flown before, and that this period of waiting atop a rocket is nothing new. I am just as tense this time, but the tenseness comes mostly from an appreciation of the enormity of our undertaking rather than from the unfamiliarity of the situation. [...] I am far from certain that we will be able to fly the mission as planned. I think we will escape with our skins, or at least I will escape with mine, but I wouldn't give better than odds on a successful [Moon] landing and return. There are just too many things that can go wrong.

—Mike Collins, July 16, 1969 (Carrying the Fire: An Astronaut's Journeys. New York, NY: Farrar, Strauss, Giroux, 1989)

I was finishing up this book when I learned about the Space Shuttle *Columbia* accident of 1 February 2003 and the loss its seven astronauts. STS-107 was a life and material sciences mission, so this was clearly a tragic event in the history of human spaceflight and for space research in particular.

Although I am convinced that space research is and will be helping in many ways our daily life, the price of giving the lives of brave men and women for science is too much to pay. I had personally met one of the STS-107 crewmembers, when he kindly volunteered to participate in one of my experiment, so I am even more saddened by this tragedy. *Columbia* has also a special place in my own memories. It was the ship that carried aloft the Spacelab from 1983 until 1998. I was the principal or co-investigator of several experiments that flew aboard this remarkable spacecraft (IML-1, LMS, and more recently, Neurolab). This Space Shuttle in particular was the proud focus of not only my life but that of hundreds of scientists around the world for many years. *Columbia*, you served us well!

This book is dedicated to the astronauts and cosmonauts, without whom the work reported here would have been impossible.

I would like to give my sincere thanks to the 73 astronauts and cosmonauts who have participated as test subjects in my own space experiments aboard Salyut-7, Mir, and the Space Shuttle, and the hundreds of volunteers who served as ground-based control subjects for these experiments.

I would also like to take this opportunity to thank my space mentors (Millard Reschke, Francis Lestienne), my long-term friends and collaborators in these experiments (Scott Wood, William Paloski), and my former students (Corinna Lathan and Claudie Haigneré—the latter is now the French Minister of Research and High Technologies), with whom I have shared exciting moments of space exploration. On a broader scale, acknowledgments are due to CNES, CNRS, ESA, and NASA for their continuous support in my research activities.

As mentioned earlier, my knowledge in certain areas of the book is significantly due to the contributions of Doug Hamilton. His experience in the Medical Division Branch at the NASA Johnson Space Center as a flight surgeon (the second best job at NASA, just after the astronaut one) was priceless. As my partner in crime at ISU since many years, Doug has helped me to put together all this material in an easily accessible format. He should be recognized as the author of the PowerPoint lectures on the "Cardio-Vascular System in Space" and on "Operational Space Medicine".

Thanks to many people for their help in providing content for this book, and in particular Elisa Allen, Oliver Angerer, Oleg Atkov, Mike Barratt, Alain Berthoz, Sheryl Bishop, Susanne Churchill, Guillaume Clément, Bernard Comet, Jan Cook, Susan Doll, Elena Grifoni, Michel Imbert, Sonya Japenga, Corrie Lathan, Patricia Santy, Barbara Ten Berge, and Michel Viso.

I am grateful to Mr. Philippe Tauzin, from the Service Commun Multimédia of the Université Paul Sabatier in Toulouse, for his help with some of the illustrations, including the book cover. I also appreciate the great work done at NASA to document, catalogue, and give access to its archives, and video and pictures gallery. Many photographs in this book come from this gallery, and most of them were taken by the astronauts themselves during their space missions.

My deep appreciation to those who have kindly accepted to review this book, and to Drs. Didier Schmitt and Douglas Hamilton for writing the Preface.

Last but not least, I thank Dr. Harry (J.J.) Blom, my editor, who offered me the opportunity to disseminate the legacy of space life sciences research in the Space Technology Library series.

Gilles Clément, Ph. D.

Toulouse, 28 May 2003

Chapter 1

INTRODUCTION TO SPACE LIFE SCIENCES

This first chapter describes the hazards the space environment poses to humans (radiation, vacuum, extreme temperatures, microgravity), and how spaceflight affects the human body (where we are). We will then review the historical context of human spaceflight (how we got there), and end with the challenges facing humans in space (where do we go from here)

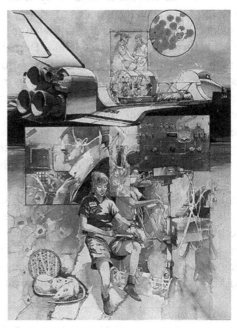

Figure 1-01. Diagram illustrating the research on physiological responses to weightlessness performed on board the first space laboratory dedicated to life sciences research (Spacelab Life Sciences-1, STS-40) in 1991. (Credit NASA)

1   SPACE LIFE SCIENCES: WHAT IS IT?

1.1   Objectives

Life Sciences are specifically devoted to the working of the living world, from bacteria and plants to humans, including their origins, history, characteristics, habits, you name it.

The study of life on Earth ranges from elucidating the evolution of the earliest self-replicating nucleic acids to describing a global ecology comprising over three million species, including humans. However, throughout its evolution, life on Earth has experienced only a 1-g environment. The influence of this omnipresent force is not well understood, except that there is clearly a biological response to gravity in the structure and functioning of living organisms. The plant world has evolved gravity sensors; roots grow "down" and shoots "up". Animals have gravity sensors in the inner ear. Many fertilized eggs and developing embryos (amphibians, fish, birds, and mammals) also have clear responses to gravity. For example, the amphibian egg orients itself with respect to gravity within a few minutes after fertilization. During that short time the dorso-ventral and anterior-posterior axes of the future embryo are established. Do we conclude therefore that the gravitational input is a required stimulus for the establishment of these axes?

To better understand a system, the scientific method consists in studying the consequences of its exclusion. This approach has led considerable advances in the knowledge of human physiology, thanks to the 19th century physiologist Claude Bernard who set out the principles of experimental medicine. Clearly, the removal of gravity is a desirable, even necessary, step toward understanding its role in living organisms. In a sense, removal of gravity for studying the gravity-sensing mechanisms is like switching the light for studying the role of vision. Transition into weightlessness abolishes the stimulus of gravity by a procedure physiologically equivalent to shutting off the light. What can be accomplished in such an elegant fashion aloft can never be done in Earth-based laboratories.

Space physiology is of basic scientific interest and deals with fundamental questions concerning the role of gravity in life processes. Space medicine is another, albeit more applied, research component concerned with the health and well fare of the astronauts. These two objectives complement one another, and constitute the field of *Space Life Sciences*. In short, space life sciences open a door to understanding ourselves, our evolution, the working of our world without the constraining barrier of gravity.

Biology
- Improve overall health of people of all ages
- Improve crop yields using less nutrients and smaller surface and volume
- Advance understanding of cell behavior
Biotechnology
- Provide information to design a new class of drugs to target specific proteins and cure specific diseases
- Culture tissue for use in cancer research/surgery and bone cartilage and nerve injuries
Biomedical Research
- Enhance medical understanding of the role of force on bone in disease processes including osteoporosis (bone loss)
- Advance fundamental understanding of the brain and nervous system and help develop new methods to prevent and treat various neurological disorders (e.g., multiple sclerosis)
- Develop methods to keep humans healthy in low-gravity environments for extended time periods
Education
- Use science on orbit to encourage and strengthen science education on Earth

Table 1-01. Major applications of space life sciences research.

More specifically, space life sciences are dedicated to the following three objectives:

- Enhance fundamental knowledge in cell biology and human physiology. Access to a space laboratory where gravity is not sensed facilitates research on the cellular and molecular mechanisms involved in sensing forces as low as 10^{-3} g and subsequently transducing this signal to a neural or hormonal signal. A major challenge to our understanding and mastery of these biological responses is to study selected species of higher plants and animals through several generations in absence of gravity. How do individual cells perceive gravity? What is the threshold of perception? How is the response to gravity mediated? Does gravity play a determinant role in the early development and long-term evolution of the living organism? These studies of the early development and subsequent life cycles of representative samples of plants and animals in absence of gravity are of basic importance to the field of developmental biology.

- Protect the health of astronauts. As was amply demonstrated by Pasteur, as well as countless successors, investigations in medicine and in agriculture contribute to and benefit from basic research. Understanding the effect of gravity on responses of humans and of plants has enormous practical significance for human spaceflight. For example, the process of bone demineralization seen in humans and animals as a progressive phenomenon occurring during spaceflight is not only a serious medical problem. It raises the question of abnormalities in the development of bones, shells, and special crystals (such as the otoconia of the inner ear) in species developing in absence of gravity. The study of such abnormalities should provide insight into the process of biomineralization and the control of gene transcription.

- Develop advanced technology and applications for space and ground-based research. In addition to the scientific need to study basic plant and animal interactions with gravity, there is a practical need to study their responses. They are also essential to our ultimate ability to sustain humans for a year or more on the surface of extraterrestrial bodies or in spaceflight missions of long duration where re-supply is not possible, and food must be produced in situ. Experiments during long-term space missions will determine which plants and animals are most efficient and best suited for our needs (for instance, can soybeans germinate, grow normally, produce an optimum crop of new soybeans for food and new seed for ensuring crops?). All of this biological cycling, plus the development of equipment for water and atmosphere recycling, and management of waste will also bring important benefits for terrestrial applications. Also, the absence of gravity is used to eliminate microconvection in crystal growth, in electrophoresis, and in biochemical reactions. The resulting products can be used for both research and commercial application.

Space life sciences include the sciences of physiology, medicine, and biology, and are linked with the sciences of physics, chemistry, geology, engineering, and astronomy. Space life sciences research not only help gaining new knowledge of our own human function and our capacity to live and work in space, but also explore fundamental questions about gravity's role in the formation, evolution, maintenance, and aging processes of life on Earth (Table 1-01).

1.2   The Space Environment

The space environment (radiation, microgravity, vacuum, magnetic fields) as well as the local planetary environments (Moon, Mars) have been extensively reviewed in Peter Eckart's book "Spaceflight Life Support and Biospherics" (Space Technology Library, Kluwer Academic Publishers, Dordrecht, 1996). In this section, I will mainly focus on microgravity. The medical issues related to space radiation will be developed in Chapter 7 (Section 3.3).

1.2.1   Microgravity

The presence of Earth creates a gravitational field that acts to attract objects with a force inversely proportional to the square of the distance between the center of the object and the center of Earth. When we measure the acceleration of an object acted upon only by Earth's gravity at Earth's surface, we commonly refer to it as *1-g* or one Earth gravity. This acceleration is approximately 9.8 m/sec^2.

We can interpret the term *microgravity* in a number of ways, depending upon the context. The prefix micro- derives from the original Greek *mikros,* meaning "small". By this definition, a microgravity environment is one that imparts to an object a net acceleration that is small compared with that produced by Earth at its surface. We can achieve such an environment by using various methods including Earth-based drop towers, parabolic aircraft flights, and Earth-orbiting laboratories. In practice, such accelerations will range from about one percent of Earth's gravitational acceleration (on board an aircraft in parabolic flight) to better than one part in a million (on board a space station). Earth-based drop towers create microgravity environments with intermediate values of residual acceleration.

Quantitative systems of measurement, such as the metric system, commonly use micro- to mean one part in a million. By this second definition, the acceleration imparted to an object in microgravity will be 10^{-6} of that measured at Earth's surface.

The use of the term *microgravity* in this book corresponds to the first definition: small gravity levels or low gravity.

Microgravity can be created in two ways. Because gravitational pull diminishes with distance, one way to create a microgravity environment is to travel away from Earth. To reach a point where Earth's gravitational pull is reduced to one-millionth of that at the surface, we would have to travel into space a distance of 6.37 million kilometers from Earth (almost 17 times farther away than the Moon). This approach is impractical, except for automated spacecraft.

However, the act of free fall can create a more practical microgravity environment. Although planes, drop tower facilities, and small rockets can establish a microgravity environment, all of these laboratories share a common problem. After a few seconds or minutes of low-g, Earth gets in the way and the free-fall stops. To establish microgravity conditions for long periods of time, one must use spacecrafts in orbit. They are launched into a trajectory that arcs above Earth at the right speed to keep them falling while maintaining a constant altitude above the surface.

Newton (1687) envisioned a cannon at the top of a very tall mountain extending above Earth's atmosphere so that friction with the air would not be a factor, firing cannonballs parallel to the ground. Newton demonstrated how additional cannonballs would travel farther from the mountain each time if the cannon fired using more black powder. With each shot, the path would lengthen and soon the cannonballs would disappear over the horizon. Eventually, if one fired a cannon with enough energy, the cannonball would fall entirely around Earth and come back to its starting point. The cannonball would begin to orbit Earth. Provided no force other than gravity interfered with the cannonball motion, it would continue circling Earth in that orbit (Figure 1-02).

This is how the Space Shuttle stays in orbit. It launches into a trajectory that arcs above Earth so that the orbiter ravels at the right speed to keep it falling while maintaining a constant altitude above the surface. For example, if the Space Shuttle climbs to a 320-km high orbit, it must travel at a speed of about 27,740 km/hr to achieve a stable orbit. At that speed and altitude, due to the extremely low friction of the upper atmosphere, the Space Shuttle executes a falling path parallel to the curvature of Earth. In other words, the spacecraft generates a centrifugal acceleration that counterbalances Earth's gravitational acceleration at that vehicle's center of mass. The spacecraft is therefore in a state of free-fall around Earth, and its occupants are in a microgravity environment. Gravity *per se* is only reduced by about

10% at the altitude of low Earth orbit, but the more relevant fact is that gravitational acceleration is essentially cancelled out by the centrifugal acceleration of the spacecraft.

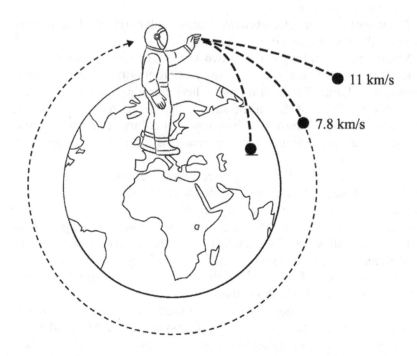

Figure 1-02. Artificial satellites are made to orbit Earth when their velocity is equal or higher than 7.8 km/sec. The rocket carries the satellite at the desired altitude and communicates to the satellite its horizontal velocity, also called "ejection velocity".

1.2.2   Other Factors of the Space Environment

Beside microgravity, during spaceflight living organisms are also affected by ionizing radiation, isolation, confinement, and changes in circadian rhythms (24-hr day/night cycle). In plants for example, spaceflight offers the unique opportunity to separate the gravitational input from other environmental stimuli known to influence plant growth, for example, phototropism (Figure 1-03), watertropism, and the circadian influences of the terrestrial environment. Spaceflight thus provides the opportunity to distinguish between the various tropic responses and to investigate the mechanisms of stimulus detection and response.

The absence of natural light in spacecraft may have significant effects on humans, too. A typical person spends his days outdoor, exposed to light provided by the Sun's rays (filtered through the ozone layer), including a small but important amount of mid- and near-ultraviolet light, and approximately equal portions of the various colors of visible light. Indoor lighting in most offices and in spacecrafts is of a much lower intensity and, if emitted by fluorescent "daylight" or "cool-white" bulbs, is deficient in ultraviolet light (and the blues and reds) and excessive in the light colors (yellow-green) that are best perceived as brightness by the retina.

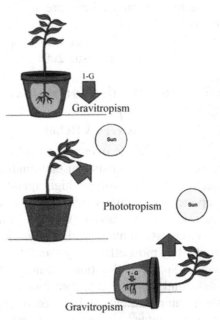

Figure 1-03. Gravitropism is the way plants grow in response to the pull of gravity. When placed near a window, plants exhibit phototropism (bending toward the light source). This behavior can be easily observed by placing a plant on its side; within minutes the roots and stem begin to reorient themselves in response to both gravity and light. (Credit NASA)

If the only effect of light on humans was to generate subjective brightness, then this artificial light spectrum might be adequate. It has become clear, however, that light has numerous additional physiological and behavioral effects. For example, light exerts direct effects on chemicals near the surface of the body, photoactivating vitamin D precursors and destroying circulating photoabsorbent compounds (melanin). It also exerts indirect effects via the eye and brain on neuroendocrine functions, circadian rhythms, secretion from the pineal organ, and, most clearly, on mood. Many people exhibit major swings in mood seasonally, in particular toward depression in the fall and winter when the hours of daylight are the shortest. When pathological, the "Seasonal Affective Disorder Syndrome" is a disease related to excessive secretion of the pineal hormone, melatonin, which also may be treatable with several hours per day of supplemental light. While not yet proved, it seems highly likely that prolonged exposure to inadequate lighting (that is, the wrong spectrum, or too low an intensity, or too few hours per day of light) may adversely affect mood and performance.

The effects of spaceflight on biological specimens might also be related to other factors. Even the gentlest of launch vehicles produces enormous amounts of noise and vibration, plus elevated g forces, until orbital velocity is achieved, or during the reentry into Earth atmosphere. Once in orbit, machines and astronauts continue to produce vibrations that are difficult to control. The space environment also exposes animals and individuals to high-energy radiation unlike anything they experience on Earth (see Chapter 2, Section 5.1). To control these and other external factors (for example, fluctuations in atmospheric pressure as astronauts enter and exit a spacecraft), the biologists studying the effects of microgravity *per se* ideally need onboard centrifuges that can expose control specimens to the level of gravity found on Earth's surface (Wassersug 2001).

1.3   Justification for Human Spaceflight

1.3.1   Humans versus Robots

The debate over space exploration is often framed as humans versus robots. Some scientists fear that sending humans to the Moon and Mars might preclude the pursuit of high quality science. On the other hand, some proponents of human exploration are concerned that doing as much science as possible using robots would diminish interest in sending humans. Nevertheless, humans will always be in command. The question is where would they most effectively stand?

Space exploration should be thought of as a partnership to which robots and humans each contribute important capabilities. Opposing robotics versus human crews is like comparing apples and oranges. The discussion must be framed in terms of relative strengths of humans and robots in exploring the Moon and Mars. For example, robots are particularly good at repetitive tasks. In general, robots excel in gathering large amounts of data and doing simple analyses. Hence, they can be designed for reconnaissance, which involves highly repetitive actions and simple analysis. Although they are difficult to reconfigure for new tasks, robots are also highly predictable and can be directed to test hypotheses suggested by the data they gather. However, robots are subject to mechanical failure, design and manufacturing errors, and errors by human operators. Also, before robots can explore and find evidence of life on Mars, their functional capabilities, particularly their mobility, need to be radically improved and enhanced. In addition, the delay in communication between Mars and Earth (in the order of 40 minutes round trip) poses a serious problem for tele-operation maneuvers.

People, on the other hand, are capable of integrating and analyzing diverse sensory inputs and of seeing connections generally beyond the ability of robots. Humans can respond to new situations and adapt their strategies accordingly. In addition, they are intelligent operators and efficient end-effectors. They may easily do better than automated systems in any number of situations, either by deriving a creative solution from a good first hand look at a problem or by delivering a more brainless kick in the right place to free a stuck antenna.* Either may be mission saving. Finally, only humans are adept at field science, which demands all of these properties. Obviously, humans would have a clear role in doing geological field work and in searching for life on Mars.

On the other hand, humans are also less predictable than robots and subject to illness, homesickness, stress from confinement, hunger, thirst, and other human qualities. They need protective space suits and pressurized habitats. Hence, they require far greater and more complicated and expensive support than robots.

1.3.2   Space Science

There is often criticism that human missions are costly out of proportion to their scientific yield when compared with automatic (unmanned) platforms such as are designed for solar system exploration or Earth observation. A direct comparison is unjust, however. Automatic probes have indeed returned spectacular results, but it is wrong to compare these directly with human flights. Historically, space life sciences are a rather recent discipline. In most space agencies, at least until recently, the term "Space Science" refers to space physical sciences, such as astrophysics or search for life on other planets. Perhaps as a reminiscent of this past, human spaceflight critics often discount the value of Space Life Sciences on the "Discovery Ledger (Big Book)".† This point of view is often due to the following fundamental differences: physical sciences lead to more concrete discoveries in a relatively unexplored sphere (once a new star is discovered, it is easy to confirm its presence), whereas space life sciences is an inherently inexact science, which must take into account background physiological variability and requires repeated measurements. For instance, large clinical trials are needed to determine the efficacy of a new drug. It may be obvious that space life sciences suffer from the small number of subjects studied and the many

---

* I was in the Mission Control Center in Moscow when Jean-Loup Chrétien and Alexander Volkov, during their space walk from Mir in 1998, were kicking into an antenna which refused to deploy automatically. They did not tell the ground until their "solution" finally worked. Before the relief, the tension was very high in Mission Control, especially when looking at the signals of the accelerometers mounted on the antenna!

† In a February 2003 interview to the Chicago Tribune, a physics professor at the University of Maryland and a director at the American Physical Society, a professional organization of physicists, said: "The International Space Station is not exploration; it's going in circles closer to the Earth than Baltimore is to New York." He added: "It is the single greatest obstacle of continued exploration of the solar system—it's blocking just about everything".

confounding factors that are difficult to control for. But with all this, it is likely that the life sciences data obtained in low Earth orbit studies will be practically utilized for going further (such as establishing a Mars base) or for improving our knowledge of clinical and aging disorders on Earth, long before information on the magnetic field of Neptune (Barratt 1995).

It is true that the cost of human-based space infrastructures, such as the International Space Station (ISS), is much higher than unmanned missions. However, the primary purpose for the ISS was a political one. The ISS is a major accomplishment for all countries involved even in its current incomplete state. It is the largest on-orbit structure ever built and the largest multi-national cooperative project in history. In building the ISS infrastructure and research equipment, aerospace companies are acquiring unique capabilities that make them recognized world players in areas such as space structures, automation, robotics, avionics, fluid handling, advanced life support systems and medical equipment. Both in view of the need to develop advanced technologies and by virtue of the research carried out on board, the ISS can have a significant impact on the competitiveness of aerospace industry. In the same way that one would not charge the cost of a road-system to a single car (or even the first dozen cars), the cost of the ISS cannot be endorsed by the scientific return of its first experiments.

The opportunities for in-depth studies in space life sciences have indeed been sparse. This is simply the nature of the current space program, with much to do and a few flight opportunities that must be shared. Experiments that might take weeks on Earth take years to plan and execute in space. Limitations of the spaceflight environment also have limited control experiments and often kept the number of specimens studied far from statistically ideal. Often space studies are paralleled by Earth-based simulation studies using centrifuges or clinostats, but results in actual microgravity are somewhat different.

Another argument often proposed against space life sciences is that no Nobel prizes have been given in this field of research. Although a true statement, there are several instances, however, of Nobel Prizes formerly delivered in life sciences related fields which would presumably not have been presented based on the recent results obtained in space. For example, Robert Bàràny, a Viennese ENT specialist, received the Nobel Prize of Medicine in 1906 for his discovery of a clinical test aimed at evaluating the functionality of the balance organs in the inner ear (see Chapter 3). During this test, irrigation of the external auditory ear with water or air above or below body temperature generates rhythmic eye movements (nystagmus) and the subject experiences slight vertigo. Barany's theory was that the caloric irrigation of the ear canal generated eye movements (the so-called caloric nystagmus) because of the heat, gravity-driven convection within the canal fluid. A space experiment carried out on board Spacelab in 1983 proved this theory to be wrong since caloric nystagmus was also observed in microgravity, where no heat current convection is generated. Later studies revealed that it is more likely the changes in pressure or temperature that are at the origin of the eye movement response (Scherer et al. 1986).

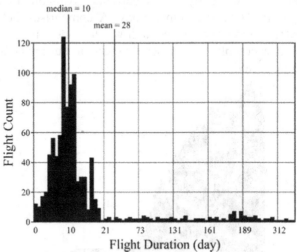

Figure 1-04. Frequency of manned spaceflight as a function of flight duration from 1961 to 2003. Most flights were of short duration, with a mean value of 28 days. The median value, however, is in the order of 10 days. (Adapted from Reschke and Sawin 2003)

1.4 Where We Are

Human spaceflight began only about 40 years ago, in 1961 with Yuri Gagarin's single orbit of the Earth on board Vostok-1. Since then, astronauts and cosmonauts have spent a considerable amount of time beyond the Earth's surface. As I write these lines, the total number of days spent in space is about 26,000 crew days. This number might sound large, but it corresponds to only about 71 years, that is less than a lifetime for a single individual.

Actually, these 71 years are the cumulative time spent by all 433 astronauts and cosmonauts who have flown in space to date. So, the average amount of time spent in space by astronauts and cosmonauts is 26,000 days / 433 = 60 days, or about 2 months. If we include the re-flights, the number of flown individuals goes up to 944. However, most of these individuals have spent less than 30 days in space, even by cumulating 3 or 4 flights. The mean duration of all spaceflights to date is about 28 days, but the median time in orbit is close to 10 days (Figure 1-04). Flight duration longer than 6 months is limited to about 40 individuals, and only four individuals have experienced spaceflights longer than one year (Figure 1-05).

Had all the astronauts and cosmonauts been the subjects of space life sciences investigations during their spaceflight, the total amount of collected data would be limited to a very small period of a human lifetime (71 years). Yet, since life sciences investigations were not conducted on all astronauts and cosmonauts, and since most of them have flown more than once, the limited number of individuals and observations makes the significance of this data very low.

This simple arithmetic is to illustrate how little research time—on how few space flyers—is currently available to determine the effects of spaceflight on the human body. A comparison between space research and extreme environment research would undoubtedly show that much more has been accomplished on Mt. Everest or during polar expeditions during the same period.[*]

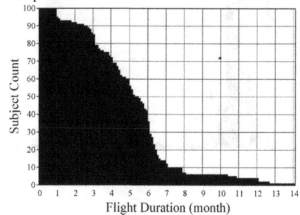

Figure 1-05. Cumulative histogram showing the astronauts and cosmonauts count as a function of flight duration. (Source John Charles, NASA)

The International Space Station (ISS) will allow extensive investigations on humans in space. However, the nominal duration of a stay in orbit for long-duration crew on ISS, or "increment", does not exceed 3-4 months. Therefore, no data is gained anymore during very long (more than 6 months) spaceflights.

The record of spaceflight duration is currently held by Dr. Valery Polyakov, a Russian physician, who spent 437 days during a single mission on board the space station Mir in 1994-1995 (see Figure 4-02). This was his second spaceflight, though. In 1989, he had already spent 242 days on board Mir, so his total time spent in space actually is 679 days, or about 22 months. But this is not the longest duration in space for a single individual. Sergey Avdeyev has logged 748 days during 3 stays on board Mir in 1993 (188 days), 1996 (179 days) and 1999 (379 days) and he currently holds the all-time cumulative total for days in space. His total time, and that of Polyakov, corresponds to nearly two years, which is just about the duration of exposure

---

[*] More than 1200 persons have reached the summit of Mount Everest between May 1953 and May 2001. About 400 expeditioners have explored successfully the Arctic or Antarctic (with or without outside assistance) between 1908 and 2002 (Source: www.adventurestats.com).

to microgravity that will be experienced during a mission to Mars. However, although we know that humans can survive to long duration in space, the data collected on two individuals is extremely limited.

There is a general perception that since a small number of cosmonauts have survived in low Earth orbit for as long as a year or so, there are no major physiological problems likely to preclude longer-term human planetary-exploration missions. One must admit that, over the years, there has been only access to anecdotal data from the Russian space program. This anecdotal information is, while interesting, not sufficiently reliable for drawing conclusions for a number of reasons. There are differences in the scientific method, the experimental protocols and the equipment, and the results are not published in peer-reviewed international scientific journals. Fortunately, the increased recent cooperative activities between Russia and its partners of the International Space Station allow a standardization of experimental procedures and better data exchange.

2 THE LEGACY OF SPACE LIFE SCIENCES RESEARCH (HOW WE GOT THERE)

2.1 Major Space Life Sciences Events

2.1.1 The Pioneers

The first powered flight in 1903 by the Wright Brothers at Kitty Hawk Beach in North Carolina is traditionally considered as the milestone in manned flight and aerospace medicine. Historically, Icarus was the first victim of a flying adventure, when he and his father Daedalus tried to escape their prison on the island of Crete by flying using waxed feathers. The legend says that Icarus, ignoring both advice and warning, flew too close to the Sun. The heat softened the wax and the feathers detached, precipitating Icarus in a dreadful fall.

However, there were no witnesses of Icarus and Daedalus flight. This was not the case for the second human flight in the history though. In June 1783, two brothers, Jacques Etienne and Joseph Michel Montgolfier, sent a large, smoked-filled bag 35 feet into the air. This first balloon flight was recorded by the French Academy of Sciences. Three months later, a duck, a rooster, and a sheep became the first passengers in a balloon, since no one knew whether a human could survive the flight. All three animals survived the flight, although the duck was found with a broken leg, presumably due to a kick from the sheep after landing. Finally on November 21, 1783, a human flight was attempted before a vast crowd that included the King and Queen of

France, and recognized scientists (Tillet et al. 1783). Pilatre de Roziers [*] and the Marquis d'Arlandes [†] piloted what became the first aerial voyage of humankind (Figure 1-06).

After this event, ballooning became quite popular for over half a century in Europe. Ten days after the first manned hot air flight, a French physicist named J.A.C. Charles made the first human flight in a hydrogen-filled balloon. When he reached an altitude of 2750 m, he began to experience physiologically some of the realities of this new environment. He complained of the penetrating cold at this altitude and a sharp pressure pain in one ear as he descended. This is the first description of symptoms experienced in aerospace medicine. In 1784 in England, after several animals were used in free flight tests, Mrs. Tible became the first woman to fly a balloon, and Jean-Pierre Blanchard became the first to cross the Channel from England to France. Feeling outdone, Pilatre de Roziers built a new balloon, using a combination of hot air envelope and a small hydrogen balloon, to fly from France to England. In January 1785, he left France, and after a few minutes in-flight the burner's flame ignited the small hydrogen balloon, creating an inferno. Ironically, the first to fly in a balloon became the first balloon casualty. The hazards of high altitude flight were demonstrated in following flights, where balloonists experienced and described for the first time the symptoms of hypoxia (altitude sickness, increase in heart rate, fatigue) (DeHart 1985).

Figure 1-06. Drawing of the first manned balloon flight taking off in front of the Château de la Muette with passengers Pilatre de Roziers and the Marquis d'Arlandes.

---

[*] The word "pilot" is derived from his name.
[†] The Marquis d'Arlandes was born in my hometown, Anneyron, a small village in the south of France.

2.1.2 Spaceflights with Animals

In the 1950s, as human spaceflight began to be seriously considered, most scientists and engineers projected that if spaceflight became a reality it would build upon logical building blocks. First, a human would be sent into space as a passenger in a capsule (Projects Vostok and Mercury). Second, the passengers would acquire some control over the space vehicle (Projects Soyuz and Gemini). Third, a reusable space vehicle would be developed that would take humans into Earth orbit and return them. Next, a permanent space station would be constructed in a near-Earth orbit through the utilization of the reusable space vehicle. Finally, lunar and planetary flights would be launched from the space station using relatively low-thrust and reusable (and thus lower cost) space vehicles.

Just like for balloon flights, animals were sent up in rockets before humans to test if a living being could withstand and survive a journey into space (Figure 1-07). The first successful spaceflight for live creatures came on September 20, 1951, when the former Soviet Union launched a sounding rocket with a capsule including a monkey and eleven mice. A few attempts to fly animals had been made before (in fact, since 1948 in the nose cones of captured German V-2 rockets during U.S. launch tests) but something always went wrong. These attempts were made with one main purpose: to study the effects of exposure to solar radiation at high altitude, and to determine the effects, if any, of weightlessness (Lujan and White 1994).

Figure 1-07. Rats and cats were the first living passengers of a suborbital flight in a French rocket in the 60's. (Credit CNES)

Orbital flight then began in 1957 (October 4) when the former Soviet Union sent the Sputnik-1 satellite into space. This was an unmanned satellite, but before the end of the year a second satellite, Sputnik-2, was launched carrying the first living creature into orbit, a dog named Laika. Laika had been equipped with a comprehensive array of telemetry sensors, which gave continuous physiological information to tracking stations. The cabin conditioning system maintained sea-level atmospheric pressure within the cabin, and Laika survived 6 days before depletion of the oxygen stores caused asphyxiation. Laika's flight demonstrated that spaceflight was tolerable to animals. Twelve other dogs, as well as mice, rats and a variety of plants were then sent into space for longer and longer duration between 1958 and 1966. In 1996, a Soviet biosatellite Cosmos mission carried two dogs in orbit for 23 days. The dogs were observed via video transmission and biomedical telemetry. Their spacecraft landed safely.

In 1959, one rhesus and one squirrel monkey rode in the nose cone of a U.S. missile during a non-orbital flight, successfully withstanding 38 times the normal pull of gravity and a weightless period of about 9 minutes. Their survival of speeds over 18,000 km/hr was the first step toward putting a human into space. Although one of the monkeys died from the effects of anesthesia given to allow the removal of electrodes implanted for the spaceflight, a subsequent autopsy revealed that the monkey had suffered no adverse effects from the flight. Between 1959 and 1961, three other monkeys made suborbital flights in Mercury capsules, and one monkey flew two orbits around Earth in a Mercury capsule in preparation for the next, human flight (Figure 1-08). These experiments paved the way for human expeditions.

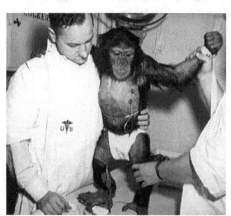

Figure 1-08. Chimpanzee "Ham" with biosensors attached is being prepared for his trip in the Mercury-Redstone 2 on January 31, 1961. (Credit NASA)

While these animals were in space, instruments also monitored various physiological responses as the animals experienced the stresses of launch, reentry, and the weightless environment. The results of these animal flights showed that:
- Pulse and respiration rates, during both the ballistic and the orbital flights, remained within normal limits throughout the weightless state. Cardiac function, as evaluated from the electrocardiograms and pressure records, was also unaffected by the flights;
- Blood pressures, in both the systemic arterial tree and the low-pressure system, were not significantly changed from preflight values during 3 hours of the weightless state;
- Performance of a series of tasks of graded motivation and difficulty was unaffected by the weightless state;

- Animals trained in the laboratory to perform during the simulated acceleration, noise, and vibration of launch and reentry were able to maintain performance throughout an actual flight.

On the basis of these results, it was concluded that the physical and mental demands that the astronauts would encounter during spaceflight "would not be excessive", and the adequacy of the life support system was demonstrated (Henry 1963).

2.1.3   Humans in Space

Earlier, on October 7, 1958, the new National Aeronautics and Space Administration (NASA) had announced Project Mercury, its first major undertaking. The objectives were threefold: to place a human spacecraft into orbital flight around Earth, observe human performance in such conditions, and recover the human and the spacecraft safely. At this early point in the U.S. space program, many questions remained. Could a human perform normally as a pilot-engineer-experimenter in the harsh conditions of weightless flight? If yes, who were the right people (with the right stuff) for this challenge?

In January 1959, NASA received and screened 508 service records of a group of talented test pilots, from which 110 candidates were assembled. One month later, through a variety of interviews and a battery of written tests, the NASA selection committee brought down this group to 32 candidates. Each candidate endured even more stringent physical, psychological, and mental examinations, including total body X-rays, pressure suit tests, cognitive exercises, and a series of unnerving interviews. Of the 32 candidates, 18 were recommended for Project Mercury without medical reservations. At a press conference, NASA introduced the seven Mercury astronauts to the public.

The following year, the Soviet Union announced that 20 fighter pilots had been selected for its space program. Physiological studies and special psychophysiological methods "permitted the selection of people best fitted to discharge the missions accurately and who had the most stable nerves and emotional health", according to the Soviet report of April 26, 1961. In 1962, 5 female parachutists joined this first group of 20 male cosmonauts.

On April 12, 1961, Yuri Gagarin became the first human to orbit the Earth. According to the press release, Gagarin felt "perfectly well" throughout the orbiting phase and also during the period of weightlessness. It was noted, however, that "measures" had been taken to protect the spacecraft from the hazards of space radiation.

Six weeks later, U.S. President Kennedy would announce as a national objective an accelerated space program to accomplish a landing on the Moon in the 1969-70 period. However, after the suborbital flights of Alan

Shepard and Gus Grissom in May and July 1961, respectively, observations made during U.S. orbital spaceflights with monkeys raised some concerns. Variations in cardiac rhythm had been recorded in one chimpanzee during a three-orbit mission (Stringly 1962). It was found that the problem came from faulty instrumentation, and that the data were therefore invalid. Accordingly, it was recommended that John Glenn's orbital flight proceeded as scheduled.

On August 6, 1961, after Grissom's suborbital flight in July, the USSR launched Cosmonaut Gherman S. Titov into orbit. The following day, August 7, Titov successfully landed after 17 orbits in 25 hours and 18 minutes. This was the first human flight of more than one orbit, and the first test of human responses to prolonged weightlessness. Two years later, Valentina Tereshkova became the first woman in space (Figure 1-09). She remained in space for nearly three days and orbited the Earth 48 times. Unlike earlier Soviet spaceflights, Tereshkova was permitted to operate the controls manually. After her spacecraft reentered the Earth's atmosphere, Tereshkova parachuted to the ground, as was typical of cosmonauts at that time. Although her spaceflight was announced as successful, it was 19 years until another woman flew in space, Svetlana Savitskaya, aboard Soyuz T-7 in 1982. Apparently, something went so wrong during Tereshkova's flight that no further flight included women. Savitskaya must have turned out all right, since she flew twice, and during her second mission on board Soyuz T-12 in July 1984, was the first woman to ever perform a space walk.

Soviet Cosmonaut Aleksei Leonov made the first space walk during the Voskhod-2 mission on March 18, 1965. He was followed by Astronaut Edward White who stepped out of Gemini-4 for 20 minutes. White propelled himself away from the spacecraft with a special gun that gushed out compressed oxygen to move him in any direction. However, because his propulsion gun ran out of fuel, he had to pull on his life support system umbilical line to maneuver around and reenter the spacecraft.

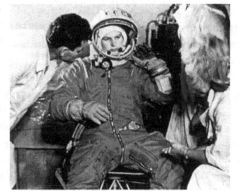

Figure 1-09. Russian cosmonaut Valentina Tereshkova after returning from a three-day spaceflight. (Credit CNES)

2.1.4   Space Life Sciences Investigations

The Mercury flights had made it clear that the body undergoes some real changes during and after spaceflight, such as measurable weight loss. A more complex set of in-flight medical studies was carried out during the

Gemini missions, which served as precursors to the lunar missions (Figure 1-10). Among those missions, Gemini-7 (December 1965) primary objective was to conduct a two-week mission and evaluate the effects of long-term exposure to weightlessness on its crew. Many medical experiments were conducted in-flight, including on vision and sleep. Extensive testing, for example on balance, was also performed just after landing. Blood and urine samples were collected throughout the mission for analysis, and astronauts exercised twice daily using rubber bungee cords.

Of particular interest was the visual acuity experiment, which was driven by earlier observations of Mercury astronauts who thought their capability to identify small objects on the Earth's surface was enhanced in weightlessness. This experiment used a visual acuity goggle combined with measured optical properties of ground objects and their natural lighting, as well as the atmosphere and spacecraft window. The results failed to show that the visual acuity was improved while in space.

Also interesting is the Gemini-11 flight, where artificial gravity was (involuntarily) first tested in space. The Gemini spacecraft was tethered to an Agena target vehicle by a long Dacron line, causing the two vehicles to spin slowly around each other. According to the Gemini commander, a TV camera fell "down" in the direction of the centrifugal force, but the crew did not perceive any changes.

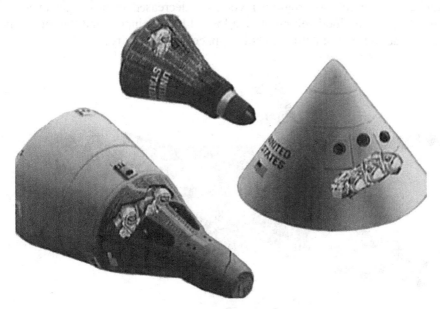

Figure 1-10. Comparison between the Mercury, Gemini, and Apollo manned spacecrafts. Mercury could accommodate only one crewmember, Gemini two, and Apollo three crewmembers. (Credit NASA)

Significant orthostatic hypotension and weight loss were observed in the crewmembers of Gemini-3, -4, -5, and -6 immediately after flight (see Chapter 4). Also, red blood cell mass losses in the order of 20% were noted after the 8-day Gemini flight. Scientists were concerned that spaceflight might affect the balance of body fluids and electrolytes since fluid losses can contribute to both of these symptoms. This led to a series of ground-based studies to simulate some of the conditions of spaceflight. These studies utilized bed rest and water immersion as a means of simulating microgravity. In addition, Biosatellite-3 was launched in 1969, three weeks before the first men were to land on the Moon, with a monkey passenger. The flight was planned for a full month, but the monkey was brought down, ill from loss of body fluids, after only nine days. It died shortly after landing. Despite the concern that the same problem could occur to humans, the Apollo missions to the Moon proceeded as planned.

During the Apollo missions, a medical program was developed which would make provision for emergency treatment during the course of the mission in case a serious illness occurs. Indeed, during the orbital flights of Mercury and Gemini, it was always possible to abort the mission and recover the astronaut within a reasonable time should an in-flight medical emergency occur. This alternative was greatly reduced during Apollo. The events of Apollo-13 showed that this medical program proved effective. Biomedical findings of the Apollo program revealed a decreased in postflight exercise capacity and red blood cell number, a loss of bone mineral, and the relatively high metabolic cost of extra-vehicular (space walk) activity.

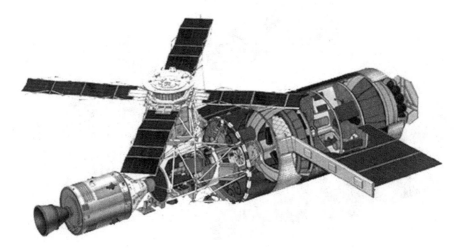

Figure 1-11. Drawing of the Skylab workshop showing the Orbital Module laboratory ("transparent" walls, right) and the Apollo crew return vehicle (left). (Credit NASA.)

In addition, symptoms of space motion sickness such as nausea and vomiting, earlier described by Soviet cosmonauts, were experienced. These observations raised another concern for future human spaceflights, and therefore constituted the starting point of detailed life sciences investigations in the Skylab program in the 1970's.

The U.S. Skylab (Figure 1-11) and the Soviet Salyut space stations allowed scientists to conduct investigations on board large orbiting facilities during missions lasting up to 3 months. They gave a basic picture of how the body reacts and adapts to the space environment. The number of subjects was, however, still limited.

The Space Shuttle (or *Space Transportation System*, STS), which began flying in 1981, provided the opportunity to test more crewmembers. Also, as the first spacecraft that could be used again and again, the Space Shuttle has provided space life scientists with a more regular opportunity to conduct experiments, and to repeat and refine those experiments.

However, with the Space Shuttle, other concerns appeared. The Space Shuttle was remarkably different from the previous spacecrafts because it returned to Earth by landing on a runway (Figure 1-12). Critical issues existed concerning the ability of crews to perform the visual and manual tasks involved in piloting and landing the Shuttle, and their capacity to achieve unaided egress after long exposure to weightlessness. It was later found that the astronauts-pilots were able to pilot and manually land the Space Shuttle, as long as the flight duration did not exceeded two weeks. Such critical achievement was in part due to the development of special simulators built to train crews to fly and land the Shuttle, in what is now popularly termed a "virtual reality" setting. In fact, astronauts returning from Shuttle missions reported that the simulations were so accurate they felt they had flown the mission many times.

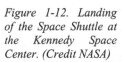

Figure 1-12. Landing of the Space Shuttle at the Kennedy Space Center. (Credit NASA)

2.1.5   Today's Access to Space

In more than 110 flights to date since its maiden mission in 1981, the Space Shuttle has repeatedly demonstrated unique capabilities as space transporter, repair ship, scientific platform, and research center. It first accomplished its role of "shuttle" by rendez-vous and docking with Mir in 1995, a few months after the end of Valery Polyakov's 14-month mission. From February 1994 to June 1998, NASA Space Shuttles made 11 flights to the Russian space station Mir, and U.S. astronauts spent 7 residencies, or "increments", on board Mir. Space Shuttles also conducted crew exchanges and delivered supplies and equipment. The Space Shuttle was then the first spacecraft to dock with the International Space Station in May 1999. Since the permanent crew occupation of the ISS in November 2000, the Space Shuttle has ensured most of the crew transport, with the Soyuz, between Earth and ISS.

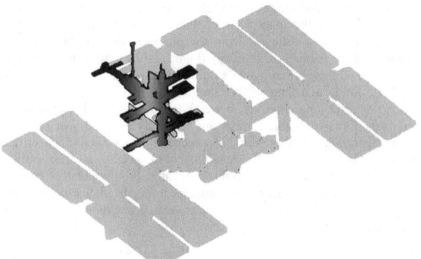

Figure 1-13. Comparison between the sizes of the Russian Mir space station (dark gray) and the International Space Station in its final configuration (light gray).

More than four times as large as the Russian Mir space station (Figure 1-13), the completed International Space Station will have a mass of about 450 tons and more than 1200 m^3 of pressurized space in six state-of-the-art laboratories (Table 1-02). The United States will provide two laboratories (*United States Laboratory* and *Centrifuge Accommodation Module*) and a habitation module for four crewmembers. There will be two Russian research modules; one Japanese laboratory referred to as the *Japanese Experiment Module* (JEM) named "Kibo" (Hope); and one European Space Agency (ESA) laboratory called the *Columbus Orbital Facility* (COF). The ISS internal volume will be roughly equivalent to the passenger cabin volume of two Boeing 747 jets.

All six laboratories together will provide 37 International Standard Payload Racks (ISPR). An ISPR, about the size of a home refrigerator, holds research equipment and experiments. Additional research space will be available in connecting nodes and the Russian modules. The JEM also has an exterior "back porch" with 10 spaces for mounting experiments that need to be exposed to space. The experiments will be set outside using a small robotic arm on the JEM. There are also four attached payload sites on the truss and two spaces on the COF for mounting external experiments.

Today, the three-person crew allowed on board the ISS[*] is so small that the astronauts spend the vast majority of their time on maintenance, leaving little room in their schedule for actual experiments. Once completed, the ISS will house an international crew of up to seven for stays of approximately three months. Emergency crew-return vehicles will always be docked with the ISS while it is inhabited, to assure the return of all crewmembers. A Russian Soyuz spacecraft, which has a crew capacity of three, is presently used. Later, a higher capacity vehicle called the Crew Return Vehicle (CRV) will allow up to seven people to return at once to Earth.

It is important to realize that the ISS is far more than a science platform alone. The ISS constitutes a highly visible signature in the sky for human endeavor, courage, spirit, and international peaceful collaboration, and it is the greatest technological challenge the human race has tackled so far. To a large part this was the early political motivation that led to its conception. In addition, and looking more towards the future the ISS provides the gateway for human exploration of the solar system. The ISS has also the potential for becoming an ideal tool to support educational activities. In particular, educational programs encouraging and supporting the study of science, mathematics, technology and engineering can be implemented on board the ISS, making use of its facilities and resources. Other education projects can be implemented that focus not only on science and technology, but also on a larger variety of subjects, such as languages, composition and art.

In April 2001, an American engineer and millionaire flew on a Soyuz and spent eight days on board the ISS. His trip erupted in a controversy when NASA and the other ISS partners objected to a tourist visit in the middle of a critical series of assembly operations at the ISS. The ISS partners reluctantly gave their approval for a visit that was going to come off with or without their

---

[*] Since the *Columbia* accident and the grounding of all the Shuttle fleet, the re-supply of the ISS is entirely dependent upon the Progress and Soyuz vehicles. For this reason, the crew on board ISS is currently limited to two persons.

approval, in return for a promise by the Russians to meet new standards for paying visitors in the future.

Actually, unofficial "tourists" had already flown on several occasions both on U.S. (a senator, a congressman, a teacher, and a prince from Saudi Arabia) and Soviet/Russian (a reporter, an engineer from a chocolate company, and guests from allied countries) space missions. Europe took this opportunity of a paying visitor on the Soyuz to allow its astronauts to have a regular access to the ISS for one week at a time, the so-called "Taxi" missions. No doubt about it, space tourism is a reality, and it's a good and necessary development for the future of commercial work in orbit.[*]

Total residents and visitors since start of assembly	121
Men	102
Women	19
Flights since start of assembly	35
American	16 Shuttle
Russian	2 Proton
	7 Soyuz
	10 Progress
Space walks since start of assembly	51
Shuttle-based	25
ISS-based	26
Expedition crews mission duration	
Expedition-1	140 days
Expedition-2	167 days
Expedition-3	128 days
Expedition-4	195 days
Expedition-5	184 days
Expedition-6	161 days

Table 1-02. International Space Station (ISS) facts and figures (as of May 2003).

---

[*] As I write these lines, 24 teams of entrepreneurs from seven countries have entered in a competition for winning a $10 million cash prize to become the first private venture to finance, build, and launch a manned space vehicle. The objective is to launch a spacecraft carrying three or more people to a height of at least 100 km and then repeat the process within two weeks. Like Charles Lindbergh, who flew nonstop across the Atlantic in 1927 in a race to claim $25,000, these modern entrepreneurs are chasing both a prize and prestige. In addition to the $10 million purse, to be awarded by the nonprofit X-Prize Foundation of St. Louis, the competitors are seeking to open the door to space travel and other commercial uses that don't depend on government backing.

Note added for this 2nd revision: In its preliminary attempt to claim the X-Prize, the first private manned space vehicle flew to the edge of space and back on June 21, 2004. The SpaceShipOne craft, built by aviation pioneer Burt Rutan, went over space's 100 km boundary, and landed safely after a 90-minute flight.

2.2 Surviving the Odyssey

Early predictions of the response of humans to spaceflight assumed that space adaptation would be analogous to human disease processes rather than to normal physiology. The predictions made by scientists about the ability of humans to endure spaceflight were indeed dire. Despite ground-based studies proving the contrary, there was true concern that the g forces of launch and reentry (6 to 8 g for the earliest spacecraft) would render human passengers unconscious, severely impaired or even dead. The mystique of this alien environment was so great that many feared a psychotic breakdown when humans would find themselves disconnected from and looking down on mother Earth.* Some physicians voiced concerns that body functions in weightlessness might suffer from a long list of calamities: swallowing, urination, and defecation would be impaired or impossible in the absence of gravity (although anyone who has ever swallowed while standing on their head hanging upside down could have proven otherwise); the bowels would not work without gravity; the heart might cavitate like a pump or beat so irregularly as to cause problems; sleep would be impaired; and muscles, including the heart, would become so weakened as to prohibit return to Earth (Churchill 1999).

The first space missions showed, however, with the proper protection, humans could survive a journey into space. Biomedical changes have been observed during spaceflight, due to the effects of microgravity, but also to other phenomenon, such as high launch and reentry gravitational forces, radiation exposure, and psychological stress.

To illustrate of what *we do know* at this point, my colleague and friend Susanne Churchill, in one of her lectures at the International Space University, used to describe the space journey of an hypothetical space traveler who experiences all of the known problems. I will use the same approach below.

So, let us take a journey with our hypothetical astronaut. She is in excellent health and fully trained for the rigors of her 3-month increment on board ISS. Launch occurs as anticipated: a couple of hours before launch she had joined the others lying down in the seats of the Space Shuttle, strapped in, feet above head, as in the early Mercury, Gemini, or Apollo launches. But there, the similarity ends. For during Shuttle lift-off she does not undergo the unpleasant gravity load, which went as high as 8 g on earlier flights. Instead, she experiences 3 g only twice. The first time comes and goes quickly near the two-minute mark, just before the two solid rocket boosters burn out and drop by parachute into the Atlantic Ocean. The final 3-g load comes five

---

* This was one reason why the earliest spacecrafts were totally automatic, with no controls for a disoriented or "crazed" pilot to use independently.

minutes later and lasts for a minute. Less than ten minutes from lift-off, she finds herself floating in the weightlessness of space.

Without warning, however, she suddenly vomits and is overwhelmed with intense symptoms of motion sickness: nausea, a sense of dizziness, and disorientation. Her symptoms become worst when she moves about in the cabin or sees one of her fellow crewmembers floating upside-down. She is unable to keep food down and rejects even water to drink, so she quickly dehydrates. She is concerned that she could not help the rest of her crew with the rendez-vous procedures of the Shuttle with the ISS, since looking out of the windows triggers more symptoms. She takes some pills and is getting ready for sleep. However, when looking in the mirror above the wash basin, she realizes that her eyes seem smaller, her face is round and puffy (Figure 1-14), and her neck veins bulging. The good news is that her wrinkles have disappeared and she looks younger. When undressing, she notices that her legs look like sticks. She tries to sleep but has a persistent backache, a definite feeling of sinus congestion, and keeps waking to discover that her arms are floating above her head. So disconcerting!

Figure 1-14. An example of "puffy face". The normal face of the astronaut of Earth (left) is contrasted with the swollen-looking face of the astronaut in space (right). (Credit NASA)

When she wakes up and dresses, she finds her clothes too short. Because of the absence of perceived gravity, her vertebrate disks are less compressed, making her height increase by 5-6 cm (Figure 1-15) and causing continuing back pain. Also, as a result of the fluid shift (which is also responsible for her puffy face and "chicken" legs) her waistline has shrunk about 4 cm and she must tighten the bands of her pants. Her shoes have also become too loose.

Within a couple of days, the motion sickness symptoms begin to subside, through her face and legs remain changed. Her posture too, is different, but not for the better. Joints go to their midpoint in zero gravity so that the hips and knees are bent into a slight crouch. Her arms tend to float in front of her unless she consciously holds them down. When she sits at a workbench, she has to strap herself in place. Even so, her seated posture is to lean back. Nevertheless, she learns to move around in weightlessness by gently pushing and pulling her body with her fingertips.

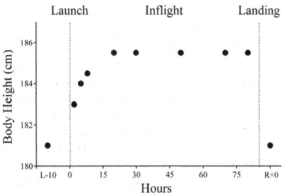

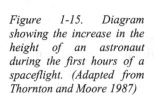

Figure 1-15. Diagram showing the increase in the height of an astronaut during the first hours of a spaceflight. (Adapted from Thornton and Moore 1987)

Rendezvous and transfer to the ISS occur without incident and she starts to settle for a 3-month stay on board with her two fellow crewmates. Personal hygiene is limited to "sponge" bathing; food becomes bland tasting and she must add spices for interest. There are experiments to monitor and several hours of exercise daily on the treadmill or cycle ergometer. After a few weeks, however, the routine is boring and it becomes harder and harder to keep up with the exercise. The more she looks out of the window, the more she longs for the sounds of rain and wind, and the smells of flowers. The objects outside the station look "unreal in clarity". However, when she closes her eyes, she experiences light flashes, especially when the ISS flies over the South Atlantic Anomaly. The crew starts to argue about the smallest things. One planned space walk has to be cancelled because of a persistent irregular heartbeat in one of the crewmember. Since that incident, this crewmember seems to be withdrawing from the others. The weekly videoconferences with family and friends are eagerly anticipated, but she wonders why there has been no communication from her youngest child for several weeks. Has something happened? Anxiety arises and she has a persistent pain in her lower abdomen, which, if it continues, might prompt an emergency evacuation to Earth.

But at last the time to return approaches. An interesting mixture of excitement and anxiety pervades the crew. Visions of favorite foods and what to do first are the main topics of conversation. Yet, the group has become so firmly a part of each other that the thoughts of reintegrating into Earth society are intimidating. But at last the crew is on its way home. When donning her reentry space suit, she realizes it is too tight because she has grown up of a few centimeters. During the reentry into the Earth atmosphere our traveler experiences disorientation again when she tilts or rolls her head. After landing, she reports an unbelievable sense of "heaviness" and finds herself unable to stand up unassisted from her seat, much less walking down the stairs. Her heart is beating fast; she sweats and almost faints. Even after several days of rehabilitation, balance is poor and walking uncoordinated.

Muscle weakness is very evident; she quickly feels short of breath and is constantly thirsty. Weight loss that occurred in space is rapidly disappearing, but her physician tells her that she had lost much of bone density in her hips and that her immune system seems to be impaired. Now she is concerned because she remembers that the various bacterial colonies they were studying on board the ISS laboratories showed explosive growth rates! Several months later though, all her body functions seem to have readapted to Earth gravity.

This story is not meant to discourage anyone from wanting to be an astronaut. In reality, not all people experience all of the adverse effects of spaceflight. It is rather meant to show how little we really know about the human body's response to spaceflight and how very dangerous this new environment can be. The interpretations for the observed physiological and psychological changes during spaceflight will be detailed in Chapters 3, 4, 5, 6, and 7 of this book.

2.3 Life Support Systems

Spaceflight includes conditions such as vacuum, extreme temperatures, noise (mostly due to the life support systems), and radiation. Protecting humans from these harmful conditions requires the use of life support equipment and technologies such as space suits, pressurized and isolated living quarters, and radiation shielding.

In addition, certain basic physiological needs must be met in order for human beings to stay alive. On Earth, these needs are met by other life forms in conjunction with chemical processes that effectively use human waste products in conjunction with energy from the Sun to produce fresh supplies of food, oxygen and clean water. In the artificial environment of a spacecraft, these materials must be provided, and human wastes removed, without relying on the natural resources of the Earth's biosphere.

To date, space missions have used a simple "open" system, bringing along all necessary food, water and air for the crew and venting waste products to space or collecting and storing them for return to Earth. When the point is reached where it is no longer cost effective or logistically possible to re-supply the spacecraft or habitat with water, atmosphere, and food, ways must be found to recycle all these components. This recycling of material is referred to as a "closed" system, and can be achieved using physical-chemical systems, or better, using biological systems (Figure 1-16).

Think of the human body as a sealed box with one pipe in and one pipe out. In go oxygen (O_2), water and food; out come solid and liquid wastes, bacteria, and carbon dioxide (CO_2). The outlet pipe is fed into a second sealed box, the *controlled ecological life support system* (CELSS). The CELSS must be as "magical" as the first, for it must transform these by-products of the body into fresh supplies and pipe them back (Collins 1990).

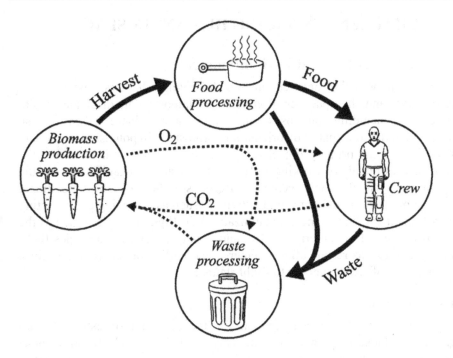

Figure 1-16. A controlled ecological life support system (CELSS) employs biological components and uses higher plants. Higher plants are easily digestible and are customary sources of human food. Besides producing food they also remove carbon dioxide from the atmosphere, produce oxygen, and purify water through the process of respiration.

Trying to recreate the cycles of nature in a relatively small volume is a great technical challenge. Plants "breathe" CO_2 and "exhale" O_2, so in a broad sense human wastes are used by plants, and vice versa. But in nature the nutrients, air, water, and energy are freely available. In a CELSS system all of these elements must be imported and carefully managed in a closed cycle. There are critical questions being addressed for CELSS during human missions. For example: How far can we reduce reliance on expendables? How well do biological and physic-chemical life support technologies work together over long periods of time? Is a "steady state" condition ever achieved with biological systems? How do various contaminants accumulate and what are the long-term cleanliness issues? Eventually, in the case of planetary missions, is it possible to duplicate the functions of the Earth in terms of human life support, without the benefit of the Earth's large buffers—oceans, atmosphere, land masses? How small can the requisite buffers be and yet maintain extremely high reliability over long periods of time in a hostile environment?

Eckart (1996) has addressed most of these questions. We will summarize them in Chapter 7 (Section 5.2).

3   CHALLENGES FACING HUMANS IN SPACE

3.1   Astronauts' Health Maintenance

As it will be detailed in the following chapters, exposure to microgravity and the space environment has important medical and health implications, including bone loss (matrix and minerals), increased cancer risk from space radiation, spatial disorientation, orthostatic hypotension, and many others. One of the primary objectives of space life sciences is to ensure the health of crewmembers working on board the spacecraft and in the hostile environment outside their vehicles. Responsibilities of the operational medicine program include preflight activities (such as screening and selecting new astronaut candidates, health stabilization), in-flight activities (such as the administration of countermeasures and medical care), and postflight procedures (such as rescue after an emergency landing or rehabilitation for a prompt return of crewmembers to flight status).

3.1.1   Preflight

The minimal medical criteria for the selection of astronauts are different for those astronauts who actually pilot the vehicle (*Pilots*), those who support onboard operations and perform extra-vehicular activities (*Mission Specialists*), those who perform specific onboard experiments (*Payload Specialists*), and those who just participate either as politicians, journalists, or tourists (*Participants*) (for more details see Chapter 7).

Based on the knowledge of specific health risk factors associated with spaceflight, appropriate and proven tests are utilized in selecting the astronauts. Annual medical evaluations are then performed to identify and correct medical risks to maintain health, provide certification for flight duties, and ensure career longevity. These tests may include further clinical evaluation (e.g., using state-of-the-art imagery techniques) or fitness assessments in order to prescribe individualized exercise programs and provide one-on-one preflight and postflight conditioning activities. Both selection and periodic medical evaluations rely on the accepted ground-based standards of preventive medicine, health maintenance, and medical practice. These standards are revised on a periodic basis to ensure that they are fair and appropriate to meet the needs of human spaceflight.

During preflight training, the primary emphasis of medical support is on *prevention*. For example, the purpose of the Crew Health Stabilization program is to prevent flight crews from exposure to contagious illness just before launch. A preflight quarantine limits access to flight crew during seven days just prior to launch. Even before this period, the health of an active duty crewmember family is of critical importance, and factors such as infectious disease and stress affecting a crewmember family may have serious adverse effects on the crewmember health and performance, as well as the health and performance of other crew members. Medical and dental care is provided by an onsite flight medicine clinic to the crewmember's immediate family, as long as the crewmember is eligible for assignment to a spaceflight mission.

Crewmembers are also trained in the use of special countermeasures to spaceflight physical deconditioning and in medical monitoring and clinical practice procedures. Medical training for the crew, medical supervision of mission planning, schedules, payloads, exercise training, and conditioning, and other health maintenance activities are all part of the preflight period.

Figure 1-17. Astronaut during a spacewalk or extra-vehicular activity (EVA) in the vacuum of space. (Credit NASA.)

3.1.2 In-flight

The primary emphasis of in-flight medical support is on *health maintenance*. Health monitoring and medical intervention, countermeasures to body functions deconditioning, and environmental monitoring enable a comprehensive program tailored to crew and mission needs and for the periodic assessment of crew medical status, including the identification of potential and unexpected health risks.

Among these potential health risks are the levels of acceleration, vibration, and noise during launch, the exposure to toxic substances and pressure changes, and the risk due to radiation. With the possible exception of the immune system, body changes that occur after entering microgravity represent normal homeostatic responses to a new environment. The body's control systems recognize the lack of gravity and begin to adapt to this unique situation, not realizing that the ultimate plan is to return to 1 g after a transient visit to microgravity. In-flight, typical adaptive and patho-physiological changes occur in the heart and blood vessels (dysrhythmias), muscles (strength), bones (fractures, renal stones), nervous system (disorientation and nausea), and in the immune system (infection). Space-walks or extra-vehicular activity (EVA) (Figure 1-17) can also be responsible for strain on muscles and bones, and decompression-related disorders.

Psycho-sociological issues become increasingly more important as space missions become longer, and spaceflight teams become larger and more heterogeneous. The isolated, confined, and hazardous environment of space create stress beyond that normally encountered on Earth, even when training for a space mission. Extended duration missions place an even greater stress on individuals, interpersonal, and group relations for astronaut crews, between astronaut crews and ground control, and on astronaut families. Current countermeasures focus primarily on the individual, mission crew, and to some extent the families of mission crews, by providing psychological training and support through in-flight communications.

Finally, for spaceflight missions, emphasis is not only on health maintenance, disease prevention, and environmental issues, but also on the provision of medical care to manage possible illnesses and injuries.

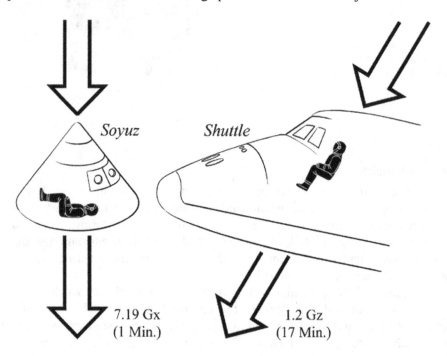

Figure 1-18. Comparison of the direction and amplitude of g forces experienced during the landing between a Soyuz capsule (left) and the Space Shuttle (right). Less g forces are tolerated when directed along the body longitudinal axis (Gz), in a direction parallel to the big blood vessels (see Chapter 4, Table 4-02).(Adapted from Nicogossian and Parker 1982)

3.1.3   Postflight

The primary emphasis of postflight medical support is *medical care*. During return to Earth, piloting tasks are challenged by the presence of g forces in deconditioned individuals (Figure 1-18). After nominal landing, astronauts often exhibit difficulties in standing, a phenomenon known as "postflight orthostatic intolerance" (see Chapter 4), and walking. These difficulties could prove dramatic in the case of a non-nominal landing where the crew may be required to egress in emergency the vehicle with no help from ground support.

Astronauts must have career longevity, normal life expectancy, with rehabilitation and recovery capabilities available upon their return from spaceflight. After landing, health monitoring and physical rehabilitation are performed to accelerate the return of crewmembers to normal Earth-based duties. An important factor to take into account is the return to flight status for pilot astronauts.

There is a large catalog of reported postflight symptoms captured in the mission medical debriefs which are collected after a space mission through interviews between the astronauts and crew flight surgeon. After every Shuttle mission, a NASA flight surgeon holds a medical debrief with each crewmember on the day of landing and then three days later. Standardized debrief forms are utilized during these meetings, at which time the physician and crewmember discuss pre-, in-, and postflight medical issues. The crewmembers are interviewed about their experiences, utilizing both open-ended and specific questions. Information from these debriefs is available in a database known as the Longitudinal Study of Astronaut Health (LSAH). NASA flight surgeons of the Flight Medicine Clinic at NASA Johnson Space Center provide the interface with the LSAH, which is a long-term program investigating whether the unique occupational exposures of astronauts are associated with increased health risks. Such studies are particularly relevant regarding the issues of radiation exposure.

3.2   Environmental Health during Space Missions

During space mission, the medical care does not only focus on health maintenance and disease prevention, but also on environmental issues. Spacecrafts are closed compartments, and therefore standards for air, water, microbiology, toxicology, radiation, noise, and habitability must be established. In-flight environmental monitoring systems are available to prevent crew exposure to toxicological and microbial contamination of internal air, water, and surfaces; to radiation sources from within and external to the spacecraft; to vibration and noise. These systems must have near real-time and archival sampling, and provide a mechanism to alert crewmembers when measured values are outside acceptable limits.

Habitability of spacecraft is vitally important to crew health, well being, and productivity, especially as mission duration increases. Habitability issues regarding human presence in space includes human factor design considerations (colors, equipment layout, and hardware design), adequate and ergonomically correct work and living volume, with similarly adequate stowage volume. Areas must be designed that allow for restful sleep and personal space, with adequate lighting and exterior views. Schedules must produce interesting work, with sufficient rest and recreation periods to avoid chronic fatigue. Ideally, each crewmember should have private time and physical space for fitness and recreation, in order to keep his/her motivation.

Time and resources are set aside for personal hygiene and sanitation (see Chapter 7, Section 5.1). In addition, healthy, palatable variety of food and beverage is provided (Table 1-03). The daily food supply totals a high 3000 calories, plus snacks. The meals also attempt to compensate for the body's tendency to lose essential minerals in microgravity, such as potassium, calcium, and nitrogen.

Thermostabilized
Heat processed foods ("off-the-shelf" items) in aluminum or bimetallic tins and retort pouches.
Irradiated
Foods preserved by exposure to ionizing radiation and packed in flexible foil laminated pouches.
Intermediate Moisture
Dried foods with low moisture content such as dried apricots. Packed in flexible pouches.
Freeze Dried
Foods that are prepared to the ready-to-eat stage, frozen and then dried in a freeze dryer that removes the water by sublimation. Freeze-dried foods such as fruits may be eaten as is while others require the addition of hot or cold water before consumption.
Re-Hydratable
Dried foods and cereals that are re-hydrated with water produced by the Shuttle Orbiter's fuel cell system. Packed in semi-rigid plastic container with septum for water injection.
Natural Form
Foods such as nuts, crunch bars, and cookies. Packed in flexible plastic pouches.
Beverages
Dry beverage powder mixes packed in re-hydratable containers.

Table 1-03. The Space Shuttle menu currently features more than 70 food items and 20 beverages. Shuttle crewmembers have a varied menu every day for six days. Each day, three meals are allowed, with a repeat of menus after six days. The pantry also provides plenty of foods for snacks and between meal beverages and for individual menu change. (Source NASA)

At the same time, the meals must be attractive, not like the early missions when astronauts had to suck their meals out of "dentifrice" tubes or plastic bags without being able to see or smell the food. Nowadays, attention is given to individual crewmember preference with regard to palatability and nutritional adequacy of food items during missions.

Medical and psychological personnel have also an opportunity to review all design considerations early in the design process to ensure that spacecraft design and support systems meet medical and psychological requirements.

3.3   Human Mars Mission

As in the ISS program, the political objectives of human missions to Mars will presumably focus on a large-scale international cooperation. Only cooperation in the realization of such a multi-billion dollar program with a corresponding distribution of tasks and costs appears to be a viable option. A human Mars mission can also be regarded as an important cultural task for humankind with the objective to globalize the view of our home planet Earth, thereby contributing to the solution of local conflicts. In any case, a human Mars mission would meet the natural human need to explore and expand to new regions.

Obviously, for a cost-effective Mars exploration, an appropriate combination of unmanned and manned activities supplementing each other in a logical way will be developed (e.g., selection of landing site by unmanned precursor missions). In this context the development and test of technologies for in-situ resources utilization (e.g., propellant from the CO_2 of the Mars atmosphere or from the water ice) will also be taken into account, which would allow even more cost-effective missions in the future.

According to conventional spacecraft configurations that have mainly been designed until 1995, the total departure mass of a manned spacecraft in low Earth orbit is around 1000 tons. That is if the mission is carried out in one-shot with a 4-stage expendable vehicle on a low energy transfer, using oxygen and hydrogen propellants for the main propulsion systems. Newer spacecraft designs (like the NASA Reference Mission V3.0 from 1997, based on Dr. Robert Zubrin "Mars Direct" Concept) lead to lower total masses of about 400 tons, based on a split mission design and on-site propellant production plants.

Using the current rocket technology, traveling between Earth and Mars will require lots of fuel and good timing. The most fuel-efficient trajectory occurs when Earth is at a six o'clock position at launch and Mars is at about four o'clock—a juxtaposition that occurs just once every 26 months. The first leg will take 256 days. Astronauts must wait on Mars for their launch toward home until Earth is in alignment. Total mission duration is then 972 days (Figure 1-19).

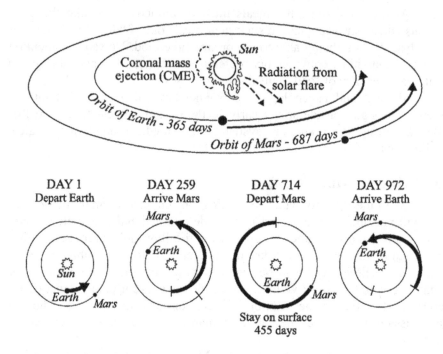

Figure 1-19. One possible scenario for a Mars mission. Top: Schematic of the orbits of Earth and Mars showing their position for a more fuel-efficient trajectory during launch. Bottom: The respective positions of the Earth and Mars determine both the duration of travel and stay on the Mars surface. (Adapted from National Geographic 2001)

Luckily, the Mars gravity of 0.38 g might act as a countermeasure to the physiological deconditioning that will take place during the trip from the Earth to Mars. However, landing maneuvers on Mars and Earth are characterized by maximum g-loads of up to 6 g due to the atmospheric drag. If the interplanetary cruise is carried out a zero gravity level (i.e., if no artificial gravity is provided within the spacecraft) such high g levels in deconditioned astronauts appear critical for the health of the crew.

Beyond the shield of Earth's magnetosphere, solar and galactic radiation can cause severe cellular damage or even cancer (see Chapter 7). The crew needs to be protected against the occasional solar flare. This can be done with a "storm shelter", e.g., with food racks and water tanks packed around the walls to absorb the radiation. Fortunately, most of a solar flare's energy is in alpha and beta particles that can be stopped with a few centimeters of shielding.

Cosmic rays are a different story. They are constantly present, coming from all directions. The radiation consists of heavy, slow moving atomic nuclei that can do far more damage to more cells than alpha and beta particles.

This radiation requires several meters of shielding for complete blockage, and since the nuclei come from all directions at all times, unlike the brief solar flares that last only a few hours or days, a storm shelter would be insufficient to protect the crew. However, the permanent habitats of the Mars base can be covered with thick layers of soil to provide full-time radiation protection, so nearly all the crew's radiation exposure would occur during the year of interplanetary travel. Even if such a system proves difficult to engineers, some scientists believe that the cosmic ray doses can simply be endured. Exposure to a thin, continuous stream of radiation does far less damage than an equal magnitude of radiation delivered in one day. There is still the possibility of cancer, but this probability is rather low.

According to the scenario shown in Figure 1-19, the astronauts would spend about nine months traveling to Mars, fifteen months on the surface, and nine months returning to Earth. Fifty rem (0.5 Sievert) per crewmember is one estimate for total exposure in that time. This dose leads to a 1% increase in probability of contracting a fatal cancer later in life, compared to an already existing 20% cancer risk for non-smokers on Earth, and would probably be acceptable to the volunteers on this mission. However, since the biological effects of cosmic radiation are poorly understood, the resulting cancer risk may conceivably be off by as much as a factor of 10, and thus jump to 10%, or drop to 0.1%.

Not much research can be done safely on Earth to investigate these radiation effects, since cosmic rays are difficult to generate, and no one would consent to being exposed to a theoretically fatal dosage. The International Space Station could provide a good testing ground, since large numbers of astronauts will be exposed to modest amounts of radiation in their six-month tours of duty, but a full investigation might require waiting decades until these astronauts retire and die either of natural causes or of cancer. Obviously Mars mission advocates have no intention of waiting that long. It actually makes the most sense to accept the radiation risk on the Mars mission, since after all this is a journey into the unknown, and the risk of radiation is mild compared to the dangers that explorers on Earth have faced, and overcome, in the past (Reifsnyder 2001).

3.4   Countermeasures

We will see in the following chapters that the changes in human physiology during spaceflight are appropriate adaptations to the space environment. They are not life threatening for at least 14 months, which is the longest period that humans have been in space. That's the good news. The bad news is that adaptation to space creates problems upon returning to Earth. Difficulty in standing, dizziness, and muscle weakness present problems after landing. Therefore, appropriate countermeasures must be developed that balance health risk against mission constraints, and particularly the limited resources regarding medical care possibilities.

Countermeasures refer to the application of procedures or therapeutic (physical, chemical, biological, or psychological) means to maintain health, reduce risk, and improve the safety of human spaceflight. The countermeasures typically aim at:
- Eliminating or preventing adverse and harmful effects on crew health. Examples include the provision of a substitute gravitational effect on orbit (artificial gravity), thus preventing microgravity from degrading the health of the astronauts;
- Mitigating the effect of adverse or harmful agents or enhancing the astronaut's ability to ward off the effects of these agents. Examples include preflight and in-flight exercise to counteract the effects of microgravity, in-flight administration of medications to prevent space motion sickness, spacecraft design changes to minimize radiation exposure;
- Minimizing the effect of adverse or harmful agents on the crew once mal-adaptation, disease, or injury has been identified. Examples include fluid loading to minimize postflight orthostatic intolerance (Figure 1-20), or a postflight rehabilitation program to reverse space mission-induced musculo-skeletal or cardio-vascular deconditioning.

Figure 1-20. A Shuttle crewmember prepares containers of drinking water and salt tablets to be consumed by his crewmates prior to reentry. Fluid loading is a standard procedure on all Shuttle flights, as an effective countermeasure to orthostatic intolerance upon return to Earth gravity. (Credit NASA)

Preflight countermeasures include activities to support appropriate crew selection and psychological training, fitness and exercise, physiological adaptive training, health stabilization program, and circadian shifting.

In-flight countermeasures include those activities necessary to maintain physiologic balance and health, mental and behavioral health, nutritional health, and physical fitness and mission performance. Typical physical exercise includes cycling, running on a treadmill, or rowing.

Postflight countermeasures include those activities necessary to assist the crewmembers in a return to preflight physical, physiological, and behavioral health baselines. Examples of countermeasures include, but are not limited to, circadian rhythm shifting, hormone replacement, and physical exercise.

For long-duration missions, the Mir and ISS experience indicates that current in-flight countermeasures are not optimal, to say the least. Evidence for this is provided by the images of the cosmonauts unable to stand immediately after returning to Earth after a long-duration stay on board Mir. They are helped from the spacecraft and "ceremoniously hauled around like nabobs in sedan chairs" (Figure 1-21). The situation with the ISS has not changed much. Astronauts on board the ISS exercise on a cycle-ergometer or a treadmill for two hours per day. Considerable muscle and bone loss is still observed after landing (see Chapter 5). We cannot ethically request that they exercise three hours a day. Consequently, it is largely admitted that using the current countermeasure methods, humans would not be operational after landing on Mars.

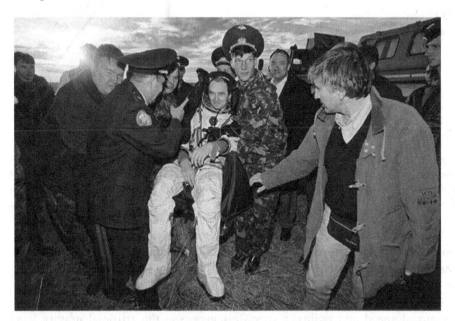

Figure 1-21. French Cosmonaut Jean-Pierre Haigneré is transported by ground personnel after returning from a six-month stay aboard the Russian space station Mir. (Credit CNES)

3.5 Artificial Gravity

One possible countermeasure to the effects of weightlessness is the use of artificially-produced gravity on board the spacecraft. Artificial gravity could be accomplished either through rotation of the entire vehicle or by the inclusion of an onboard centrifuge.

During rotation about an eccentric axis the resulting centrifugal force provides an apparent gravity vector. The centrifugal force produced by rotation is dependent upon two parameters of the rotating structure, the square of its angular velocity (ω^2) and its radius r (Figure 1-22).

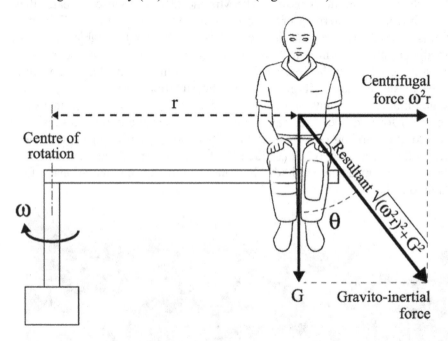

Figure 1-22. Rotating a subject at angular velocity ω, in a horizontal plane at a distance r from the center of rotation, produces a centrifugal force of $\omega^2 r$ and a resultant force of $[(\omega^2 r)^2 + g^2]^{1/2}$ acting at an angle θ to his vertical mid-body axis. Tan $\theta = (\omega^2 r/g)$. (Adapted from Howard 1982)

On Earth the centrifugal force combines with the gravitational force, and the resulting vector, the so-called gravito-inertial force, is both larger in magnitude than the centrifugal force itself, and tilted with respect to gravity. In microgravity, the subject will only be exposed to a centrifugal force, referred to as artificial gravity. Since the centrifugal force depends on both rotational speed and radius, a specific increase in the artificial gravity level can be achieved either by increasing the radius, or by increasing angular velocity. This translates to a trade-off between cost and complexity (which depends mostly on the radius of the structure) versus the physiological and psychological concerns (both of which depend mostly on its angular velocity) (Diamandis 1997).

One significant drawback of this technique is the Coriolis force that is generated every time a linear motion is attempted in any plane not parallel to the axis of rotation. The Coriolis force has a magnitude of $2\omega V$; where ω is the angular velocity for the rotating environment and V is the linear velocity of the moving object. When attempting any linear movement out of the plane of rotation, the Coriolis force combines with the centrifugal force to produce a different apparent gravity vector in magnitude alone or in both magnitude and direction. To a human in a rotating environment, this vector may be manifest in two ways. First, it adds to the apparent weight of a body moving in the direction of rotation and subtracts from the apparent weight when moving against the direction of motion. Second, when a body moves toward the center of rotation, the Coriolis force is exerted in the direction of rotation at right angles to the body's motion; when moving away from the center of rotation the force is opposite to the direction of rotation. By contrast, a motion parallel to the axis of rotation will generate no Coriolis force (Stone 1973).

The Coriolis forces affect not only whole-body movements, but also the vestibular system (see Chapter 3). Rotation of the head out of the plane of rotation generates cross-coupled angular accelerations that induce stimulation of all three semicircular canals. Such head movements in a stationary environment do normally not stimulate some of the canals, and this results in illusory sensations of bodily or environmental motion. Nausea and vomiting may result after a few head movements, particularly if the angular velocity of the centrifuge is high. Based on ground-based studies, it has been postulated that the lightest acceptable system for providing artificial gravity using a rotating spacecraft would be one having a radius of rotation of about 14 m, rotating at 6 rpm, and providing 0.58 of Earth gravity (Thompson 1965).

However, a rotating spacecraft presents serious design, financial, and operational challenges for a maneuvering station. Also, head movements and resultant Coriolis and cross-coupled accelerations on a rotating spacecraft may limit the usefulness of centrifugation for other than brief periods of intermittent stimulation. From a practical perspective, it is very likely that humans do not need gravity (or fraction of it) 24 hours a day to remain healthy. If intermittent gravity is sufficient, we won't need a permanently rotating spacecraft to produce a constant gravity force. An onboard human short-arm centrifuge presents a realistic near-term opportunity for providing this artificial gravity.

Several designs for onboard short-arm centrifuge have been proposed. I was the Principal Investigator of an experiment using a human-rated centrifuge generating 0.5 g or 1 g along the longitudinal body axis, which flew aboard the Neurolab mission of the Space Shuttle (STS-90) in 1998

(Figure 1-23). This experiment was (and remains) the first in-flight evaluation of artificial gravity on astronauts. The results of this experiment, described in more details in Chapter 3 (Section 2.1), suggested that centrifugal force of 0.5 g was well tolerated by the crew, and that cardio-vascular deconditioning was reduced in those astronauts who rode the centrifuge 20 minutes every other day during a 16-day space mission.

Figure 1-23. This 0.65-m radius centrifuge was developed by the European Space Agency for the Neurolab mission (STS-90) to investigate the adaptation of the vestibular system in detecting changes in motion and orientation in weightlessness. (Credit NASA)

Scientists at NASA Ames Research Center are currently conducting tests by having healthy volunteers exercise on a treadmill on a short-arm centrifuge (Figure 1-24). Their approach is to determine whether exercising under increased gravitational forces will decrease the amount of time required to maintain health and fitness. If the results prove positive and the amount of in-flight exercise are reduced by centrifugation, then such device could be a good candidate countermeasure for the ISS or the spacecraft en route to Mars.

3.6   A New Science is Born

Space life sciences is a young science, having come into existence with the first studies carried out on animals during the first suborbital flight less than 60 years ago. Since then, people have visited the Moon and have lived in space for about the period planned for a mission to Mars. Still, our understanding of how spaceflight affects living organisms remains rudimentary.

Today's opportunities for carrying out space research are scarce, since the Spacelab is retired and the International Space Station is not yet equipped with all the planned biological research facilities (see Chapter 2, Section 6). Some experiments can still be performed during Shuttle missions and on board ISS, but crew time is limited by the operations of construction, assembly and maintenance.

When the ISS will be fully operational (hopefully with a permanent six or seven-crew), it should prove an ideal platform for studying fundamental biological processes in microgravity. This will undoubtedly lead to the growth and development of a new science: gravitational physiology.

Figure 1-24. Platform assembly developed by NASA Ames Research Center. The platform rides on roller bearings and supports two cycle couches. The three pedaling stations, two on board at the end of each couch and the third off board, transform subjects and operators pedaling power to platform rotation. The supine subjects are oriented with their heads at 26 cm from the center of rotation. Centrifugal force of 5 g can be generated along the body longitudinal axis (at a distance of 168 cm from the center of rotation). (Greenleaf et al. 1997)

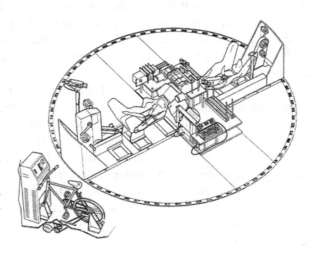

Gravity affects some materials and fluids dynamics. It is required for convective mixing and other weight-driven processes, such as draining of water through soil, and assuring that what goes up comes down. One might predict that plants would grow taller without gravity, yet the lack of gravity might facilitate increased levels of growth-inhibitory or aging environmental factors around the plants, thereby causing them to dwarf. Gravity also has a role to play in development of load-bearing structures. The scaling effect of gravity is well known: the percentage of body mass contributing to structural support is proportional to the size of a land animal (e.g., 20 g mouse = ~5%, 70 kg human = ~14%, and 7000 kg elephant = ~27%). This scaling effect in land animals would likely change in space and could result in a static scale comparable to marine mammals on Earth (~15% of mass as supporting tissues over a wide range of weights). Legs are bothersome in space and not only get in the way but also are involved in the fluid shifts that occur early in-flight. Whether legs would disappear over time without gravity (perhaps similar to the extraterrestrial ET) or become more like grasping talons is unknown. Form follows function and as function changes, so will form. How much change and what form organisms will assume over time in space is unknown (Morey-Holton 1999).

Data to date suggest that certain biological structures have evolved to sense and oppose biomechanical loads, and those structures occur at the cellular level as well as at the organism level. There is evidence that the musculo-skeletal system of vertebrates change following acute exposure to space (see Chapter 5, Section 1). What will happen over multiple generations is speculative. The "functional hypothesis" theory suggests that what is not used is lost. If this theory holds over multiple generations in space, then gravity-dependent structures may ultimately disappear or assume a very different appearance in space. The next chapter reveals that we only have short snapshots of how small living organisms actually change in the space environment.

Another example is plants. Plants are the first organisms to be raised to seeds in space, from seeds that were themselves raised in microgravity. We now know that plants can grow in space, but the Mir studies have indicated that air and water require special management in microgravity. Further studies in this area are of paramount importance if one wants to move from the current physico-chemical to ecological controlled life support systems.

Carrying out research in space often comes at a considerable cost (sometimes human, as demonstrated by the recent *Columbia* tragic event). The most striking difficulties are the small subject pool available, the lack of adequate controls, and the fact that science is (by necessity) secondary to mission safety when conducting experiments in such a hostile environment. Nevertheless, the success of the manned space program is dependent on the concomitant success of life science research in microgravity to solve the considerable dangers still faced by crewmembers on long-duration missions.

In this respect, the human Mars mission represents another fascinating challenge for space medicine. Such mission, when it will be undertaken, will probably be the longest period of exposure to a reduced gravitational environment, and probably the longest period away from Earth too. A recent report also suggested that radiation on Mars might be at much higher levels that previously believed. So high in fact, that it would make living there almost impossible. All together, these conditions make a human Mars mission a challenge from both the physiological and psychological points of view.

4   REFERENCES

Barany R (1906) Untersuchungen über den vom Vestibularapparat des Ohres reflektorisch ausgelösten rhythmischen Nystagmus und seine Begleiterscheinungen. *Monatschr Ohrenkeilkde* 40: 193-207

Barratt M (1995) A Space Medicine Manifesto. Notes of the Lecture on *Current and Future Issues in Space Medicine* presented at the Summer Session of the International Space University, Stockholm

Churchill SE (1999) Introduction to human space life sciences. In: Houston A, Rycroft M (eds.) *Keys to Space. An Interdisciplinary Approach to Space Studies*. Boston, MA: McGraw Hill, pp. 1813- 1821

Clarke A (1951) *The Exploration of Space*. London, UK: Temple Press

Collins M (1990) *Mission to Mars*. New York, NY: Grove Weidenfeld

DeHart RL (ed) (1985) *Fundamentals of Aerospace Medicine*, Philadelphia, PA: Lea and Febiger

Diamandis PH (1997) Countermeasures and artificial gravity. In: Churchill SE (ed) *Fundamentals of Space Life Sciences*, Malabar, FL: Krieger Publishing Company, Volume 1, Chapter 12, pp 159-175

Greenleaf JE, Gundo DP, Watenpaugh DE, Mulenburg GM, McKenzie MA, Looft-Wilson R, Hargens AR (1977) Cycle-powered short radius (1.9 m) centrifuge: Effect of exercise versus passive acceleration on heart rate in humans. Houston, TX: National Aeronautics and Space Administration, *NASA TM-110433*, pp 1-10

Henry JP (1963) Synopsis of the Results of the MR-2 and MA-5 Flights. In: *Results of the Project Mercury Ballistic and Orbital Chimpanzee Flights*, Houston, TX: National Aeronautics and Space Administration, NASA SP-39, Chapter 1

Howard IP (1982) *Human Visual Orientation*. Chichester, NY: John Wiley & Sons

Lujan BF, White RJ (1994) *Human Physiology in Space*. Teacher's Manual. A Curriculum Supplement for Secondary Schools. Houston, TX: Universities Space Research Association

Morey-Holton ER (1999) Gravity, a weighty-topic. In: Rothschild L and Lister A (eds) *Evolution on Planet Earth: The impact of the Physical Environment*, New York: Academic Press

Mullane MR (1997) *Do Your Ears Pop in Space?* New York, NY: John Wiley & Sons

Newton I (1687) *Principia*, Chapter VII, Book III

Nicogossian AE, Parker JF (1982) *Space Physiology and Medicine*, NASA SP-447. Washington, DC: National Aeronautics and Space Administration Scientific and Technical Information Branch

Reifsnyder R (2001) Radiation Hazards on a Mars Mission. MIT Mars Society Youth Chapter. *The Martian Chronicles* 8. Available at: http://chapters.marssociety.org/youth/mc/issue8/index.php3

Reschke MF, Sawin C (2003) *A historical review of physiological research in space flight*. Paper presented at the 14th IAA Humans in Space Symposium, Banff, Alberta, Canada, 18-22 May 2003

Rogers MJB, Vogt GL, Wargo MJ (1997) *The Mathematics of Microgravity*. Washington, DC: National Aeronautics and Space Administration, NASA Educational Brief EB-1997-02-119-HQ

Scherer H, Brandt U, Clarke AH, Merbold U, Parker R (1986) European vestibular experiments on the Spacelab-1 mission: 3. Caloric nystagmus in microgravity. *Experimental Brain Research* 64: 255-263

Stone RW (1973) An overview of artificial gravity. *Fifth Symposium on the Role of the Vestibular organs in Space Exploration*. 19-21 August 1970, Naval Aerospace Medical Center, Pensacola, FL, NASA SP-314, pp 23-33

Stringly LE (1962) *Countdown and Procedures for Project Mercury. Atlas-5 Flight (Chimpanzee Subject). Final Technical Documentary Report*. Holloman AFB: Aeromedical Research Laboratory Technical Documentary Repository 62-17

Thompson AB (1965) Physiological design criteria for artificial gravity environments in manned space systems. *First Symposium on the Role of the Vestibular organs in the Exploration of Space*, 20-22 January 1965, US Naval School of Aviation Medicine, Pensacola, FL, NASA SP-77, pp 233-241

Thornton WE, Moore TP (1987) Height changes in microgravity. In *Results of the Life Sciences DSOs conducted aboard the Space Shuttle 1981-1986*. Space Biomedical Research Institute, Houston, TX: NASA Johnson Space Center, pp 55-57

Tillet, Brisson, Cadet, Lavoisier, Bossut, de Condorcet, Desmarest (1783) *Rapport Fait à l'Académie des Sciences sur la Machine Aérostatique de Mrs. De Montgolfier*

Wolfe T (1979) *The Right Stuff*. New York, NY: Farra, Straus & Giroux

Zubrin R, Wagner R (1996) *The Case for Mars: The Plan to Settle the Red Planet and Why We Must*. New York, NY: A Touchstone Book, Simon and Schuster

Chapter 2

SPACE BIOLOGY

Gravity provides a directional stimulus which plays an important role in basic life processes in the cell, such as biosynthesis, membrane exchange, and cell growth and development. Likely, growth and development of plants are determined by hormones whose transport is also influenced by gravity. Will these functions develop normally when deprived of the gravitational stimulus? This chapter will review the fundamental questions raised by the space environment in the areas of gravitational biology, developmental biology, plant biology, and radiobiology.

Figure 2-01. One application of space research is to improve overall health of people of all ages.

1. WHAT IS LIFE?

It is generally admitted that, for scientific purposes, an object must meet six criteria to be considered alive: movement (even plants move: stems shoot upward, flowers open and close, and leaves follow the movement of the Sun); organization (animals and plants have organs, whose structure is nearly identical within the same species); homeostasis (the ability to maintain constant conditions within the body); energy (all living things absorb and *use* energy); reproduction, and growth (during the growth process, cells not only increase in number but they also develop into different types of cells that are needed to form the organs and tissues of the new individual) (DuTemple 2000).

1.1 Life on Earth

The planet Earth is thought to be 4.6 billion years old. The first life form appeared about 4 billion years ago by the spontaneous aggregation of molecules that rapidly evolved into microscopic, relatively simple cells. Over the following years, these primitive cells evolved into at least 10 million different species, which represent Earth's existing biological diversity. All organisms, including animals, plants, fungi, and an untold collection of microbial species, have their common ancestral roots within these earliest life forms (Figure 2-02).

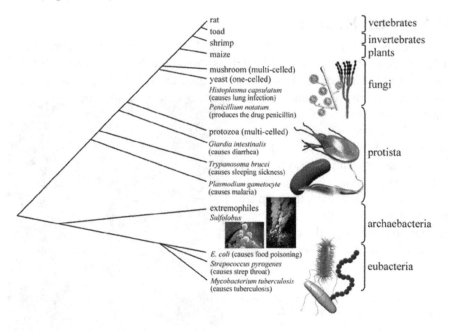

Figure 2-02. Evolution of organisms deduced from their gene sequences.

Chemical and fossil evidence indicates that life on Earth as we know it today evolved by natural selection from a few simple cells, called *prokaryotes* because they lacked nuclei. The earliest prokaryotes probably already had mechanisms that allowed them to replicate their genetic information, encoded in nucleic acids, and to express this information by translation into various proteins. Typical prokaryotic cells are bacteria (Figure 2-03). They are small, with relatively simple internal structures containing desoxyribo nucleic acid (DNA), proteins, and small molecules. They replicate quickly by simply dividing in two (a single cell can divide every 20 minutes and thereby give rise to 5 billion cells in less than 11 hours). Their ability to divide quickly (growth rate) enables these cells to adapt rapidly to changes in their environment. Bacteria can utilize virtually any type of organic molecule as food (including sugars, amino acids, fats, hydrocarbons) and get their energy (adenosin triphosphate, ATP) from chemical processes in absence or presence of oxygen.

About 1.5 billion years ago there appeared larger and more complex cells such as those found in "higher" organisms: the unicellular protists, fungi, plants, and animals we know today. The important organelles of energy metabolism (plastids and mitochondria) originated 2 to 1.5 billion years ago through the symbiosis of prokaryotes. In this process, bacteria having one set of specialized functions were engulfed by host cells with complementary requirements and functions. These eukaryotic cells (or *protozoa*) have a nucleus, which contains the cell's DNA and cytoplasm where most of the cell's metabolic reactions occur. They get their ATP from aerobic oxidation of food molecules (respiration) or from sunlight (photosynthesis). Consequently, more than 2 billion years ago, the biota had used the process of photosynthesis to create an oxidizing atmosphere from one previously poor in oxygen. Carbon dioxide was also removed from the atmosphere in the form of carbonate precipitates. A myriad of bacteria, mollusks, corals, and other organisms contributed to vast limestone deposits and continue to do so today. With these and other processes, the Earth biosphere has transformed a once sterile planet, intermediate in character between Venus and Mars, into the living planet we now enjoy.

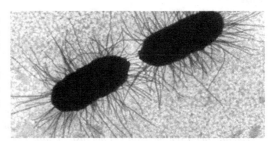

Figure 2-03. Escherichia coli is the most known bacteria. It is characterized by rudimentary chromosomes, rapid generation time, and a well-defined life cycle. Like other bacteria, E. coli is able to generate new mutants when challenged by its environment.

Bacteria have been detected or isolated from many hostile environments on Earth, including the dry, extremely cold surfaces and interstices of rocks in the dry valleys of the Antarctic, hot environments associated with submarine and terrestrial volcanoes and geothermal systems, and deep subsurface sediments and aquifers. Investigations in extreme terrestrial environments are in their infancy, and we still know little about either most of the organisms inhabiting these environments (also called *extremophiles*) or in many cases the geochemistry and geophysics of the environments themselves.

Nevertheless, in the last decade or so, a variety of novel organisms have been isolated. They include *hyperthermophiles* capable of growing at 110°C, *barophiles* capable of growing at the pressures found in the deepest ocean trenches, and *anaerobes* capable of using iron, manganese, or even uranium as electron acceptors. Similarly, a variety of strategies have been identified by which microorganisms can survive environmental conditions that do not allow growth, including low temperature and low nutrient conditions (Table 2-01).

Physiological Characteristic	Description
Temperature	
• Psychrophile	• Optimal temperature for growth is 15°C or lower (range:0-20°C)
• Psychrotroph	• Capable of growing at 5°C or below
• Mesophile	• Optimal temperature for growth is 37°C (range: 8-50°C)
• Thermophile	• Grows at 50°C or above
• Hyperthermophile	• Grows at 90°C or above (range: 80-113°C)
Oxygen	
• Aerobe	• Can tolerate 21%t oxygen present in an air atmosphere and has a strictly respiratory-type metabolism
• Anaerobe	• Grows in the absence of oxygen
• Facultative anaerobe	• Can grow aerobically or anaerobically—characteristic of a large number of genera of bacteria including coliforms such as Escherichia coli
• Microaerophile	• Capable of oxygen-dependent growth but only at low oxygen levels
pH	
• Acidophile	• Grows at pH values less than 2
• Alkalophile	• Grows at pH values greater than 10
• Neutrophile	• Grows best at pH values near 7
Salinity	
• Halophile	Requires salt for growth: classified as extreme (all are archaea) or moderate halophiles (15 to 20% NaCl)
Hydrostatic pressure	
• Barophile	• Obligate barophiles, no growth at 1 atmosphere of pressure; barotolerant bacteria, growth at 1 atmosphere but also at higher pressures. A number of deep-sea bacteria are called barophilic if they grow optimally under pressure and particularly if they grow optimally at or near their in situ pressure
(100 atm per 1,000-m depth) (0.987 atm = 1 bar = 0.1 megapascal)	
Nutrition	
• Autotroph	• Uses carbon dioxide as its sole source of carbon
• Heterotroph	• Unable to use carbon dioxide as its sole source of carbon and requires one or more organic compounds
• Chemoorganoheterotroph	• Derives energy from chemical compounds and uses organic compound
• Chemolithoautotroph	• Relies on chemical compounds for energy and uses inorganic compounds as a source of electrons
• Mixotroph	• Capable of growing both chemo-organo-hetero-trophically and chemolithoautotrophically; examples include sulfur-oxidizing bacteria
• Oligotroph	• Can develop on media containing minimal organic material (1 to 15 micrograms carbon per liter)
• Copiotroph	• Requires nutrients at levels 100 times those of oligotrophs

Table 2-01 Microorganisms with particular physiological and nutritional characteristics. (Source NASA)

An interesting, although alarming, discovery was made during the Apollo program. The Apollo-12 Lunar Module landed on the Moon about 200 m away from an unmanned probe, Surveyor-3, which had landed there 2.5 years earlier. The astronauts of Apollo-12 inspected the Surveyor spacecraft for damages and recovered an external camera for detailed analysis back on Earth. A specimen of bacteria (*Streptococcus mitis*) was found alive on the camera. Because of the precautions the astronauts had taken, it is almost sure that the germs were there before the probe was launched. This poses the problem of the contamination of other planets by Earth biotope.

These bacteria had survived for 31 months in the vacuum of the Moon atmosphere while exposed to considerable solar and cosmic radiation. They suffered huge monthly temperature swings and the complete lack of water. Just like if they had hibernated. In fact, freezing and drying, in the presence of the right protectants, are actually two ways normal bacteria can enter a state of suspended animation. And interestingly, if the right protectants are not supplied originally, the bacteria that die first supply them for the benefit of the surviving ones!

Likewise, spores of the Bacillus bacteria were found during the summer of 2000 in salt crystals buried 600 meters below ground at a cavern in New Mexico. When they were extracted from the crystals in a laboratory and placed in a nutrient solution, the microorganisms revived and began to grow. These bacteria had survived in a state of suspended animation for 250 million years. Until now, the world's oldest living survivors were thought to be 25-40 million-year-old bacteria spores discovered in a bee preserved in amber. Traditionally, endospore and cyst development were considered the principal mechanisms for long-term survival by microorganisms, but it is now clear that many microorganisms have mechanisms for long-term survival that do not involve spore or cyst formation.

1.2 Life on Mars

Without exception, life in Earth's biosphere is carbon-based and is organized within a phase boundary or membrane that envelops reacting biomolecules. Every documented terrestrial cellular life form is a self-replicating entity that has genetic information in the form of nucleic acid polymers (DNA) coding for proteins. Biologically active systems require at a minimum liquid water, carbon, nitrogen, phosphate, sulfur, various metals, and a source of energy either in the form of solar radiation or from chemosynthetic processes.

The conditions that nurtured early self-replicating systems and their transition into microbial cells are speculative. In contrast, it is much easier to model the early stages of evolution. Origins-of-life experiments have outlined the synthesis of the basic building blocks of life, including amino acids, nucleotides, and simple polypeptides and polynucleotides. Yet creation of self-sustaining, self-replicating biological entities capable of evolution has not yet been achieved in the laboratory, and even if successful would not necessarily mimic how life started on Earth or in other parts of the Universe.

For life to originate, the presence of liquid water and a source of utilizable free energy are necessities. The synthesis and polymerization of basic organic building blocks of life on Earth eventually led to self-replicating nucleic acids coding for proteins, but the earliest replicating systems were not necessarily composed of amino acids and nucleotides. If extraterrestrial biological systems exist, their modes of information storage, retrieval, and processing and their enzymatic activity may not be identical to those of biological entities on Earth. Understanding this prebiotic evolution is one of the major goals of the *astrobiology* program, that is the biology of the early Earth and elsewhere in the Universe.

In the search for extra-terrestrial life, microbes are far more likely than multicellular organisms to retain viability on small solar system bodies because they can adapt to a much wider range of environmental conditions. As mentioned above, single-cell organisms such as bacteria have infiltrated virtually every corner of Earth's biosphere and still constitute the bulk of the Earth's biomass. They grow in temperate marine and terrestrial settings, within other microbial or multicellular organisms, in deep subsurface niches, and in extreme environments that would be lethal for other life forms. They often influence geochemical reactions within the biosphere and frequently play key roles in food chains and complex ecosystems.

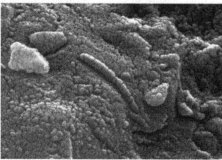

Figure 2-04. Left: ALH84001 is by far the oldest Martian meteorite, with a crystallization age of 4.5 billion years. Right: The small amount of carbonate in ALH84001 is the center of attention concerning the possibility of life on Mars. (Credit NASA)

Figure 2-04 (left panel) shows a 4.5 billion-year-old rock which is a portion of a meteorite (ALH84001) that was dislodged from Mars and that fell to Earth in Antarctica about 16 million years ago. It is believed to contain fossil evidence that primitive life may have existed on Mars more than 3.6 billion years ago. The small grains on the right panel in Figure 2-04 appear to have formed in fractures inside this rock in the presence of liquid water or other fluid. There is considerable debate about the origin of these carbonates. These grains are the sites of the three types of evidence that McKay and his colleagues (1996) suggest represent fossil life on Mars.

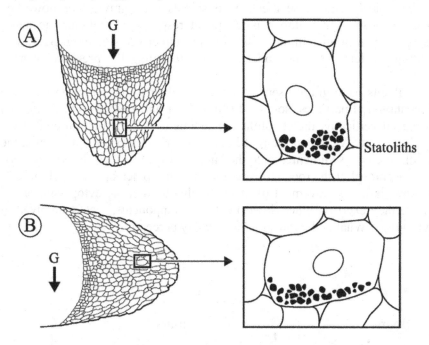

Figure 2-05. Schematic of the microscopic view of the top of a of a Zea maize root cap showing the sedimented statoliths (black particles) at the bottom of the cells (A) and their migration onto the lateral walls of the cells in the direction of the gravitational force on Earth (B). (Adapted from Wilkins 1989)

2 GRAVITATIONAL BIOLOGY

Throughout its entire evolution, life on Earth has experienced only a 1-g environment. The influence of this omnipresent force is not well understood, except that there is clearly a biological response to gravity in the structure and functioning of living organisms. Gravitational biology aims to understand the molecular mechanisms whereby a cell detects gravity and converts this signal to a neuronal, ionic, hormonal, or functional response.

2.1 Questions

How cells, as single unicellular organisms or as the basic units of multicellular organisms, are sensitive to gravity (*gravitropism*)? How do plant cells detect the gravity vector and transform this force into hormonal and non-hormonal signals?

Changes in the physical environment surrounding cells, in vivo or in vitro, can lead indirectly to changes within the cell. Little is known about if or how individual cells sense mechanical signals (i.e., gravity) or how they transduce those signals into a biochemical response. A cellular mechano-sensing system might initiate changes in numerous signaling pathways. Spaceflight offers a unique opportunity for revealing the presence of such system.

Plants have gravity-sensing organs in their roots, which involve the sedimentation of particles (so-called *statoliths* or *amyloplasts*). On Earth, in a root placed vertically, the statoliths are sedimented at the bottom end of the cell (Figure 2-05A). When the root is placed horizontally for three hours, the statoliths are now sedimented onto the lateral walls of the cell (Figure 2-05B).

Removal of the root abolishes the capacity to detect gravity (Figure 2-06). Now, is it the movement of the statoliths through the cytoplasm, or the pressure they exert on other (lower) cellular components, that is involved in graviception? What is the threshold for gravity perception?

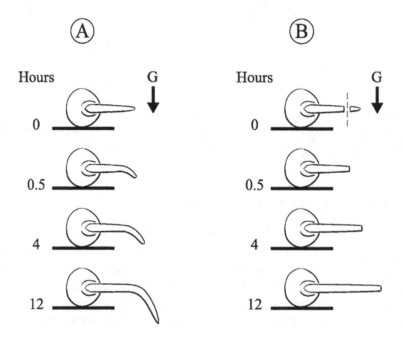

Figure 2-06.A. Time sequence showing the gravitropic response on Earth of the seedling root of Zea maize placed horizontally at time zero (top) and photographed after 0.5 hr, 4 hr, and 12 hr. B. After removal of the root cap, the seed growths straight for at least 12 hr (bottom) after having been placed in the horizontal position. (Adapted from Wilkins 1989)

It is now known that this gravitropic response utilizes growth-stimulating hormones, such as gibberellin and indoleacetic acid, and an inhibiting hormone, abscisic acid. These same hormones, along with electric current, are also involved in phototropism; thus, the interactions between the two responses are complex. Only spaceflight conditions can help distinguish between the two mechanisms.

The threshold of gravity sensing systems in plants and animals is unknown, but there are reasons for believing that, in certain organisms, it is in the range of 10^{-5} to 10^{-6} g. It is calculated that, at these gravity levels, sedimentation and thermal convection are no longer operative. These two physical factors are believed to be the principal intracellular phenomena in effecting the gravitational input within the cell, a testable hypothesis at the appropriate microgravity level. The body (*protoplasm*) of all living cells is composed of a large variety of particles and aggregates of particles, suspended in a heterogeneous matrix. The normal force of gravity makes these particles tend to float or sink with respect to the other cell components, depending on their relative densities (Figure 2-07).

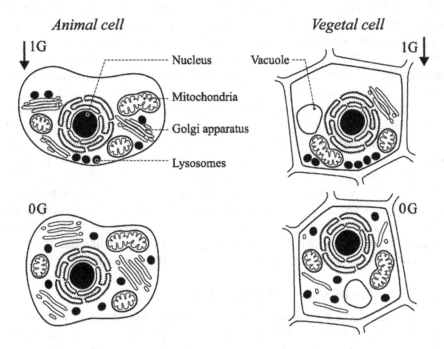

Figure 2-07. Decrease in pressure and loss of sedimentation in microgravity compared to 1 g. Above (1 g), in both animal and plant cells, the denser items have sedimented to the lower part of the cell. Below (0 g), the cell elements are almost evenly distributed in the cell volume. In addition, the physical pressure and strain on structures is reduced. (Adapted from Cogoli et al. 1989)

However, the turbulence within every living cell (a movement of the cell contents termed protoplasmic streaming) and the binding of particles to cellular skeletons, tend to obscure the visible effects of gravity-dependent stratification of cell particles. The principal characteristic of low-gravity environments is that the net gravitational body forces are reduced substantially. This leads to decreases in hydrostatic pressure, buoyancy-driven flows, and rates of sedimentation. Reduction of the gravitational force permits other forces, such as that due to surface tension, to become important, if not dominant. An immediate consequence is that under microgravity conditions, other system behaviors (e.g., mixing, separation, and interfacial phenomena) are significantly altered. Many fluid transport phenomena involving heat and mass transfer are also influenced by the reduction of gravity.

The balanced exchange of ions and molecules through cell membranes may be potentially sensitive to gravity. The same may hold for membrane turnover, a basic process in cell life, and for intercellular diffusion of substances of varying molecular weight.

To investigate these phenomena, the research program in the biological sciences and biotechnology has focused on three primary areas of interest: a) separation physics aimed at providing improved resolution and sensitivity in preparative and bioanalytical techniques; b) cell biology, cell function, and cell-cell interactions; and c) physical chemistry of biological macromolecules and their interactions, including studies of protein crystal growth directed at supporting crystallographic structure determinations.

In the field of biotechnology, for example, the absence of convection and sedimentation can help the separation and isolation of biological specimens. The increase in surface tension will improve transport processes, and consequently secretion and growth. The objective is to cultivate proteins (hormones, enzymes, antibodies) and cells that secrete a medically valuable substance. The purified product would be returned to Earth for medical use, product characterization, or improvement of ground-based separation techniques. However, this process is now challenged by ground-based computer graphics models, and by genetic-engineering techniques (such as the cloning process) that are much less expensive than experiments in space.

2.2 Results of Space Experiments

When gravity is altered, biological changes are observed even when cells are isolated from the whole organism and grown in culture (*in vitro*). Physical scientists predicted this would not occur because gravity is an extremely weak force compared with the other fundamental physical forces acting on or within cells. However, spaceflight results suggest that microgravity may alter the characteristics of cultured cells. Most cells flown in space have either been suspended in an aqueous medium or attached to an extracellular matrix bathed by an aqueous medium.

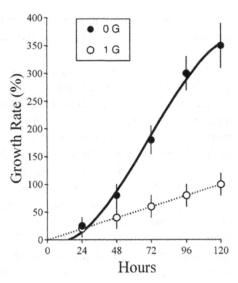

Figure 2-08. Diagram showing the increase in growth rate of bacteria (Bacillus subtilis) cultured aboard Skylab (0 g) by comparison with the same ambient conditions as Skylab, but on Earth (1 g).

2.2.1   Suspended Cultures

Many space missions have flown bacteria in experimental cultures. During spaceflight, bacteria (the most studied were *Escherichia coli* and *Bacillus subtilis*) exhibit an increased duration of exponential growth, and an approximate doubling of final cell population density compared to ground controls (Figure 2-08). Other experiments suggest that space conditions make bacteria (*E. coli*) more resistant to antibiotics. These results suggest that humans are at greater risk in space, given that there may be larger populations of bacteria in a confined environment, which are, moreover, less sensitive to antibiotics.

These effects may simply reflect the fact that when challenged with a new environment, the first response of bacteria is to increase growth rate until new mutants appear that are better adapted. However, these differences may also be related to the lack of convective fluid mixing and lack of sedimentation, both processes that require gravity. We already mentioned that the major effect of reduced gravity environments is a reduction in gravitational body forces, thus decreasing buoyancy-driven flows, rates of sedimentation, and hydrostatic pressure. Under such conditions, other gravity-dependent forces, such as surface tension, assume greater importance. These alterations in fluid dynamics in a reduced gravity environment have significant implications. For example, in cell culture experiments, the diffusion of nutrients, oxygen, growth factors, and other regulatory molecules to the plasma membrane, as well as the diffusion of waste products and CO_2 away from the cell, will be reduced in the near absence of convection unless countered by stirring or forced flow of medium.

It is thought that, in reduced gravity, the more uniform distribution of suspended cells may initially increase nutrient availability compared to the 1g sedimenting cells that concentrate on the container bottom away from available nutrients remaining in solution. This phenomenon would increase growth rate. However, if waste products build up around cells in the absence of gravity, then after some time they could potentially form a pseudo-membrane that decreases the availability of nutrients or directly inhibits cell metabolism. It is suggested that inhibitory levels of metabolic byproducts, such as acetate, may be formed when glucose is in excess within the medium. Therefore, although perhaps somewhat counter-intuitive, a reduction in glucose availability actually may be beneficial to cell growth. Also, local toxic byproducts could become concentrated on the bottom of the 1-g container with cells in increased proximity to each other. Such a process could limit cell growth. Thus, changes in bacteria and possibly other cells during spaceflight may be related to alterations in the microenvironment surrounding non-motile cells, e.g., the equilibrium of extracellular mass-transfer processes governing nutrient uptake and waste removal.

The current view is that a "cumulative" response resulting from reduced gravity may be responsible for the observed effects at the level of the single cell. Earlier predictions suggesting that no effect of spaceflight should be expected were more focused on the physical inability of gravity to elicit an immediate or "direct" response from organisms of such small mass. Rather than a "direct" response, reduced gravity is suspected to initiate a cascade of events: the altered physical force leads to an altered chemical environment, which in turn gives rise to an altered physiological response (Klaus 1998).

2.2.2   Attached Cells

Early results with cultured cells from muscles or bones suggest that spaceflight induces a wide variety of responses. For example, delayed differentiation and changes in the cytoskeleton, nuclear morphology, and gene expression have been reported for bone cells (Hughes-Fulford et al. 1996). Muscle fibers cultured in space were 10-20% thinner (i.e., atrophied) compared with ground controls due to a decrease in protein synthesis rather than an increase in protein degradation (Vandenburgh 1999). Interestingly, the atrophy of isolated muscle fibers in culture was very similar to the amount of muscle atrophy reported in flight animals (see Chapter 5, Section 3.3). These data from bone and muscle cells suggest that spaceflight affects adherent cells and tissues even when isolated from systemic factors. The same results were obtained during ground-based studies using clinostats (Figure 2-09).

Changes in the physical environment surrounding cells, *in vivo* or *in vitro*, can lead indirectly to changes within the cell. Cellular structures that might oppose mechanical loading are only beginning to be defined. Exciting research on the interaction of the cell cytoskeleton with membrane components and the extracellular matrix is shedding light on possible "force sensors" at the cellular level that might be essential for the differentiation process (Wayne et al. 1992). Ingber (1998, 1999) has applied the concept of "tensional integrity", which is a tension-dependent form of cellular architecture that organizes the cytoskeleton and stabilizes cellular form, to cells. This architecture may be the cellular system that initiates a response to mechanical loading as a result of stress-dependent changes in structure that alter the mechanical load on extracellular matrix, cell shape, organization of cytoskeleton, or internal pre-stress between cell and tissue matrices.

The consensus of physical chemists prior to this decade was that forces exerted between molecules within a cell were far greater than gravitational forces. Thus, they concluded that gravity should not be perceived at the cellular level (Brown 1991). However, at that time very little was known about how cells interacted with components of the extracellular environment. These interactions might function to either suppress or amplify signals generated by gravitational loading. Defining the cellular connections that might sense and transduce mechanical signals into a biochemical response may also shed light on the events initiating cell maturation. As a cell matures, it stops dividing and begins to express characteristics of a mature cell type. However, if a cell does not mature, it will continue to divide. This is the definition of a cancer cell. The maturation process may be triggered by multiple factors, including loads placed on the extracellular matrix during different phases of development.

In summary, flight experiments suggest that gravity, quite likely, is perceived by cells through physical changes both in the aqueous medium surrounding cells in culture and in cellular structures that oppose or sense mechanical loads. Exactly how the gravity signal is then transduced to cellular functions is yet to be determined. The answer to this question is not only relevant to understanding the fundamental processes in normal cell physiology, but also in the patho-physiology of certain diseases, such as age-related bone loss, cancer, or immune disorders (Bouillon et al. 2001).

Figure 2-09. The clinostat is a simple device that places a plant, a small organism, or cell growing in culture on a rotating platform. The rotation causes the biosystem under test to be subjected to the gravity vector from all directions. From the system's point of view, the rotation cancels the gravity vector by continuous averaging, thus approximating the highly reduced vector found in the actual space environment. (Credit CNES)

2.2.3   Human Blood Cells

Although the reports to date are conflicting, some indicate that a microgravity environment may compromise the immune system function. These investigations are carried out on cultures of lymphocytes prepared on the ground and tested in space, and with whole-blood samples taken from the crew and tested in-flight, respectively (Figure 2-10). Cogoli et al. (1980) reported that cultures of human lymphocytes subjected to microgravity responded to concanavalin A, a lymphocyte stimulating agent, 90% less than ground-based controls. This is a standard test used to evaluate the competence of peripheral blood lymphocytes to multiply when stimulated with this agent. Studies on the astronauts of the first four Shuttle flights revealed that the lymphocyte responses to photohemagglutinin, another lymphocyte stimulating agent, were reduced from 18 to 61% of normal following spaceflight. It has been suggested that the above changes were due to stress-related effects, but this should be studied further.

These studies are important because, as discussed earlier, the concentrations of microorganisms in space vehicles may be significantly higher than normal. The conditions associated with space travel, space stations, and planetary colonies raise many new and important problems concerned with host-parasite interactions involving humans and animals. Rotation of crewmembers on the ISS will introduce different strains of fungi, bacteria, and viruses that could contribute to the emergence of "new" strains of opportunistic pathogens through mutation and genetic exchange.

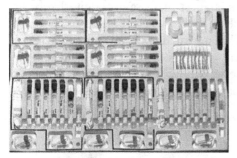

Figure 2-10. Blood draw kit used in-flight by the astronauts. (Credit NASA)

Clearly, spaceflight is associated with a significant increase in the number of circulating white blood cells, including neutrophiles, monocytes, T-helper cells, and B cells. In contrast, the number of natural killer cells is decreased. Plasma norepinephrine levels are increased at landing and were significantly correlated with the number of white blood cells (Mills et al. 2001). These data suggest that the stress of spaceflight and landing may lead to a sympathetic nervous system–mediated redistribution of circulating leukocytes, an effect potentially attenuated after longer missions. Whether *hematopoesis* (or maturation of lymphocytes) is compromised is yet to be established. The multiple stresses of spaceflight may also lead to hormonal imbalances, and corticosteroid release may lead to immuno-suppression. Oogenesis and spermatogenesis (i.e., the formation of female and male sexual gametes) may also be compromised. In any case, additional research is required to confirm or reject the presence of these problems.

On the other hand, there is a significant reduction in the percent of whole blood that is comprised of red blood cells (hematocrit) in some astronauts. The hematocrit is a compound measure of red blood cells number and size. This reduction in the number of red blood cells in astronauts is often referred as the *space anemia*. This reduction may be due to several factors. While in space, the overabundance of fluids in the upper part of the body causes the kidneys to remove this excess fluid, part of which is plasma (see Chapter 4, Section 3.3). This reduction in plasma volume causes an over-abundance of oxygen-carrying capability, which, in turn, would reduce the production of erythropoietin and consequently decrease red blood cell production. This process would be favored by the fact that muscles lose mass (see Chapter 5, Section 1.1) and thus require less oxygen. However, it is also possible that the over-abundance of oxygen-carrying capacity in the blood is responsible for an increase in the destruction rate of red blood cells. Finally, as we will see in Chapter 5, as astronauts lose calcium in their bones, the structure and function of the bone and its marrow may change and may result in a decrease in red blood cell production.

2.3   Bioprocessing in Space

Research in biotechnology relies on the manipulation of cells of living organisms. The purpose of these manipulations is to produce useful molecules, natural or artificial, in useful quantities, to develop new organisms or new biological molecules for specific uses, or to improve yields of plant and animal products through genetic alteration. Recombinant techniques, for example, make it possible to produce natural or artificially mutated versions of proteins exhibiting a wide range of activities and uses, scientific and medical, in large quantities. The techniques essential to these manipulations are applied in aqueous environments and are subject to fluid dynamics and transport processes.

Gravity affects biological systems through its influence on the transfer of mass and heat, particularly in the area of fluid dynamics and transport, as well as its impact on cell structure and function (Figure 2-11). Consequently, microgravity may lead to new knowledge about biological systems, to improvements in current experimental techniques, and to the development of new experimental approaches.

Examples include fermentation processes, compartmental targeting of expressed products within the cell, and the ultimate purity, structural integrity, and activity of a protein product. Particle sedimentation under the influence of gravity, for example, can interfere with aggregation processes such as those mediating cell-cell interactions, cell fusion, cell agglutination, and cellular interactions with substrates.

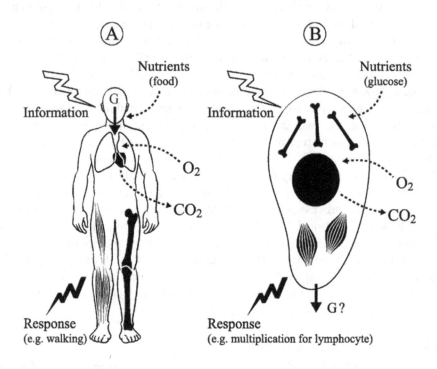

Figure 2-11. Schematic comparison between body (A) and cell (B) functions, showing that biological processes that occur at cellular level are similar to those at organism level.

A detailed knowledge of the three-dimensional architectures of biological macromolecules is required for a full understanding of their functions, and of the chemical and physical effects that they manage to achieve these functions. To be able to synthesize new proteins, whether for medical uses or as complex biomaterials, it is necessary to be able to relate molecular structure and function. Protein crystallography, currently the principal method for determining the structure of complex biological molecules, requires relatively large, well-ordered single crystals of useful morphology. Crystals with these qualities may be difficult to produce for a variety of reasons, some of which may be influenced by gravity, through density-driven convection and sedimentation. Protein crystal growth experiments conducted on board the Space Shuttle (Figure 2-12) have provided persuasive evidence that improvements can, in fact, be realized for a variety of protein samples.

There are two types of biological materials for which commercial bioprocessing in space could offer advantages over production on Earth:

proteins and cells. The proteins include hormones, enzymes, antibodies and vaccines. The cells with medical prospects are: a) those that when cultivated, secrete a medically-valuable substance that can be isolated either in space or on Earth; b) those that can be implanted in man for therapeutic purposes; and c) those that, through cell fusion, can yield antibody-producing hybrid cells (Bonting et al. 1989).

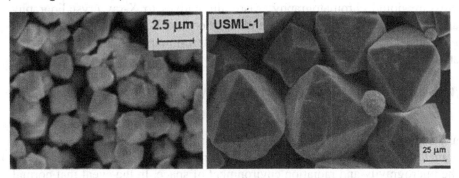

Figure 2-12. Zeolites have a rigid crystalline structure with a network of interconnected tunnels and cages, similar to a honeycomb. Zeolites have the ability to absorb liquids and gases such as petroleum or hydrogen, making them the backbone of the chemical processes industry. Industry wants to improve zeolite crystals so that more gasoline can be produced from a barrel of oil. The zeolite crystals grown on the ground (left) are smaller than ones grown in space (right). The Zeolite Crystal Growth Furnace Unit aboard the ISS will allow to grow zeolite crystals and zeo-type materials in space. (Credit NASA)

How does space bioprocessing work? The raw material, whether a protein mixture or a mixture of living cells, is brought into space and separated in microgravity; the purified product is then returned to Earth for medical use, product characterization, or improvement of a ground-based processing technique. Table 2-02 lists some of the medical products that could be obtained through bioprocessing in space.

Materials	Condition
Alpha-1-antitrypsin	Emphysema
Antihemophilic factor	Hemophilia
Beta cells pancreas	Diabetes
Epidermal growth factor	Burns
Erythropoietin	Anemia
Granulocyte stimulating factor	Wound healing
Growth hormone	Growth problems
Immunoglobulins	Immune deficiency
Interferon	Viral infections
Transfer factor	Multiple sclerosis
Urokinase	Thrombosis

Table 2-02. Some candidate biological materials for space processing and their medical prospects. (Source Bonting et al. 1989)

However, the continuous production of such biological materials on a commercial scale in space proved not compatible with the cost for access to space, and space bioprocessing remains marginal today. Furthermore, ground-based genetic engineering in mammalian (or human embryo) cells is now a very strong alternative to space bioprocessing, together with purification methods such as affinity or immuno-affinity chromatography and high-pressure liquid chromatography. Also, alternatives to X-ray crystallography are emerging, using physical and mathematical models and computer graphics, that are equally useful in determining the three-dimensional structure of proteins.

3   DEVELOPMENT BIOLOGY

The major goal for developmental biology is to determine whether any organism can develop from fertilization through the formation of viable gametes (reproductive cells) in the next generation, i.e., from egg to egg, in the microgravity and radiation environment of space. In the event that normal development does not occur, the priority is to determine which period of development is most sensitive to microgravity.

3.1   Questions

Can higher plants and animals be propagated through several generations in the space environment? Although many embryos orient their cleavage planes relative to the gravity vector, we do not understand whether gravity, per se, is essential to gametogenesis, fertilization, implantation in animals, organogenesis, or development of normal sensory-motor responses. Given the effects of microgravity exposure on bone, muscle, and vestibular function, there is some doubt whether vertebrates can develop normally in space.

The amphibian has been used as a model for many experiments on embryonic development in space (Souza et al. 1995). In *Xenopus laevis* (the South African three-clawed frog) for example, the unfertilized egg has a polarized structure because of an unequal distribution of the yolk: the animal pole is poor in yolk, whereas the vegetative pole contains large quantities. Before fertilization, the egg, surrounded by a layer of jelly, is oriented randomly. After fertilization, the whole egg detaches itself from this layer and rotates, so that the heavier vegetative pole moves downwards, in the direction of gravity. Very roughly, the animal pole corresponds to the head, and the vegetal pole corresponds to the dorsal side (Figure 2-13). An hour or so after fertilization, a second rotation occurs: the cortex rotates by 30 deg relative to the cytoplasm. This rotation establishes the dorso-anterior axis of the animal. The egg then begins to divide and form the embryo that, after an appropriate time, emerges from the jelly-like egg as a tadpole.

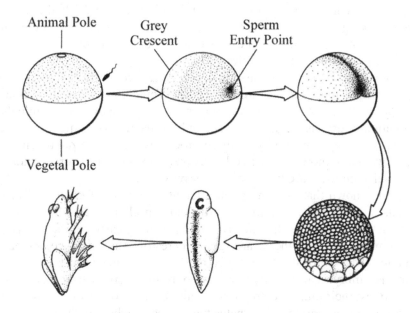

Figure 2-13. The fertilized egg of a common amphibian is shown as it develops from single cell to larva and adult. Cells constituents of the egg are segregated by density—the dark, less dense material rises to the upper half of the sphere, while the denser light-colored material settles to the bottom. Continued development of the embryo follows this orientation. (Credit NASA)

The cortex rotation depends on a transient array of parallel microtubules at the vegetal cortex. A kinesin-like protein is associated with the microtubules and is thought to move the cortex along the microtubules, anchored in the cytoplasm (Elinson et al. 1990). The cortex rotation can be influenced by gravity in several ways. First, extremes of gravity, caused by centrifugation, can overcome the microtubule mechanism and produce a dorso-anterior axis on the centripetal side (Black and Gerhart 1985). Second, gravity alone can produce a dorso-anterior axis in the absence of the microtubule mechanism (Scharf and Gerhart 1980). Third, gravity alone can orient the microtubules prior to their formation, thereby directing where the dorso-anterior axis will form (Zisckind and Elinson 1990). Gravity in these cases acts by moving the heavy yolk-rich cytoplasm downward, producing a cytoplasmic rearrangement.

These gravity effects have led to repeated attempts to place frog eggs in space in order to see how they develop in microgravity. In the most successful of such experiments, there was little or no perturbation of the dorso-anterior axis (Souza et al. 1995). A normal head formed, indicating that some form of cytoplasmic rearrangement had occurred. This arrangement was likely due to the functioning of the parallel microtubule mechanism. One possibility is that gravity-induced rearrangement is an evolutionarily primitive mechanism, which substitutes for the microtubule mechanism. If there were any frogs lacking the microtubule mechanism, their eggs would be interesting objects to put in space: the hypothesis is that the dorso-anterior axis would be altered in the resulting space tadpoles.

Developmental biology also includes all aspects of the life span of an organism, from fertilization through aging. Topics of research includes gamete production, fertilization, embryogenesis, implantation (in mammals), the formation of organs (organogenesis), and postnatal development (changes after birth). The role of gravity in these processes is entirely unknown. For example, we don't know whether cell division (mitosis), nor the orientation of the bilateral symmetry, are influenced by gravity.

It is known that at some point after fertilization, different in diverse organisms, cells become committed to developing along a certain pathway. This restriction in fate is called *determination*. During early cell divisions on most animal embryos, there are gradual restrictions in developmental potentiality. This is not the case in plants. Sooner or later in all animals the cells in the embryo can usually give rise only to a certain tissue or organ. They have lost their plural potentialities. This second process of development is *differentiation*, a term that designates the processes whereby the differences that were "determined" become manifest. The mechanisms by which the determination and differentiation occur at the right time to produce the normal organisms is called the formation of pattern (i.e., not only do they realize their fates, but they do so in the correct place at the correct time).

The formation of the various tissues and organs (organogenesis) not only spans several developmental stages, but also continues after birth or hatching and into the natal period. For each organ system, there appears to be a *critical period* during which development can be disrupted by relatively small environmental stresses. The systems affected by weightlessness in the adult (e.g., vestibular apparatus, bone metabolism, formation of blood cells) might suffer more severe and more permanent effects if the gravity stimulus were withdrawn during the appropriate stage of organogenesis. This would be similar to the results of the experiments showing that the visual system (including the receptors, their neural connections, and the visual cortex) develops abnormally in animals raised in complete darkness (Imbert 1979).

Further, the transition from the neonatal period to adulthood is marked by fundamental developmental events, such as cell specialization, cell-cell interactions, the development and integration of many physiological and biochemical functions, and growth (Figure 2-14). For example, radical changes in the structure and connections of neurons occur during the development of the nervous system. From tissue layers found in embryonic animals, cells increase in number and eventually differentiate and migrate to their appropriate function and position in the developing nervous system. In all, up to 75% of neurons are lost by the process of apoptosis, or programmed cell death during development. Those that remain must form synapses with communicating neurons. Because these processes are regulated by both chemical and mechanical factors, gravity may play a crucial role as a stimulus for proper development.

Regenerative processes are also fundamental developmental responses to postnatal tissue loss and injury. In many situations, these processes are simply responses to changes in the environment to which the individual is exposed. Understanding the role of gravity not only in ontogeny (the development of the individual) but also in phylogeny (the evolution of species) justifies the studies on various species in space for successive generations, over long period of time. By acting on this external factor, would it then be possible to modify the blueprints contained in the genome and change some characters of the species? In other words, would we all become boneless, jellyfish-like organisms, after many generations in space?

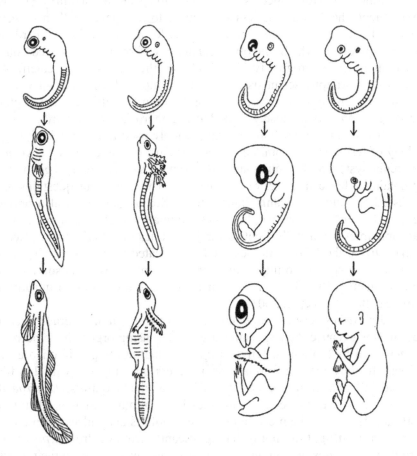

Figure 2-14. Embryonic development. The early stages are closely similar among species (drawn to scale). The later stages are more divergent (not drawn to scale).

3.2   Results of Space Experiments

Diverse organisms have been subjected to microgravity for varying periods of time. The results of these studies have been inconsistent. Both normal and abnormal developments have been observed, dependent on the organism and the stage of development at which the material was subjected to microgravity. Moreover, in the study of embryonic material in particular, most experiments have by necessity been performed with eggs that were fertilized on the ground, well before orbital flight, so that the critical g-sensitive time period immediately after fertilization was spent at 1-g. Also, in many experiments, the other environmental factors (such as launch and entry forces, atmosphere, and radiation level) were not adequately controlled.

3.2.1   Invertebrates

Since aquatic species normally live in a neutrally buoyant environment, they should be less susceptible to microgravity than terrestrial species. However, it has been shown that the formation of skeletal hard parts (shells, spicules) which involve calcium carbonate is altered during development in microgravity. By studying the sea scallop calcification process, for example, scientists hope to learn more of the mechanics behind bone density loss in humans during long-duration spaceflight (see Chapter 5, Section 1.2), a problem closely related to osteoporosis here on Earth.

Sea urchins are a long-standing, widely used model for studying the biology of fertilization. Common genetic origins, or homologies, between the sea urchin system and mammalian systems make the sea urchin a good model for obtaining basic information that can point to important questions to be addressed by studying mammalian systems. Sea urchin sperm also provides the added benefit of survivability; these animals are able to tolerate delays that sometimes occur with flight research. A series of experiments carried out in space using the ESA's Biorack facility indicated that microgravity caused an increase in sperm motility. However it has not been demonstrated if this increase in motility allows the sperm to get to the eggs more quickly and fertilize better (Tash et al. 2001).

Jellyfishes serve as excellent subjects for research on gravity-sensing mechanisms because their specialized gravity-sensing organs have been well characterized by biologists. Jellyfish *Ephyrae* that developed in microgravity had significantly more abnormal arm numbers as compared with 1-g flight (centrifuged) and ground controls. As compared to controls, *Ephyrae* that developed in space showed abnormalities in swimming behavior when tested postflight. However, the mean numbers of statoliths and pulses per minute as determined postflight did not differ significantly from controls. *Ephyrae* that were flown after developing on Earth tended to show changes in their gravity-sensing organs too. Studies on gravity threshold conducted in the onboard centrifuge revealed that more than 50% of the animals convert to Earth-like swimming behavior upon exposure to 0.3 g. The swimming behavior of both *Ephyrae* hatched on Earth and in microgravity showed that they had difficulty orienting themselves in space (Souza et al. 2000).

Experiments on fruit fly revealed that mating is possible without gravity, and that developmental processes and morphogenesis were normal in microgravity. Studies on gypsy moth have been performed to study the effect of microgravity on the diapause cycle. Diapause is the dormant period in an insect life cycle when it is undergoing development into its next phase. Results showed that microgravity may shorten the diapause cycle of gypsy moths and lead to the emergence of larvae that are sterile. The capability to produce sterile larvae may lead to the development of a natural form of pest control.

According to the laws of aerodynamics, insects cannot produce enough lift pressure to fly. The mechanism whereby they achieve flight must involve unsteady flows interacting with the dynamically changing wing surfaces. Interestingly, experiments carried out on insects in space have shown that larvae of fruit fly that developed in space learned not to fly and preferred to float without wing beat. Wing abnormalities and mutations have also been reported in floor beetle when examined after spaceflight. Similarly, honeybees were unable to fly normally and tumbled in weightlessness with no wing beat.

Perhaps the most famous space experiment using invertebrates is the one carried out on Skylab, in 1973 to study if two common cross spiders (*Arachnous diadematus*) spin webs differently in microgravity. Since the spider senses its own weight when constructing the web to determine the required amount of silk to make the web, gravity plays an important role in the construction of the web (Figure 2-15). Therefore, the objectives of this experiment were to observe how microgravity affects the weight sensing mechanism for web construction. During their first attempt in space, the webs were different from ground controls, but later webs were nearly identical (Summerlin 1977).[*] However, although the spiders did not spin their web patterns differently, it seems that the threads themselves were different.

This experiment should fly again soon on the ISS. Indeed, it has since attracted interest from the industry in advanced composite material. Spider silk is an ultra-lightweight fiber that combines enormous tensile strength with elasticity. Each fiber can stretch 40% of its length and absorb a hundred times as much energy as steel without breaking. Spiders have specialized rear legs which are capable of applying the sticky silk without adhering to it, and engineers would like to mimic the action of these legs, which are known in

---

[*] A spider spins a web in about 2 hours. Since Skylab was completing a full Earth orbit in about 90 minutes, one can say that this was the first "world wide web" !

engineering as an "end-effector". Coincidentally, NASA refers to space-age technology used in "down to Earth" applications as "spin-offs"!

An experiment selected by ESA for the ISS will use scorpions onboard the ISS. It is known that the circadian patterns in animals and humans are also influenced by activities such as food intake and locomotion. The exposure of scorpions to microgravity will help to analyze entraining and coupling mechanisms of biological clocks and will contribute to the analysis of disturbances of clock systems in humans, by fully automatic measurement of physiological parameters with circadian patterns (i.e., locomotion, eye movements, O_2 consumption and cardio-vascular activity). Scorpions represent an interesting animal model since they tolerate a complete lack of food and water for more than six months without nutritional care. The animals will be connected to sensors and electrodes and exposed to microgravity, 1-g, and different light regimes (Wilson 2003).

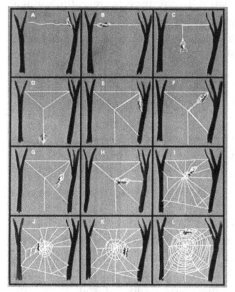

Figure 2-15. The common cross spider produces a web of nearly concentric circles each day at approximately the same time. The web is constructed in a very orderly fashion, starting with a bridge and frame (A-D), and axial threads (E-I). Spiral emanating from the hub is constructed next (J-L). Since the spider senses its own weight to determine the required thickness of the material (D), microgravity provides a new stimulus to the spider's behavioral response (Summerlin 1977).

3.2.2   Lower Vertebrates

No vertebrates have been raised from conception to sexual maturity in the absence of gravity. No birds or reptiles have bred on orbit, although fertilized chicken and quail eggs have flown on several occasions. Young chick embryos have survived. Quail eggs that were fertilized on the ground have hatched on the Mir space station, but yielded hatchlings that were disoriented and would not or could not spontaneously feed (Jones 1992).[*]

---

[*] When a cosmonaut took a hatchling from its habitat, the chick appeared content as long as it was held. But once released, the bird first flapped its wings for orientation

Studies of sea urchin, fish, frog, and newt (Dournon et al. 2001, Moody and Golden 1999,) indicate that fertilization can occur in space, but in these cases the gametes had been developed while the organism was on Earth. One investigation has suggested that gametes formed in space, however, are normal (Ijiri 1997). In this experiment, *Medaka* fishes mated freely in microgravity and the resulting larvae swam normally both in space and after return to Earth.

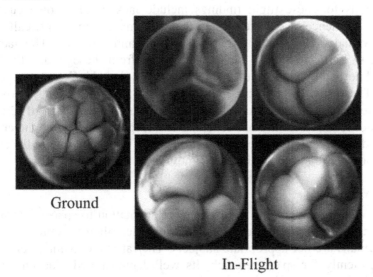

Figure 2-16. Division in amphibian (pleurodele newt) eggs, showing some abnormalities (larger sillons, odd number of cells) in the flight specimens by comparison with ground controls. (Adapted from Gualandris-Parisot et al. 2002)

Female frogs were sent into space and induced to shed eggs that were then artificially inseminated. As mentioned above, the eggs did not rotate (but the cortex did) and yet, surprisingly, the tadpoles emerged and appeared normal. After return to Earth, the tadpoles metamorphosed and matured into normal frogs. Development appeared normal during spaceflight, yet some morphological changes occurred in frog embryos and tadpoles (Figure 2-16). The embryo had a thicker blastula roof that should have created abnormalities

and began to spin like a ballerina, then kicked its legs causing it to tumble like a spinning ball. The cosmonaut noted that the chick would fix its eyes on the cosmonaut while trying to orient in space. When placed in their habitat, the chicks had difficulty flying to their perch to eat, and, unlike the adults, had difficulty grasping the perch for stability when eating. The hatchlings ate normally only when held by the crew and, thus, did not survive. By contrast, adult quails adapted quickly to the space environment. They soared, rather than flapped, their wings and held onto their perch for stability when eating.

in the tadpole, but no deformations appeared, suggesting plasticity of the embryo (Souza et al. 1995, Duprat et al. 1998).

Another interesting finding was that the tadpoles did not inflate their lungs during spaceflight. Earth or 1-g space (centrifuged) tadpoles swam to the surface, gulped air, and expanded their lungs within 2-3 days of hatching. Air bubbles were present in the tadpole aquatic habitat on orbit, yet the tadpoles did not inflate their lungs while in microgravity. Two possible explanations for these flight findings include lack of directional cues and increased influence of surface tension that may make it more difficult for an orbit-born tadpole to burst through a bubble and gulp air. The tadpoles returned to Earth within 2-3 days of emerging from the egg, and the lungs appeared normal by the time the tadpoles were 10-day old (Wassersug 2001).

These studies produced multiple important findings. They show that vertebrates can be induced to ovulate in space and that rotation of fertilized eggs is not required for normal development in space. The vertebrate embryo is very adaptive and the system is plastic, yet the long-term fate of the animal throughout its life in space remains unknown.

3.2.3 Mammals

When investigations address human adaptation to spaceflight and its health implications, the use of mammalian species often becomes necessary when humans are not appropriate subjects. The rat is the mammal employed most frequently for space research. Its well-demonstrated biochemical and structural similarity with human makes the rat an appropriate subject with which to test new drugs and investigate many disorders experienced by astronauts during and after spaceflight. Within a two-week period, which corresponds to a Space Shuttle flight, the rat neonates go through a critical development period where rapid neural and motor development occur (Figure 2-17). Also, because of their phylogenetic proximity to humans, non-human primates, such as rhesus monkeys, have occasionally served as research subjects in space biology, but only when the need has been clearly demonstrated (Souza et al. 2000).

Figure 2-17. A rat born on the ground before launch will go through a large development period within the two-week duration of a Space Shuttle space mission. (Credit NASA)

Fertilization events have been studied in several species for which fertilization occurs externally, such as newt or fish. As discussed above, the data indicate that for these animals, production of a zygote and early cleavages are mostly normal in the space environment. Fertilization events in mammals have not been studied, primarily because they occur internally. In several occasions, however, flown pregnant rats gave birth to normal neonates after flight. It was observed that during postflight delivery, flight dams have twice as many abdominal contractions as the ground controls, suggesting that more extended exposure to spaceflight could still have a detrimental effect on pregnancy, or at least the birthing process (Ronca and Alberts 2000).

The force of gravity may influence events underlying the postnatal development of motor function in rats, similar to those noted in hatchling quail. Such effects most likely depend on the age of the animal, duration of the altered gravitational loading, and the specific motor function.

Walton (1998) also reported differences in righting reflex, swimming behavior, and locomotion in neonatal rats when the musculo-skeletal system did not bear weight during critical times of development (Figure 2-18). The results from the 17-day Neurolab Shuttle Mission showed that neonatal rats flown in space exhibited altered locomotor behavioral development that persisted for the 1-month recovery period, and that righting reflex strategies were still abnormal 5 months after return to Earth. Other results indicated delayed development of certain nerve connections to muscles. The connections returned to normal after return to Earth, yet fibers in hindlimb muscles did not reach normal size even after a month back on Earth. The data suggest that biomechanical loading of limbs during early development may be essential for innervation of muscles. Another mechanism, however, may be at work: besides the lack of loading during critical times, there is also the possibility that adaptive changes in the vestibular system, particularly the reduction in descending otolith input required to maintain muscle tone (see H-reflex data in Chapter 3), modify the nerve-to-muscle connections.

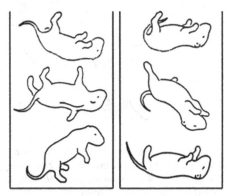

Figure 2-18. Righting reflex of neonates rats in water. Trajectories of a control pup (left) and a flight pup (right) during release in water. Righting responses were absent in flown animals. (Adapted from Ronca and Alberts 1997)

To date, relatively little neurobehavioral research has been done in microgravity with vertebrates, juvenile or adult. This is partly because the habitats for raising them in space have not existed and because the study of vertebrates up to sexual maturity requires longer exposure to microgravity, both capabilities that ISS will soon provide (Wassersug 2001). Physiological experiments using implanted electrodes in fish, rats, or rhesus monkeys have given interesting data on the adaptive changes in the neuro-vestibular system, for example (see Chapter 3). However these experiments are limited to constrained animals or caged animals, which do not experience motion in microgravity, which makes it difficult the comparison with adaptive changes in astronauts. The response of animals to free-fall is astonishingly diverse, as shown by the observations made on frogs, lizards, and snakes in parabolic flight (Wassersug 2001). Other observations of animals placed in microgravity after vestibular lesions prove interesting for understanding the role of gravity in the process of recovery of balance function (Figure 2-19).

In conclusion, short-term exposure to the space environment does not significantly affect the embryonic development, although interesting, unexplained changes occur during embryogenesis and early development. However, because the animals in most of these studies were only partially adapted to the space environment due to the short duration of the flight, it is formally possible that long-term exposure will have more significant effects. The opportunity to conduct development studies onboard the ISS will leave room for much investigation.

Figure 2-19. Drawing showing the posture of a rat 6 months after a unilateral lesion of the vestibular system in normal gravity (left) in microgravity during parabolic flight (middle) and in hypergravity (1.8 g) following parabolic flight. Vestibular compensation is disrupted when the animals are placed in microgravity, suggesting a continuous role of gravity in the recovery process. (Adapted from Réber et al. 2003)

4   PLANT BIOLOGY

4.1   Questions

On Earth, plant roots as a rule grow downward toward gravity, while stems grow up and away from gravity, a phenomenon known as *gravitropism* (Figure 1-03). By studying plants in microgravity on board spacecraft, biologists seek to understand how plants respond to gravity. Also, plants respond to environmental stimuli such as light, temperature, and magnetic or electric fields. These responses are masked on Earth by the overriding response of plants to gravity. In addition, any strategy that visualizes a long-term sustained human presence in space absolutely requires the ability to continuously grow and reproduce various plant species over multiple generations, for food and controlled environmental life support systems.

4.2   Results of Space Experiments

A large variety of plants with short life spans have flown in space: algae, carrots, anise, pepper, wheat, pine, oat, mung beans, cress, lentils, corn, soybeans, lettuce, cucumbers, maize, sunflowers, peas, cotton, onion, nutmeg, barley, spindle trees, flax, orchids, gladiolas, daylilies, and tobacco. Due to this variety, for the most part, observations on plants exposed to microgravity have been anecdotal. It has been demonstrated repeatedly that plants do grow in microgravity. However, whether plants can grow and develop normally over several generations remains to be determined.

4.2.1   Graviception

For research purposes, the gravitational response, as with any stimulus response, has been divided into three steps: a) stimulus perception: how a plant senses gravity; b) signal transduction: how the plant transfers this knowledge into action; and c) the response or resulting action: differential cell elongation or differential growth that results in the root or shoot bending in a new direction..

Where does the response occur? In roots we have already seen that perception occurs in the root cap (Figure 2-05). The sensing mechanism underlying a plant's ability to orient its organs in a gravitational field seems to involve the sedimentation of intracellular particles known as *statoliths*. The statoliths each consist of a number of starch grains surrounded by two membranes, the structure being termed an *amyloplast*. With the movement of statoliths, the cell receives a mechanical stimulus. How a cell transfers this mechanical stimulus into a chemical signal is still of great debate. One hypothesis is that as the statoliths "fall", they come to rest on other organelles such as the endoplasmic reticulum or plasma membrane, thus exerting pressure on the organelle that results in the opening of ion channels and the release of ions such as calcium that initiate the signal transduction pathway. Unfortunately, little is known of these early events.

Microgravity experiments using sounding rockets (lasting 6 minutes) or the Space Shuttle (lasting from 20 hours to 10 days) have shown that the amyloplasts do not move freely in the cytoplasm. Therefore there is also the possibility that the amyloplasts are connected with the cytoskeleton (through an actin network) and that their movement results in tension in the cytoskeleton, again, resulting in initiation of a signaling pathway (Perbal et al. 1997).

Onboard centrifuge experiments have demonstrated that seedlings grown in space required a dose of 20-30 g sec (gravity time seconds), or less than 0.1g for 200-300 sec, to elicit a gravitropic (root bending) response (Perbal and Driss-Ecole 1994). Other studies revealed that the minimum force that is sensed by plant organs is in the range of 5×10^{-4} g for roots and 10^{-3} g for shoots. The root is able to perceive its orientation with respect to a linear acceleration vector and to generate a signal of curvature in less than 30 seconds. However, the microgravity environment encountered to date (10^{-3} to 10^{-4} g) has not been low enough or long enough to definitely answer the question of threshold and response mechanisms. Cells seem to develop normally, although growth patterns are adversely affected even at these levels (see below). Thus, while considerable experience has been gained in orbital flights of days to weeks in duration, much lower gravity levels (10^{-5} to 10^{-6} g) and much longer flight duration (months) are needed.

Even on Earth, the shoot apex of a plant may not grow directly upwards, but it may exhibit continuous helical and spiral movements as it extends so that seen from the side it appears to oscillate. This "circumnutational" movement may be a constant seeking of the apex for perfect alignment along the line of the gravitational force, and be determined by constant adjustments in the levels of hormones produced in response to gravity perception. Experiments using sunflower seedlings, 4-5 days old, grown in space on a centrifuge revealed that circumnutation takes place in microgravity, albeit with some reduction in the amplitude of oscillation, and therefore that gravity is not essential (Brown et al. 1990).

4.2.2   Development of Plants

The questions raised by the space environment are the following: How will space affect seed viability and germination? Is plant development normal in space? Will plants be able to reproduce in space? Significantly, results of studies on the German Spacelab-D1 mission, which incorporated onboard 1-g centrifuge controls, indicated that single plant cells behave normally or even exhibited accelerated development. In contrast, the roots of seedlings germinated in microgravity grew straight out from the seed, and the same roots contained statoliths that were more or less randomly distributed in their cells. Control roots centrifuged at 1 g in-flight, were normally gravitropic (Figure 2-20).

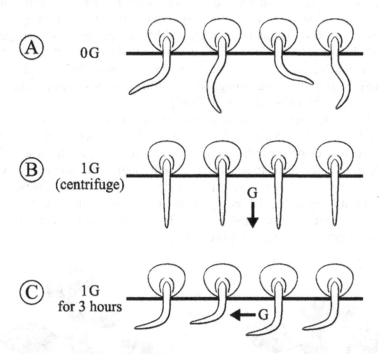

Figure 2-20. Experiments carried out on the German D1 mission showing that roots may grow randomly in microgravity (A), but can be reoriented uniformly on exposure to 1 g on a centrifuge (B and C) for as little as 3 hours. (Adapted from McLaren 1989)

Many species of plants have grown in microgravity. In 1972 the first plant, *Arabidopsis thaliana*, was successfully grown from seed to flowering plant and to next generation of seeds on Salyut-7. This was repeated for the first potential crop plant, super dwarf wheat, on Mir in 1996. It appears that the absence of gravity has no real effect on germination. For example, 12 million tomato seeds remained in space for 6 years, as part of an experiment embarked on a satellite the size of a school bus that was placed in orbit by the Space Shuttle and retrieved by another crew 6 years later (Long Duration Exposure Facility, LDEF). Postflight measurement of germination showed no difference with ground controls, indicating that seeds remain viable in space.

Cytological studies of roots flown under a variety of conditions in space have consistently revealed reduced cell divisions as well as a variety of chromosomal abnormalities. Reduced amounts of cellulose and lignin were also found in space-grown mung bean, oat, and pine. Early space experiments exhibited poor plant growth and altered development: plants died in transition from vegetative to flowering stage or plants flowered, but were abnormal.

Beginning in 1993, a series of experiments on the Space Shuttle and then on Mir was initiated to examine this problem. Experiments using the plant *Arabidopsis* indicate that at least this plant develops normally through the flowering stage. However, fruit set was decreased and seeds brought back to Earth germinated less efficiently than ground-based controls. It was suggested that early abnormalities in plant reproduction could be caused by the toxic effect of ethylene, rather than spaceflight factors, and that seed size was diminished possibly because storage and utilization of reserves are modified in absence of gravity (Musgrave et al. 1997)

Plants continue to conduct photosynthesis in space. However, some studies indicate a decreased ability to do so. Chloroplasts can have their internal structure disorganized, and their starch stores depleted. When plants were able to produce seeds in space, either additional light was needed versus Earth or it was noted that oxygen production (from photosynthesis) was reduced. In microgravity, by comparison with 1-g control on Earth, the growth of the primary root and its apical (up-down) dominance over the secondary roots were reduced (Figure 2-21).

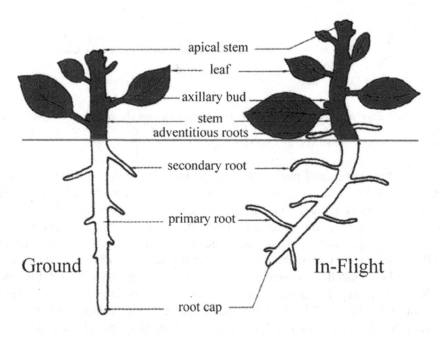

Figure 2-21. Summary of the effects of microgravity on the differentiation, development and direction of growth of plant organs. (Adapted from Perbal 2001)

Also, the growth of plant organs in space seems characterized by changes in the orientation of stem and leaves and secondary roots, more adventitious roots, and faster growth of secondary roots.

In summary, these studies suggest that the reproductive phase is complete in microgravity when the culture conditions (gas and liquid exchanges) are adequate. However, whether or not a seedling growing from the beginning in microgravity can flower and produce normal seeds, which can lead to normal plants, remains a matter of debate. Long-term flight experiments are required to determine if a variety of plant species can grow normally in microgravity and, in particular, if they can produce viable seeds.

5   RADIATION BIOLOGY

The broad spectrum of radiation encountered in space goes from extreme ultraviolet radiation, X-rays and high-energy particles as electrons, neutrons, protons, and heavy ions up to iron and even higher charges. This section will be limited to a description of the ionizing radiation encountered in low Earth orbit and its biological effects as revealed by the biology experiments performed during space missions to date. A more complete description of the space radiation environment can be found in Eckart (1996). The issues of radiation from the medical perspective will be discussed in greater details in Chapter 7 (Section 3.2).

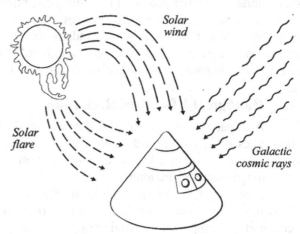

Figure 2-22. The three main sources of ionizing radiation in space.

5.1   Ionized Radiation in Space

Cosmic radiation includes galactic radiation from supernova explosions and radiation of solar origin associated with solar flares (Figure 2-22). The primary galactic radiation present outside the Earth's atmosphere is composed of about 85% protons (with a hydrogen nucleus), 12% alpha particles (with a helium nucleus), and 1.5% of heavy ions particles with high charge and energy. These high-energy particles interact with the nuclei of the nitrogen and oxygen atoms of the atmosphere, resulting in a highly complex secondary radiation, which irradiates the whole surface of the globe.

The solar particle radiation consists of 95% protons. High surface doses may be experienced, but dosage levels rapidly decline with the depth of material penetrated.

Space missions that include travel within or through the Van Allen belts also add a third source of radiation. The Van Allen belts consist of protons and electrons trapped by the geomagnetic field. A phenomenon of special importance for missions in near Earth orbit is the South Atlantic Anomaly, in which the particles are drawn closer to the Earth than at other regions of the globe due to the asymmetry of the geomagnetic field.

Finally, additional radiation is created in high-energy collisions of primary particles with spacecraft materials (Figure 2-23).

The first evidence of the effects of space radiation on human crew are the "light flashes" first observed by Apollo and Skylab astronauts.[*] During most of the Skylab-4 mission, these flashes averaged 20 per hour; however, flashes increased to 157 per hour when the Skylab orbit passed over the center of the so-called "South Atlantic Anomaly". The flashes are believed to be due to high-energy heavy particles of cosmic rays and have been reproduced in humans on Earth by exposure to high-energy ionizing particles. Three explanations have been proposed to explain the phenomenon of light flashes seen by crewmembers: a) an emission of photons by high-energy particles slowed by fluid in the eye (Cerenkov radiation); b) a light generated by these particles ionizing fluid in the eye; or c) an artificial light stimulus caused by these particles impacting retinal sensors in the eye (Pinsky et al. 1975).

5.2   Biological Effects of Radiation

All space agencies agree to the resolution that the effects of space radiation, especially on non-dividing cells of the retina and central nervous system, must be assessed before long-duration human missions beyond Earth's magnetosphere are attempted.

However, most of the biological effects of radiation remain largely unknown. The mechanisms of ionizing radiation impacts on cells are either direct, particles impacting a vital target molecule and directly transferring

---

[*] During the debriefing of the Apollo-12 mission, Pete Conrad reported the following: "We all did see these corona discharges. [...] Most of the time (we saw them) during our sleep periods when we were lying in our bunks. [...] They appeared as either a bright round flash or a particle streaking rapidly across your eyeball in a long thin illuminated line. I could determine whether it was my left eye or my right eye that did it at the time." Alan Bean also reported: "If I was thinking about watching for them, I would see one every minute or somewhat less. One of them would be a flash, and about one minute later there would be a line. It did not appear to make any difference whether we were in lunar orbit, translunar, transearth, or anything else. If you just wanted to look for them, you could see them going by" (Godwin 1999).

their energy, or indirect, particles impacting other molecules (e.g., water) to yield longer-lasting, very-reactive free radicals (with unpaired electron).

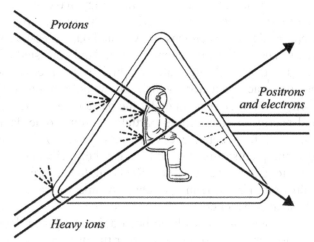

Figure 2-23. This diagram illustrates the secondary radiation generated by the collision of various high-energy primary particles with the spacecraft materials.

The biological effects of protons are fairly well understood. Through bombardment of spacecraft material, protons produce neutrons. These neutrons, upon colliding with a hydrogen nucleus, liberate their energy. Since living organisms contain many hydrogen-rich compounds, such as proteins, fat, and water (70%), they are most likely to be affected. However, the half-life of neutrons is only 11 minutes, after which they decay naturally to protons and electrons. Early dosimetry performed on Skylab and Russian space stations indicates that the flux of neutrons is probably not significant. The major space hazard comes from the highly charged energetic particles.

From ground-based experience (such as radiotherapy or nuclear explosion) it is known that the early, acute effects of radiation include skin effects (burns), eye lens opacification (cataract), graying of hair, immune system suppression (higher risk of infection), and the loss of non-dividing cells. Late effects include cancer in blood-forming organs (bone marrow, thyroid, lung, stomach, colon, and bladder) and genetic effects, which arise from, cell transformation (chromosome aberrations and translocations).

At the cellular level, when DNA strands break, non-rejoined breaks can lead to cell death, whereas incorrectly rejoined breaks can lead to mutation. The temporal and spatial characteristics of the radiation energy determine the quantity and quality of damage. Single-strand breaks can normally repair. However, double-strand breaks with close single hits or a high-density energy hit do not repair. Cells in mitosis are the most vulnerable. High-energy particles, with their large capacity to transfer their energy along the path, can generate a high percentage of double-strand DNA breaks. The effects are widespread and can lead to the death of numerous cells along the track. In addition, it is difficult to protect the vehicle and its inhabitants (shielding) from these particles (Tobias 1974).

Most results in space radiology have been obtained so far during short-term space missions in low Earth orbit. However, some observations were made on specimens flown on board free-flyers satellites for several years. The biological systems investigated consisted of bacterial spores, plant seeds and animal eggs, all of which were alternately sandwiched between nuclear track detectors which allow both a precise determination of the biological region penetrated by the particles and a determination of the charge and energy of each hitting particle. Chromosome damage and abnormalities are seen. In general, seeds are less sensitive than developing embryos or growing plants; this may be because their cells are not actively dividing. It is hard to determine if these effects in lower organisms will lead to tumor induction, life shortening, or chromosome aberration in organisms with longer life spans (Planel et al. 1994).

Several studies have been done to try and determine which radiation is most damaging, or even whether the damage was solely due to radiation at all. Some studies showed that standard radioprotectant chemicals like cysteine, aminoethyliothiourea, and 5-methoxytryptamine didn't stop the damage. This might indicate that the low-energy, indirect radiation is not at primary fault. However, some of the flights on which damage was found were short enough that galactic cosmic radiation dosages were low. Of course high-energy particles remain a possible threat, but some of the chromosomal damage and abnormalities could also be due to other environmental factor, like microgravity. On the other hand, experiments with protozoa and bacteria suggest that small doses of radiation may actually be beneficial because small doses elicit stress responses and have been shown to increase DNA repair (Planel et al. 1987).

Microlesions of cultured retina cells induced by single heavy ions were first discovered via spaceflight experiments. These findings initiated biological investigations using particle accelerators on Earth. However, the results of ground-based and space studies are often conflicting. For example, a higher number of mutations were observed in biological systems (e.g., larvae of *Drosophila*) exposed to an "artificial" radiation source while in orbit compared to ground controls. This difference suggests the possible existence of a combined effect of radiation and microgravity, the repair of radiation-induced lesions being modified in space (Planel et al. 1985).

This illustrates the difficulty of differentiating between the effects of the several factors inevitably present during spaceflight. For this reason, some biological effects, and their protection, can be studied only in space. For this purpose, a *Phantom Torso* was developed by NASA and ESA to study the effects of radiation at tissue level both during activities within the space vehicle and outside. The Phantom Torso is a part-dummy, part-dosimeterimbedded mock-up of a human's upper body, minus a set of arms, built to determine the effects of radiation on the human body (Figure 2-24). Dosimeters are mounted at critical organ-tissue locations within the Torso where critical organs are located: the head, the heart, the liver and kidneys, and they record the level of radiation received as a function of time.

Other instruments mounted on the outside of the ISS measure the spectrum of particles that first hit the ISS shielding. As they go through the station wall and the Torso, the radiation gets modified. The secondary radiation may have a different effect on tissue than the primary radiation. So the radiation spectrum is measured before and after it hits the Phantom Torso. The information gained from the Torso experiment will help determine the best types of materials and methods for shielding human crew.

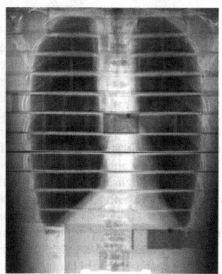

Figure 2-24. Phantom Torso mounted on the space vehicle (left panel) to monitor actual radiation level at various depths within the tissues (right panel). (Credit NASA)

For human mission to Mars, considerably better quantitative data on low-energy transfer radiation dose rates beyond the magnetosphere are still required. In particular, better predictability of the occurrence and magnitude of energetic particles from solar flares is needed, since radiation from solar flares can be life threatening in relatively short time periods.

6   ISS FACILITIES FOR SPACE BIOLOGY

Both the ESA's Biolab (Figure 2-25) and the NASA/NASDA's Centrifuge Accommodation Facility (Figure 2-26) planned for flying on board the ISS should provide unique opportunities for space biology research during long-term exposure to microgravity. These facilities provide a laboratory or platform that has an environment of 10^{-5} to 10^{-6} g. Appropriate incubators and growth chambers are provided for cells, simple organisms, plants, and animals.

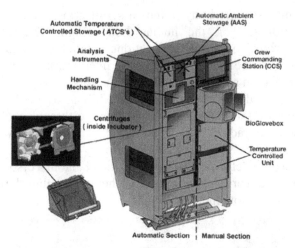

Figure 2-25. Artist's view of the of ESA Biolab facility for use in the ISS. (Credit ESA)

Units have the capability to maintain and monitor microbial, animal, aquatic, and plant cell and tissue cultures for up to 30 days. An aquatic habitat will accommodate small fresh water organisms, such as *Zebrafish,* for up to 90 days to support egg generation studies for examination at all life stages. Animal habitats will be capable of housing up to 6 rats or a dozen mice. These habitats will be compatible with another compartment accommodating pregnant mice and subsequently their offspring from birth through weaning. A plant unit will be able to support plant specimens up to 8-cm total height (root and shoot) through all stages of growth and development. An insect habitat will support drosophiles and other insects for multigenerational studies and for radiation biology. Egg incubators will support the incubation and development of small reptilian and avian eggs prior to hatching.

In order to investigate the problem of gravity thresholds in living systems, several on-board centrifuges are provided. These centrifuges provide a 1-g control for microgravity experiments, as well as the capability to explore a range of gravity levels between 10^{-6} and 2 g in order to study gravity thresholds for certain phenomena. Some habitats will have the experimental capability of selectable gravity levels of up to 2 g by being mounted on the 2.5-m diameter centrifuge. Other habitats will be equipped with internal centrifuges, which will provide selectable gravity levels from 0.01 to 1.5 g.

The laboratories will also be equipped with a –80°C freezer, a 4°C refrigerator, a cryogenic freezer, passive dosimeter system, and dissecting and observation microscope.

The centrifuge and the freezers have special "damping" mechanisms that ensure that this microgravity level is undisturbed by human or machine activity. Periodic sampling is performed automatically or by the onboard crew, in such a way as to leave the remaining material undisturbed. Sampling is obviously carefully planned and minimized to preclude vibrations and other unwanted gravitational forces. "Glove boxes" attached to the habitats provide an enclosed workspace used for performing experiments and handling research organisms.

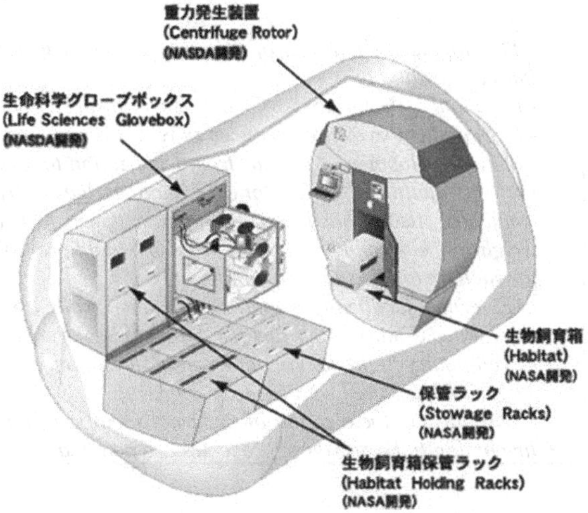

重力発生装置
(Centrifuge Rotor)
(NASDA開発)

生命科学グローブボックス
(Life Sciences Glovebox)
(NASDA開発)

生物飼育箱
(Habitat)
(NASA開発)

保管ラック
(Stowage Racks)
(NASA開発)

生物飼育箱保管ラック
(Habitat Holding Racks)
(NASA開発)

Figure 2-26. Gravitational Biology Facility and the Centrifuge facility developed by NASA and NASDA for use on the ISS. (Credit NASA and NASDA.)

For a comprehensive view of the current opportunities in space biology research, the following table (Table 2-03) lists the advantages of the animal models that could be studied on board the ISS.

Drosophila
- *Genome completely sequenced*
- *Rich information about genetics and developmental mutations*
- *Extensive homology to vertebrates at the molecular level*
- *Several indicator lines and mutants are available*
- *Hardy and small; an incubator has been built for their embryonic development*
- *Short life cycle: development from egg to adult is only 220 hours*
- *Development is external, and therefore can be videotaped*
- *Prolific producers of offspring: can create an enormous database in one experiment*
- *Can put embryos into "cold storage" to slow development; allows one to manipulate the developmental period of exposure to altered gravity*
- *Identified cells and connectivity maps in both central and peripheral nervous systems are useful for issues of neural maps*

Worm (Caenorhabditis elegans)
- *Genome completely sequenced (in Genbank)*
- *Complete development is known, tracking each cell division from egg to adult*
- *Extensive homology to vertebrates at the molecular level*
- *Developmental patterns of a large number of genes are known*
- *Hardy and small; suitable flight hardware currently exists*
- *Short life cycle: a full generation cycle takes 4-8 days*
- *Development is external, and therefore can be videotaped*
- *Large database of mutants available*
- *Can put embryos into "cold storage" to slow development (see Drosophila)*
- *Identified cells and connectivity maps in nervous system (see Drosophila)*

Amphibian
- *Genome project has begun*
- *Some mutations are available*
- *Extensive homology to mammals at the molecular level and for the mechanisms of tissue induction*
- *Developmental patterns of a large number of genes are known and GFP-marker lines are available*
- *Eggs and embryos are very hardy; several species have already flown*
- *Eggs and embryos are produced in very large number*
- *Short developmental period: from fertilization to larva in a few days. However, life cycle to adulthood is too long in Xenopus laevis, but comparable to mouse in X. tropicalis. Also, it has been shown that X. laevis larvae fail to inflate their lungs in a weightless environment. Such a problem would prevent a complete life cycle in space for such an air-breathing amphibian.*
- *External fertilization and development; all aspects of development can be videotaped. Embryos are opaque, but tadpoles of some species are semi-transparent*
- *Preliminary flight data about external development have been collected*
- *Eggs are stratified with regards to gravity*
- *Cytoplasmic localization of maternal factors necessary for formation of the germ line and initial axes; these processes are potentially affected by gravity*

Avian
- *Genome project underway by the USDA*
- *Some mutants are available*
- *Extensive homology to mammals at the cell, tissue and molecular level*
- *Developmental patterns of a large number of genes are known*
- *Flight data are available; there are some sensitive periods during which these embryos do not do well in the flight environment*
- *Short developmental cycle: eggs hatch in 21 days for chick, 16 days for quail. The life cycle is relatively long, however, in comparison to other models*
- *Can be studied in large numbers*
- *Can store early embryos at cool temperatures, and then restart development at desired times by bringing them to appropriate incubation temperature*

Fish (attributes cited are mostly for Zebrafish, but in some cases apply to other species as well)
- *Genome project well underway for Zebrafish*
- *Rich information about genetics and developmental mutations in some species*
- *Extensive homology to mammals at the molecular level*
- *Developmental pattern of a large number of genes are known and indicator lines are being developed*
- *Eggs and embryos are hardy; several species have already flown*
- *Short life cycle*
- *External fertilization and development; all aspects of development can be videotaped; Embryos are transparent*
- *Eggs and embryos are produced in very large numbers*
- *Significant flight data for vestibular system development*

Mouse
- Genome project well underway
- Rich information about genetics and developmental mutations
- Many models for human congenital defects
- Several indicator lines and mutants are available
- Short developmental cycle: 21 days gestation; Life cycle relatively short (4 months)
- Small adults; habitats take less space than other mammals

Rat
- Genome project has begun
- Some well-developed models for human disease and pathophysiology
- Significant database of maternal fetal behavior
- Some flight-based data
- Developmental cycle comparable to mouse

Table 2-03. Appropriate animal models for space studies. (Source Moody and Golden 1999)

7 REFERENCES

Black SD, Gerhart JC (1985) Experimental control of the site of embryonic axis formation in Xenopus laevis eggs centrifuged before first cleavage. *Developmental Biology* 108: 310-324

Bonting SJ, Brillouet C, Delmotte F (1989) Bioprocessing. In: *Life Sciences Research in Space*. Oser H, Battrick B (eds) Paris: European Space Agency, ESA SP-1105, Chapter 9, pp 109-117

Bouillon R, Hatton J, Carmeliet G (2001) Space biology. Cell and molecular biology. In: *A World Without Gravity*. Seibert G (ed) Noordwijk: European Space Agency, ESA SP-1251, pp 111-120

Brown A, Chapman DK, Lewis RF, Vendetti AL (1990) Circumnutations of sunflower hypocotyls in satellite orbit. *Plant Physiology* 94:233-238

Brown AH (1991) From Gravity and the Organism to Gravity and the Cell. *ASGSB Bulletin* 4: 7-18

Cogoli A, Valluchi-Morf M, Müller M, Briegleb W (1980) The effect of hypogravity on human lymphocyte activation. *Aviation Space Environmental Medicine* 51: 29-34

Cogoli A, Iversen TH, Johnsson A, Mesland D, Oser H (1989) Cell biology. In: *Life Sciences Research in Space*. Oser H, Battrick B (eds) Paris: European Space Agency, ESA SP-1105, Chapter 5, pp 49-64

Dournon C, Durand D, Tankosic C, Membre H, Gualandris-Parisot L, Bautz A (2001) Effects of microgravity on the larval development, metamorphosis reproduction of the urodele amphibian Pleurodeles waltl. *Development, Growth & Differentiation* 43: 315-326

Duprat AM, Husson D, Gualandris-Parisot L (1998) Does gravity influence the early stages of the development of the nervous system in an amphibian? *Brain Research Reviews* 28: 19-24

Dutemple L (2000) *The Complete Idiot's Guide to life Sciences.* Indianapolis, IN: Alpha Books

Eckart P (1996) *Spaceflight Life Support and Biospherics.* Dordrecht: Kluwer Academic Publishers, Space Technology Library

Elinson RP, Del Pino EM, Townsend DS, Cuesta FC, Eichorn P (1990) A practical guide to the developmental biology of terrestrial-breeding frogs. *Biological Bulletin* 179: 163-177

Gualandris-Parisot L, Husson D, Bautz A, Durand D, Aimar C, Membre H, Duprat AM, Dournon C (2002) Effects of space environment on the embryonic development up to hatching of salamander eggs fertilized and developed during orbital flights. *Biological Science in Space*

Godwin R (1999) *Apollo 12 NASA Mission Reports.* Burlington, Canada: Apogee Books, CG Publishing Inc.

Hammond TG, Lewis FC, Goodwin TJ, Linnehan RM, Wolf DA, Hire KP, Campbell WC, Benes E, O'Reilly KC, Globus RK, Kaysen JH (1999) Gene expression in space. *Nature Medicine* 5: 359

Hughes-Fulford M, Lewis M (1996) Effects of microgravity on osteoblast growth activation. *Experimental Cell Research* 224:103-109

Ijiri K (1997) Explanations for a video version of the first vertebrate mating in space: A fish story. *Biological Science in Space* 11: 153-167

Imbert M (1979) Development of the visual system: Role of early experience. *Journal of Physiology* 75: 207-217

Ingber DE (1998) The architecture of life. *Scientific American* 278: 48-57

Ingber DE (1999) How cells (might) sense microgravity. *FASEB Journal* 13: S3-S15

Jones TA (1992) Gravity and the ontogeny of animals. *The Physiologist* 35: S77-79

Klaus DM (1998) Microgravity and its implication for fermentation technology. *Trends in Biotechnology* 16:369-373

McKay DS, Gibson EK, Thomas-Keprta KL, Vali H, Romanek CS, Clemett SJ, Chillier XDF, Maechling CR, Zare RN (1996) Search for life on Mars: Possible relic biogenic activity in Martian meteorite ALH84001. *Science* 273: 924-930

McLaren A (1989) Developmental biology. In: *Life Sciences Research in Space.* Oser H, Battrick B (eds) Paris: European Space Agency, ESA SP-1105, Chapter 3, pp 31-36

Mills PJ, Meck JV, Waters WW, D'Aunno D, Ziegler MG (2001) Peripheral leukocyte subpopulations and catecholamine levels in astronauts as a function of mission duration. *Psychosomatic Medicine* 63: 886-890

Moody SA, Golden C (1999) *Developmental Biology Research in Space: Anticipating the International Space Station.* Recommendations from the September 1999 Developmental Biology Workshop sponsored by the International Space Life Sciences Working Group in Woods Hole, MA

Musgrave ME, Kuang A, Porterfield DM (1997) Plant reproduction in spaceflight environments. *Gravitational Space Biology Bulletin* 10: 83-90

Perbal G (2001) The role of gravity in plant development. In: *A World Without Gravity.* Fitton B, Battrick B (eds) Noordwijk: ESA Publications Division, ESA SP-1251, pp 121-136

Perbal G, Driss-Ecole D (1994) Sensitivity to gravistimulus of lentil seedling roots grown in space during the IML-1 mission of Spacelab. *Physiology Plant* 70: 119-126

Perbal G, Driss-Ecole D, Tewinkel M, Volkmann D (1997) Statocyte polarity and gravisensitivity in seedling roots grown in microgravity. *Planta* 203: S57-S62

Pinsky LS, Osborne WZ, Hoffman RA, Bailey JV (1975) Light flashes observed by astronauts on Skylab 4. *Science* 188: 928-930

Planel H, Tixador R, Richoilley G, Gasset G, Templier J (1985) Respective role of microgravity and cosmic rays on Paramecium tetraurelia cultured aboard Salyut 6. *Acta Astronautica* 12 :443-6.

Planel H et al. (1987) Influence on cell proliferation of background radiation or exposure to very low, chronic gamma radiation. *Health Physics* 52: 571-578

Planel H et al. (1994) Influence of a long duration exposure, 69 months, to the space flight factors in Artemia cysts, tobacco and rice seeds *Advances in Space Research* 14: 31-32

Réber A, Courjon JH, Denise P, Clément G (2003) Vestibular decompensation in labyrinthectomized rats placed in weightlessness during parabolic flight. *Neuroscience Letters* 344: 122-126

Ronca AE, Alberts JR (1997) Altered vestibular function in fetal and newborn rats gestated in space. *Journal of Gravitational Physiology* 4: P63-P66

Ronca AE, Alberts JR (2000) Physiology of a microgravity environment selected contribution: Effects of spaceflight during pregnancy on labor and birth at 1 G. *Journal of Applied Physiology* 89: 849-854

Scharf SR, Gerhart JC (1980) Determination of the dorsal-ventral axis in eggs of Xenopus laevis: complete rescue of uv-impaired eggs by oblique orientation before first cleavage. *Developmental Biology* 79: 181-198

Souza KA, Black SD, Wassersug RJ (1995) Amphibian development in the virtual absence of gravity. *Proceeding of the National Academy of Sciences* 92: 1975-1978

Souza K, Theridge G, Callahan PX, eds. (2000) *Life Into Space*. Space Life Sciences Experiments. Ames Research Center, Kennedy Space Center, 1991-1998. Life Sciences Division, NASA Ames Research Center, Moffetts Field, CA: NASA SP-2000-534

Summerlin LB (ed) (1977) *Skylab, Classroom in Space*. NASA Marshall Spaceflight Center, Huntsville, AL: NASA SP-401

Tash JS, Kim S, Schuber M, Seibt D, Kinsey WH (2001) Fertilization of sea urchin eggs and sperm motility are negatively impacted under low hypergravitational forces significant to spaceflight. *Biology of Reproduction* 65: 1224-1231

Tobias CA, Todd P (1974) *Space Radiation Biology and Related Topics*. New York: Academic Press

Vandenburgh H, Chromiak J, Shansky J, Del Tatto M, Lemaire J (1999) Space travel directly induces skeletal muscle atrophy. *FASEB Journal* 13: 1031-1038

Walton K (1998) Postnatal development under conditions of simulated weightlessness and spaceflight. *Brain Research Reviews* 28: 25-34

Wayne R, Staves MP, Leopold AC (1992) The contribution of the extracellular matrix to gravisensing in characean cells. *Journal of Cellular Science* 101: 611-623

Wassersug RJ (2001) Vertebrate biology in microgravity. *American Scientist* 89: 46-53

Wilkins MB (1989) Plant biology. In: *Life Sciences Research in Space*. Oser H, Battrick B (eds) Paris: European Space Agency, ESA SP-1105, Chapter 4, pp 37-48

Wilson A, ed. (2003) *European Utilisation Plan for the International Space Station*. Nordwijk: ESA Publications Division, ESA SP-1270

Zisckind N, Elinson RP (1990) Gravity and microtubules in dorsoventral polarization of the Xenopus egg. *Development and Growth Differentiation* 32: 575-581

Additional Documentation:

A Strategy for Research in Space Biology and Medicine in the New Century (1998) National Academy of Science. National Research Council Committee on Space Biology and Medicine. Mary J. Osborn, Committee Chairperson. Washington D.C.: National Academy Press. Also available at the following URL address: http://www.nas.edu/ssb/csbm1.html

Cell & Molecular Biology Research in Space (1999) *FASEB Journal*, Volume 13, Supplement

Plant Biology in Space: Proceedings of the International Workshop (1997) *Planta*, Volume 203, Supplement

Chapter 3

THE NEURO-SENSORY SYSTEM IN SPACE

To be aware of the environment, one must sense or perceive that environment.[*] The body senses the environment by the interaction of specialized sensory organs with some aspect or another of the environment. The central nervous system utilizes these sensations in order to coordinate and organize muscular movements, shift from uncomfortable positions, and adjust properly. One relevant question is "what is the relative contribution of gravity to these sensory and motor functions?" This chapter reviews the effects of microgravity on the functioning of the sensory organs primarily used for balance and spatial orientation. Disorientation and malaise so frequently encountered during early exposure to microgravity and on return to Earth are described. Theories and actual data regarding the role of the central nervous system in the adaptation of sensory-motor functions (including the control of posture, eye movements, and self-orientation) to changing environmental gravity levels are explored. (For a more comprehensive review of experimental space research conducted in this area during the past 30 years, see Clément and Reschke 1996).

Figure 3-01. Balance relies on vision and touch, as well as signals from the vestibular organs in the inner ear and from the muscles. The body of this ice skater is tilted to align with the resultant of centrifugal force and gravitational force, but his head is almost upright to keep an unbiased visual image.

---

[*] The words "sense" and "perceive" are from Latin words: "sense" means "to feel", whereas "perceive" means "to take in through", i.e., to receive an impression of the outside world through some portion of the body.

91

1   THE PROBLEM: SPACE MOTION SICKNESS

The neuro-vestibular system consists of organs which sense the acceleration environment, nerves which transmit this information to the spinal cord and brain, and to the central nervous system which integrates this information so that we can determine our position and orientation relative to the environment. The vestibular organs in the inner ear detect and measure linear and angular accelerations. These responses, already complex, are further integrated with visual and proprioceptive inputs (Figure 3-02). In microgravity, some of these signals are modified, leading to misinterpretation and non-adequate responses by the brain. One of these responses is *space motion sickness* (SMS) (Figure 3-03).

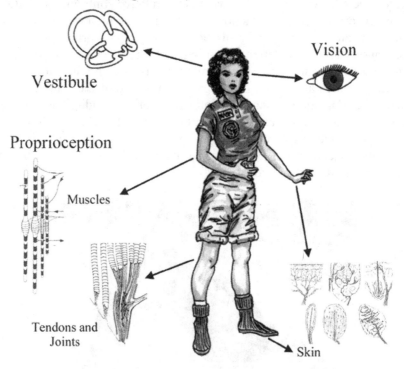

Vestibule

Vision

Proprioception

Muscles

Tendons and Joints

Skin

Figure 3-02. The eyes, the inner ear and the special receptors in the skin, muscles and joints all participate in maintaining posture and balance, and assist in our movements.

SMS is a special form of motion sickness that is experienced by some individuals during the first several days of exposure to microgravity. The syndrome may include such symptoms as depressed appetite, a nonspecific malaise, lethargy, gastrointestinal discomfort, nausea, and vomiting. As in other forms of motion sickness, the syndrome may induce an inhibition of self-motivation, which can result in decreased ability to perform demanding tasks in those persons who are most severely affected. Gastrointestinal symptoms have their onset from minutes to hours after orbital insertion. Excessive head movement early on-orbit commonly increases symptoms. Symptom resolution usually occurs between 30 and 48 hours, with a reported range of 12 to 72 hours, and recovery is rapid.

There were no reports of SMS in the Mercury and Gemini programs, while 35% of the Apollo astronauts developed symptoms. The incidence in the Skylab missions raised to 60%. About two-thirds of the Space Shuttle astronauts experienced some symptoms of SMS. There are no statistically significant differences in symptom occurrence between pilots versus non-pilots, males versus females, different age groups, or novices (first time flyers) versus veterans (repeat flyers.) An astronaut's susceptibility to SMS on his/her first flight correctly predicted susceptibility on the second flight in 77% of the cases (Davis et al. 1988). In other words, one astronaut who has been sick during his/her first flight is likely to be sick again during his/her subsequent flights.

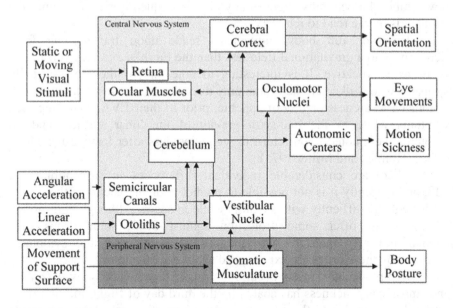

Figure 3-03. Inputs and outputs of the neuro-vestibular system. The information from the various sensory organs first reaches the lower portions of the brain. For most of our activities, we are not consciously aware of what is going on in our busy body when we sit, stand, walk, or run. However, certain sensations do eventually reach the cerebral cortex, and through them we remain consciously aware of the relative positions of our body parts. Space motion sickness is believed to be caused by a conflict among sensory information, through connections with the autonomic nervous system.

SMS affects a similar percentage of both U.S. and Russian crews. Symptom recurrence at landing, also called "Mal de Débarquement" reportedly afflicts 92% of Russian cosmonauts returning from longer missions (Gorgiladze and Bryanov 1989). No reports of *mal de débarquement* were noted in the U.S. Shuttle program. However, many astronauts returning to Earth after long-duration stay on board the ISS now experience this syndrome.

Microgravity by itself does not induce space sickness. There were no reports of motion sickness during the Mercury and Gemini spaceflights. As the volume of spacecrafts has increased (allowing more mobility) the incidence of SMS has increased as well. Movements that produce changes in head orientation seem necessary to induce SMS symptoms. In particular, many crewmembers report that vertical head movements (rotation in the pitch or roll planes) are more provocative than horizontal (yaw) head movements (Oman et al. 1990). However, once sickness has been well established, head movements in any plane are generally minimized by the affected crewmember. Indeed, movement of any kind is frequently restricted until the astronaut is on the road to recovery.

Head or full body movements made upon transitioning from microgravity to a gravitational field less than the Earth's, and vice versa, may not be as provocative. It is interesting to note that of the twelve Apollo astronauts who walked on the Moon, only three reported mild symptoms, such as stomach awareness or loss of appetite, prior to their EVA. None reported symptoms while in the one-sixth gravity of the lunar surface, and no symptoms were noted upon return to weightlessness after leaving the Moon surface (Homick and Miller 1975).

There are considerable individual differences in susceptibility to SMS, and currently it is not possible to predict with any accuracy those who will have some difficulty with sickness while in space. Although anti-motion sickness drugs offer some protection against SMS, some drugs (i.e., scopolamine) may interfere with the adaptation process, and symptoms controlled by these drugs are experienced again once treatment ceases.

Symptoms have rarely occurred during extravehicular activity (EVA). Since most space sickness has abated by the third day of flight, mission rules restrict EVA until the third mission day. Nevertheless, some astronauts medicate prior to space walking. The minimum flight duration for the Space Shuttle is also three days, to reduce the probability for astronauts, in particular the pilots, to be incapacitated by SMS symptoms prior to entry and landing (Davis et al. 1988).

2   VESTIBULAR FUNCTION

The gravity vector is a fundamental factor in human spatial orientation, which results from the integration of a complex of sensory inputs coming from the vestibular organs (in the inner ear), the eyes (mostly from peripheral retina), and tactile and proprioceptive receptors located in the skin, joints, muscles, and viscera.

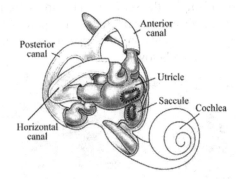

Figure 3-04. Schematic of the vestibular system in the inner ear showing the three semicircular canals (anterior, posterior, horizontal) and the two otolith organs (utricle and saccule).

2.1   The Vestibular System

2.1.1   The Vestibular End Organs

The vestibular system's main purpose is to create a stable platform for the eyes so that we can orient to the vertical—up is up and down is down—and move smoothly. The inner ear contains two balance-sensing organs: one is sensitive to linear acceleration, the other to angular acceleration (Figure 3-04).

The *linear* acceleration sensitive organ, comprised of the saccule and utricle, sends messages to the brain as to how the head is translated or positioned relative to the force of gravity. It contains two tiny sacs filled with fluid, the saccule and the utricle, and lined along their inner surface with hair cells of various lengths (Figure 3-05). Overlying the hair cells is a gelatinous matrix (the otoconia) containing solid calcium carbonates crystals (the *otoliths*, meaning "ear stone" in Greek). During linear acceleration, the crystals, being denser than the surrounding fluid, will tend to be left behind due to their inertia. It has been demonstrated that the resultant bending of the cilia causes cell excitation when the bending is toward the kino-cilium (the longest hair cell), and inhibition when away from the kino-cilium. During head motion, the weight and movement of the otoliths stimulate the nerve endings surrounding the hair cells and give the brain information on motion in a particular direction (up, down, forward, backward, right, left) or tilt in the sagittal (pitch) or the frontal (roll) plane.

The *angular* acceleration sensitive organ is comprised of three semicircular canals. It detects angular acceleration through the inertial movement of the liquid (the endolymph) within each canal and provides the brain with information about rotation about the three axes: yaw, pitch, and roll. The semicircular canals do not react to the body's position with respect to gravity. They react to a change in the body's position. In other words, the semicircular canals do not measure motion itself, but change in motion. Not surprisingly, the semicircular canals are not affected by spaceflight, as shown by the absence of changes in the perception of rotation or in the compensatory eye movements in response to rotation both in-flight and after flight (see Section 3.5 below).

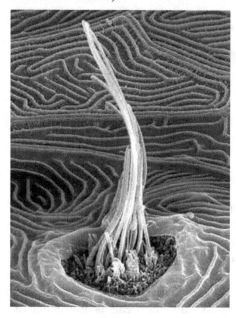

Figure 3-05. Electronic microscopy of hair cells cilia of a Zebrafish saccule. The otolith crystals that normally rest on top of the hair cells are not visible.

2.1.2 Linear Acceleration and Gravity

When our head is horizontal the hair cells in the utricles are not bent and this stimulation is interpreted as signifying "normal posture". If our head is tilted forward, the otoliths shift downward under the action of gravity, bending the hair cells. If we translate backward, again there is a shift of the otoliths forward due to the inertial forces. Thus, an equivalent displacement of the otoliths (and consequently the same information is conveyed to the central nervous system) can be generated when the head is tilted 30 deg forward, or when the body is translating at 0.5 g backward (Figure 3-06). This example simply illustrates Einstein's principle stating that, on Earth, all linear accelerometers cannot distinguish between an actual linear acceleration and a head tilt relative to gravity.[*]

---

[*] In a normal situation, the brain would easily distinguish between a tilt of the head relative to gravity and a head translation by comparing the sensory information from the otolith organs with that from the eyes or muscles proprioceptors. But in complete

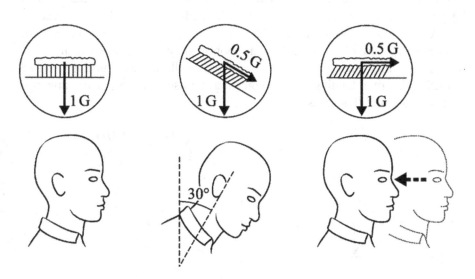

Figure 3-06. The otoliths bend the hair cells of the utricles the same way when the head is maintained at a constant tilt angle of 30deg relative to gravity and when the whole body in translated backwards at 0.5 g.

On Earth, otolith signals can be interpreted as either linear motion (translation) or as tilt with respect to gravity. Because stimulation from gravity is absent in weightlessness, interpretation of otolith signals as tilt is inappropriate (Figure 3-07). Therefore, it is possible that during adaptation to weightlessness, the central nervous system reinterprets all otolith signals to indicate translation. This hypothesis is known as the "Otolith Tilt-Translation Reinterpretation" (OTTR). This central reinterpretation would persist following return to Earth, and be at the origin of spatial disorientation, until re-adaptation to the normal gravity environment occurs (Parker et al. 1985, Young et al. 1986).

Evidence for the OTTR hypothesis comes from subjective reports by astronauts returning from spaceflight who have a sense of body translation when they voluntarily pitch or roll their head. For example, many experience a backward translation when they pitch their head forward, or a rightward translation when they roll their head to the left. The utricle and the saccule are not located at the axis of head rotation during roll or pitch head movements.

---

darkness, there could be a conflict between the proprioceptive input (e.g., signaling that the head is tilted) and the otolith input (e.g., signaling that the head translates).

Therefore, this movement must evoke otolith stimulation which could readily be perceived as translation during and immediately after landing. Such a misleading interpretation of otolith signals might be responsible for the staggering posture of the astronauts as soon as they land. The astronauts tend to lean to the outside of the turn when walking and turning corners immediately after landing, also suggesting a misevaluation of the apparent vertical from otolith signals.

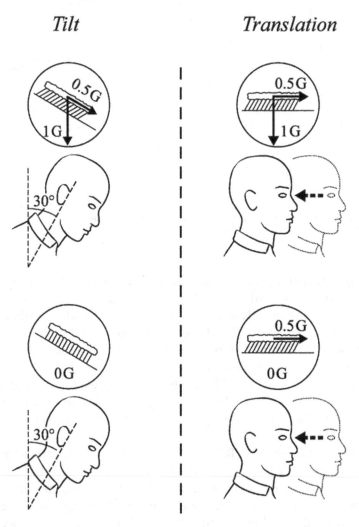

Figure 3-07. In microgravity, the otoliths are stimulated by head translation, but not by head tilt. Consequently, it is hypothesized that, after a period of adaptation, the brain reinterprets all otolith signals as signaling head translation.

The OTTR hypothesis has been the theoretical basis of much space research on the neuro-vestibular system for the past 15 years. I was fortunate enough to be able to perform a space experiment that tested this hypothesis in 1998. This experiment, which flew on board the Neurolab mission (STS-90), used a human-rated centrifuge constructed by the European Space Agency. On Earth, when an individual is rotated in a centrifuge in darkness, he senses the direction of the resultant gravito-inertial force and regards this as the vertical. If a centrifugal force equivalent to 1 g is directed sideways, the gravito-inertial force is displaced 45 deg relative to the upright body, and the subject has a sense of being tilted by 45 deg to the outside (Figure 3-08). In microgravity, however, the gravitational component is negligible and the gravito-inertial force is equivalent to the centrifugal force. This force could be interpreted either as a 90 deg tilt of the body, or a whole body translation in the opposite direction. During the Neurolab mission, four astronauts were asked to report their perceived angle of tilt during steady-state centrifugation in darkness throughout the flight and during the postflight re-adaptation period. Centrifugation was always perceived as tilt, not translation. Therefore the findings do not support the OTTR hypothesis. Despite the fact that the otoliths do not respond to head tilt in orbit, the brain continues to sense a steady-state linear acceleration applied to the otoliths as the upright in all circumstances.

The debate regarding the OTTR hypothesis is still raging. Some have proposed that the OTTR only occurs during voluntary head movements, or only during rotational head movements, or that OTTR has to be frequency dependent. Centrifugation, by applying very low frequency passive linear acceleration to the entire body, would thus not elicit OTTR. I am currently conducting a follow-up study on astronauts returning from spaceflight, by spinning them about a tilted axis (off-vertical axis rotation) at various frequencies to further address this hypothesis.

The Neurolab centrifuge experiment, however, brought another interesting result. At the beginning of the flight, during a 1-g centrifugation in darkness, the astronauts perceived a 45-deg tilt to the side, very much like on Earth. However, as the mission progressed, they felt more and more tilted, until a 90-deg tilt to the side on flight day 16. This simple result indicates that the brain does not continuously calculate the direction of gravity, but uses an internal estimate of gravity whose weighting changes during spaceflight. The internal estimate normally used on Earth (1 g) carries over to the early period of exposure to weightlessness, and therefore the astronauts continue to perceive a 45-deg tilt, despite the absence of sensed gravity. After a period of adaptation, the internal estimate declines to zero and the astronauts perceive a full body tilt to the side (Clément et al. 2001, Clément et al. 2003).

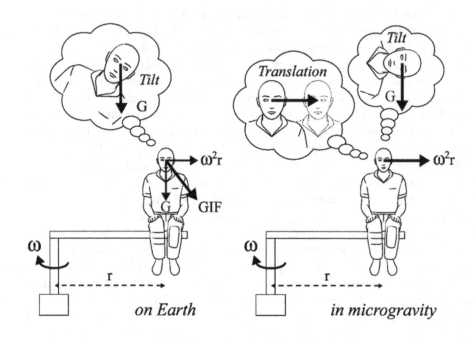

Figure 3-08. On Earth, a subject sitting at the end of a centrifuge arm in darkness adopts the direction of the gravito-inertial force (GIF) as the new direction of "gravity". In microgravity, the GIF is equivalent to the centrifugal force and the subject could perceive either a body tilt or a body translation, depending on the interpretation of the otolith signals by the central nervous system.

2.1.3   Changes in the Vestibular Receptors during Spaceflight

Although it is difficult to measure changes in the vestibular end organs directly, several attempts have been made to examine the question "Is there anatomical and physiological changes in the vestibular end organs and their primary afferents after exposure to microgravity?"

Experiments on frogs have revealed no alteration of the sensory epithelium of the vestibular organ of adults returned from an 8-day stay aboard the Russian Mir space station, or following larval development in microgravity. However, changes in the structure of the otoconia in rats had been observed during earlier missions (Cosmos-782). This degeneration of the otolith crystals could occur because of changes in body calcium, protein metabolism, and calcium exchange. In addition, it is unclear how much of these changes were due to the high accelerations experienced by the animals during take-off and landing.

More recent Spacelab experiments indicated no deleterious effects in the otoconia of the otoliths from rodents who flew as compared with the ground controls. However, an unexpected change found during the Spacelab SLS-1 mission, and later confirmed during the Neurolab mission, was an increase (by a factor of 12) in the number of synapses in hair cells from the in-flight maculae as compared with the control data. These findings suggest that mature utricular hair cells retain synaptic plasticity, permitting adaptation to an altered environment. Consistent with these results is data that show a decrease in synapse activity in centrifuged rats. These data suggest that the maculae adapt to g-forces changes in either direction by up- or down-regulation of synaptic contacts in an attempt to modulate neural inputs to the CNS (Ross and Tomko 1998).

Primary afferent fibers of the vestibular nerve are relaying the information originating at the hair cells to the brainstem (within each nerve are also efferent fibers from the CNS which provide neural feedback to modulate the activity of the peripheral organs). The resting activity of single otolith afferents and their response to centrifugal forces were found to be different in microgravity compared to the ground condition in frogs. Recently, a study recording the vestibular nerve impulse data from the oyster toadfish during the Neurolab mission confirmed these results. On the other hand, the spontaneous firing rate of single horizontal semicircular canals afferents did not change postflight relative to preflight in two flight monkeys. However, these monkeys were restrained in a laboratory chair, thus preventing any movements of the head during the flight. It is known that movement and interaction with the environment are necessary factors to drive adaptive changes. For example, vestibular patients show a faster recovery when moving around after vertigo crisis or unilateral surgery.

Few experiments on the early development of the vestibular system have been carried out in space. This is an interesting research topic since in all species the vestibular system begins to respond to stimulation (linear or angular acceleration) prior to hatching or birth, in contrast to hearing or vision, which can be postnatal in some species. Mammalian offspring emerge from the birth canal in a species-typical orientation which, for rats and humans, is head first. Fetuses typically achieve the appropriate orientation via active, in-utero behavior. Perhaps the vestibular system is employed for this early task. Indeed, many infants born in the breach position are born with vestibular disorders. Also, the so-called "righting response", by which the newborn mammals actively adjust from a supine to a prone position, is disrupted by induced vestibular disorders during development.

In the development of the visual system, activity in the retinal pathway influences the specification of those connections which determine how visual information is processed in the cerebral cortex. In every other sensory system known, especially those that make up the neural space maps in the brainstem, sensory stimulation has been implicated in the initial specifications of the connections and physiological properties of the constituent neurons. Only in the utricle and saccule gravitational pathway has it been impossible to study the role of sensory deprivation, because there is no way to deprive the system of gravitational stimulation on Earth. For this reason, experiments in microgravity should be planned to test the hypothesis that gravity itself plays a role in the development and maintenance of the components of the vestibular system. These components include both the vestibular receptors of gravity (i.e., the sensory hair cells in the utricle and saccule, vestibular ganglion cells that form synapses with vestibular hair cells, and vestibular nuclei neurons) and the motor neurons. The latter receive input from axons of the vestibular nucleus neurons composing the vestibular reflex pathways. The vestibular system also receives inputs from the proprioceptive system, involved in the control of muscle length and tension, and from the visual system, involved in the control of eye movements. Little is known about the exact nature of these interactions and virtually nothing concerning the development of these connections.

2.2   The Other Senses

In common speech, five different senses are usually recognized: sight (vision), hearing, taste, smell, and touch. Of these, the first four use special organs (the eye, ear, tongue, and nose, respectively), whereas the last use nerve endings that are scattered everywhere on the surface of the body, as well as inside the body (visceral sensations). Proprioceptive sensations arise from organs within the body, from muscles, tendons, and joints. How far these five senses are affected by spaceflight is uncertain.

2.2.1   Vision

The visual environment in space is altered in several ways. First, objects are brighter under solar illumination. Earth's atmosphere absorbs at least 15% of the incoming solar radiation. Water vapor, smog, and clouds can increase this absorption considerably. In general, this means that the level of illumination in which astronauts work during daylight is about one-fourth higher than on Earth. Second, there is no atmospheric scattering of light. This causes areas not under direct solar illumination to appear much darker and results in a transformation of normal visual intensity relationships.

Early anecdotal reports that orbiting astronauts were able to see objects such as ships, airplanes, and trucks with the naked eye suggested improved visual acuity in space. Extensive testing of Gemini astronauts was performed using a small, self-contained binocular optical device containing an array of high- and low contrast rectangles. Astronauts judged the orientation of each rectangle and indicated their response by punching holes in a record card (Figure 3-09). Another method, taking into account the particularity of the visual environment of space described above, also used large rectangular patterns displayed at ground sites in Texas and Australia. Astronauts had to report the orientation of the rectangles. Display were changed in orientation between passes and adjustments for size were made in accordance with slant range, solar elevation, and the visual performance of astronauts on preceding passes. Results with both measurement methods indicated that visual performance was neither degraded nor improved during spaceflight. The astronauts' reported ability to detect moving objects (airplanes and ships) was probably based on the detection of turbulence or waves behind the vehicles. Also, the colors contrast might improve the ability to identify features, as Astronaut William Pogue described it during his Skylab mission "We were able to see icebergs about a hundred yards in diameter quite easily because of the large contrast of white ice with the dark blue sea".

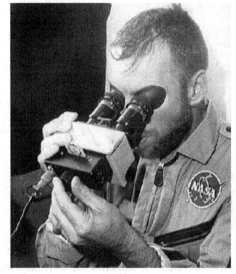

Figure 3-09. Astronaut Frank Borman performing visual acuity tests during the Gemini-7 mission. (Credit NASA)

More refined visual testing has been performed on several Shuttle flights using a specially-designed visual test apparatus to assess contrast sensitivity, phoria, eye dominance, flicker fusion frequency, stereopsis, and acuity (Figure 3-10). With the exception of reduced contrast sensitivity, no significant changes due to weightlessness were found. These changes were too small to impact operational performance. However, if contrast sensitivity continues to change during longer exposure to weightlessness, the decrement could become operationally significant.

An interesting observation is that some astronauts have described a decrease in their ability to see clearly at close range when in space. Interestingly enough, most of the astronauts experiencing this change were in their early forties and could see clearly without reading glasses when they were on the ground. One theory why this might happen is that the eye is like a water balloon. Rest it on a table and it gets longer as it flattens out (which is the normal condition on Earth). Put that balloon in space and it shortens, becoming more round. The eye could do the same thing and when it shortens it becomes farsighted, causing more difficulty seeing objects up close.

The effects of weightlessness on vision could also be attributable to the headward shift of body fluids following orbital insertion and the subsequent pooling of fluid in the trunk and lower extremities after return to Earth. Microgravity is known to produce a headward shift of 700-1400 mL of fluid (see Chapter 4, Section 3.2). Photographic studies showed a significant decrease in the size of the retinal vasculature after flight. Intraocular pressure rises during flight and drops below pre-flight levels after landing. These intraocular tension changes are probably associated with the fluid shifts. The effects of long-duration flights on these functions are unknown and remain be determined.

Figure 3-10. Near-visual acuity test performed by an astronaut on board the Space Shuttle. (Credit NASA)

It is also worth to note here the light flashes perceived by the astronauts in the absence of normal visual stimulation, which are caused by heavy ionized cosmic particles passing through retinal cells (see Chapter 2, Section 5.1). Although no performance disturbance has been associated with these light flashes, it is likely that the flashes mask transient visual stimuli.

Many astronauts have reported impairment in evaluating distances, both on the Moon [*] and during orbital flights. The collision of the Progress

---

[*] The following sentences are exerted from the postflight debriefings of the astronauts who walked on the Moon during the Appolo-12 mission (Godwin 1999):

"Everything looked a lot smaller and closer together in the air than it turned out to be on the ground. When we were on the ground, things that were far away looked a lot closer than they really were. The thing that confused me was that we were so close to the Surveyor crater. I didn't realize we were as close to it as we were." —Pete Conrad.

"In appearances, it took us a long time to convince ourselves that some of the craters which looked so close were really much farther away." —Alan Bean.

spacecraft with the Mir station in 1996 could have been due to a misevaluation by the Mir cosmonauts of the actual distance between the two vehicles. I have a personal interpretation for these changes in distance perception. I think that perception of absolute distances is altered after a long exposure in a confined environment where there is only a close distance range. It is known that distances between objects and the observer are altered because there are no objects with familiar size (such as trees, people, vehicles, etc.) in the background. This is the case on the Moon, inside a space vehicle, or any other confined place. People who spend a long time in enclosed chambers (such as divers or submariners) have trouble evaluating distance when they get out. For this reason, submarine crewmembers are not allowed to drive immediately after returning from long duties in the confined space of a submarine.

The objects seen inside a space station stay within distances of several cm to a few meters, whereas the objects outside (the Earth or the stars) are very far away. There is not an intermediate distance range. It is therefore possible that the perception of distances of objects is altered in this intermediate range. I have proposed an experiment based on virtual reality to test this hypothesis. This experiment has been accepted by ESA and NASA and is manifested to fly on board the ISS in the near future.

2.2.2 Hearing

"In space, no one can hear you [scream]". This cliché, which is commonly used in science fiction movies, has apparently not attracted the interest of scientists for study hearing during spaceflight, since very little data is available yet.

Several aspects of spaceflight can have an impact on hearing capability: a) life support equipment is continuously running (ranging from 64 dBA for the air conditioning to 100 dBA for some vent relief valves) and the noise reverberates through the spacecraft's structure; b) astronauts spend 24-hrs a day in the office, always close to noise sources; and c) there is no privacy, with a constant interaction with other crewmembers. Thus quietness periods such as on Earth do not exist: earplugs can reduce noise but not vibrations.

Spaceflight raises the spectrum of noise questions: its effect on perception and performance, adaptation effects, the fatiguing and annoying aspects of noise, and individual sensitivity differences. Because certain

---

"When we were at the ALSEP site, it looked as if we were about 450 feet west and 50 feet north of the position of the LM. It was a pretty good level site. Later when I got back to the LM and looked back, I noticed it didn't look as if the site were that far away. This was the continual problem we had, trying to judge distances." —Alan Bean.

minimum noise levels are always present, spaceflight potentially constitutes a more stressful noise environment than a simple consideration of decibel levels would imply.[*]

Although very stringent noise requirements for ISS result in a noise environment comparable to home and office, intelligibility of hearing as noise increases may vary across individuals. For example, it is known that both the lack of language proficiency and the reverberance in a room impair hearing. The performance of a native English speaker on board the ISS at 60 dBA therefore must to be compared with that of a non-native English speaker at 68 dBA! In addition, low noise levels can also be annoying and affect individual and group (communication) behavior.

The investigation of hearing in astronauts is difficult to conduct during spaceflight because classical hearing assessment techniques do not work in the noisy environments often found in spacecrafts (no soundproof laboratory). Since crew are at risk for hearing loss due to noise levels often encountered during spaceflight, techniques and investigation to track this loss are needed during and after the mission.

Nevertheless, auditory brain stem response recordings were investigated during Shuttle flights. No significant differences were observed between mean latency values for any potential on the ground or during flight, suggesting that the auditory function is not altered in microgravity (Thornton et al. 1985). Another experiment performed on Mir showed that the localization of a sound source in microgravity was within the same range as on Earth, i.e., between 1 deg and 2 deg. Since the faculty of localizing sound sources depends on normal binaural hearing, it was concluded from this study that hearing was not altered in cosmonauts.

2.2.3   Smell and Taste

It is well known that during spaceflight, astronauts ask for more spices and condiments to add taste to the prepared food. Diminished sensitivity to taste and odor could result from the passive nasal congestion reported in conjunction with the headward shift of fluid. Taste, particularly the non-volatile component mediated by the taste buds, may be susceptible to threshold shifts in microgravity, because of a reduced mechanical stimulation as a result of changes in the convection process.

Evaluation of olfactory recognition (using paper impregnated with lemon, mint, vanilla, or distilled water) and taste recognition (using solution of solutions of sucrose, urea, sodium chloride, and citric acid) demonstrated

---

[*] The threshold for hearing is defined as 0 dBA, corresponding to corresponds to 0.00000003% of atmospheric pressure (1/30 billionth). The threshold for pain is 0.03% of atmospheric pressure, or approximately 120 dBA. Even what we call "silence" on Earth is in fact a background noise of about 40 dBA.

no subjective changes in smell or taste function postflight. However, there were large differences among individuals (some of them could have been due to the reminiscence of space motion sickness symptoms!).

Materials used in spaceflight are subjected to testing for odor as well as for flammability and toxicity. Odor evaluations are made by panels of test subjects who rate materials on a scale from 0 (undetectable) to 4 (irritating) with a score of 2.5 (falling between "easily detectable" and "offensive") considering passing. Nevertheless, because particulate matter does not settle out in weightlessness, odor problems in a space habitat may be more severe than under similar Earth conditions.[*]

Also, responses to odors can be accentuated by the presence of visual cues. For example, during the earlier Spacelab missions, crewmembers complained of disturbing odors, which they attributed to the primates, and test rats which shared their facilities and which were in view (Connors et al. 1985). In later missions, the animal cages were placed in visually separated areas and no odor problems were mentioned.

2.2.4 Proprioception

The absence of gravity modifies the stimuli associated with proprioception and impact spatial orientation, including knowledge of position in the passive limb, difficulty of pointing accurately at targets during voluntary limb movement, modification of tactile sensitivity, and changes in the perception of mass. However, the nature of proprioceptive changes in microgravity has been poorly studied. There is almost no space study of neck and joint angle sensors, and on the role of localized tactile cues in the perception of body verticality.

When crewmembers point at remembered target positions with they eyes closed, they make considerable errors and tend to point low. When they are asked to reproduce from memory the different positions of a handle, the accuracy of setting the handle to a given position is significantly lower with an error towards a decrease of handle deflection angle. Also, when trying to touch various body parts, they usually note that their arms are not exactly where expected when vision is restored. The problem is that these examples

---

[*] An astronaut onboard the ISS reported: "I had the pleasure of operating the airlock for two of my crewmates while they went on several space walks. Each time, when I repressed the airlock, opened the hatch and welcomed two tired workers inside, a peculiar odor tickled my olfactory senses. At first I couldn't quite place it. It must have come from the air ducts that re-pressed the compartment. Then I noticed that this smell was on their suit, helmet, gloves, and tools. [...] The best description I can come up with is metallic; a rather pleasant sweet metallic sensation. It reminded me of my college summers where I labored for many hours with an arc-welding torch repairing heavy equipment for a small logging outfit. It reminded me of pleasant sweet smelling welding fumes. That is the smell of space" (Pettit 2003).

are suggestive of either degradation in proprioceptive function, or an inaccurate external spatial map, or both (Watt 1997, Young et al. 1993).

An elegant way to evaluate changes in the proprioceptive function is to measure the subjective sensation generated by the stimulation of proprioceptive receptors. A classic technique consists in vibrating a muscle tendon to elicit illusory limb movement. Using this technique, it was observed that the illusion of body tilt forward or backward was less pronounced in-flight than postflight during vibration of lower leg muscles. One interpretation of this result is that the utricles and saccules are unloaded in microgravity and decrease their descending modulation of alpha and gamma motoneurons, resulting in decreased tonic vibration reflexes.

A nice illustration of an alteration of proprioceptive inputs during the early exposure to microgravity is the impossibility for an astronaut to maintain a "vertical" posture, perpendicular to the foot support, in absence of visual information (Figure 3-11). The large body tilt observed in these conditions reveals an inaccuracy in the proprioceptive signals from the ankle joint (or in their central interpretation). After flight day 3, however, the astronauts are able to maintain an upright posture, suggesting that adaptive processes take place quite rapidly (Clément et al. 1988).

Figure 3-11. An astronaut with the feet attached to the floor of the Space Shuttle and placed in darkness using an occluding goggle is instructed to maintain an "upright" posture on flight day 2. In absence of gravitational and visual inputs, his body is tilted forward, suggesting a recalibration of the proprioceptive inputs from the ankle joint. (Credit NASA)

Among the somato-sensory systems projecting to the neuro-vestibular system, the position receptors of the cervical column (neck receptors) play an important role. During the Spacelab-D1 mission, the trunk of a crewmember was passively bent sidewards or forewards, while keeping his head fixed to the floor of Spacelab, thus stimulating the neck receptors. The crewmember reported an illusory rotation of a head-fixed target cross seen in the monitor of his helmet, which was entirely due to the stimulation of the cervical position receptors, since the otoliths were not stimulated.

Another interesting feature of microgravity is that it allows to separate between two distinct physical concepts, mass and weight, which both produce similar sensations of heaviness. On Earth, weight can be judged passively through the pressure receptors in the skin, if the object is placed upon a supported limb. Weight can also be judged actively, if the object is held against the force of gravity by the muscular effort, or is repeatedly lifted. Mass can only be judged actively, derived from the force required to produce a given acceleration, or from the acceleration produced by imparting a given force. Thus, active weight perception usually includes mass perception. It is therefore difficult to investigate weight without mass during active movement, except in weightlessness. Using balls of various masses that the astronauts shook up and down moving their arms, it was found that discrimination in the mass of objects in microgravity was poorer than in normal gravity. Weight discrimination was impaired for two or three days postflight, while crewmembers felt their bodies and other objects to be extra heavy. The impairment in-flight was partly due to the loss of weight information (a reduction in the pressure stimulation), and probably also to incomplete adaptation to microgravity. The increase in apparent heaviness of objects reported for static weight judgements postflight suggests that some central re-scaling of the static pressure systems had occurred (Ross et al. 1986).

3   EFFECTS OF SPACEFLIGHT ON POSTURE AND MOVEMENT

Postural activity is the complex result of integrated orientation and motion information from visual, vestibular, and somato-sensory inputs. These inputs collectively contribute to a sense of body orientation and, additionally, coordinate body muscle activities that are largely automatic and independent of conscious perception and voluntary control.

3.1   Rest Posture

Human factor studies, after investigating photographs taken during Skylab missions, have led to the NASA *Neutral Body Posture* model (Figure 3-12). This model is characterized by a forward tilt of the head (with the line of sight 25 deg lower than the body-centered horizontal reference), shoulders up (like a shrug), and arms afloat, up and forward with hands chest high.

Recent investigations, taking into account body size, gender, and mission duration suggest, however, that the neutral body posture model is too generalized, and should be modified with additional data to provide more representative spaceflight crew postures. However, it is unclear how the direction of the line of sight has been evaluated from the Skylab photographs. Also, the downward deviation of gaze in microgravity in this model is in contradiction with the results of several space experiments that actually measured the eye deviation during spaceflight (see Section 3.5).

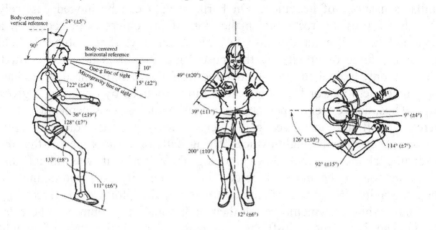

Figure 3-12. The neutral body posture model. (Credit NASA)

3.2   Vestibulo-Spinal Reflexes

Two of the more dramatic responses to orbital flight have been postural disturbances and modified reflex activity in the major weight-bearing muscles. For example, monitoring the Hoffman reflex (or H-reflex), which takes advantage of the anatomical pathways that link the otoliths and spinal motoneurons, has been selected as a method of monosynaptic spinal reflex testing to assess otolith-induced changes in postural muscles. By contrast to doctor tapping a patient's knee to produce the proverbial "knee jerk" reflex, during H-reflex the stimulus is an electrical shock to sensory fibers coming from stretch receptors in the calf (*Soleus*) muscle, and the response is the electrical activity recorded from the muscle. Each time a subject is tested, the number of motoneurons that have been excited by a standard volley of sensory impulses is counted. That number is an indicator of spinal cord excitability. Interestingly enough, this number fell in ISS crewmembers, quite quickly at first and then more gradually over many days. A return to normal was observed within days after landing (Watt 2001).

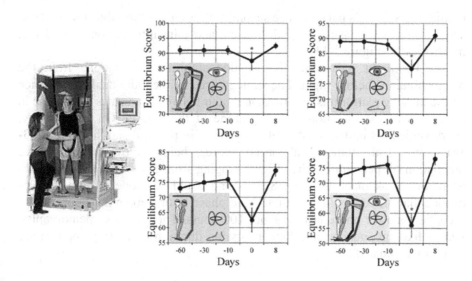

Figure 3-13. Subjects ability to stand as still as possible is investigated while standing on a platform inside a booth. The platform and the booth are designed to isolate the various sensory information used in balancing—visual, vestibular, and proprioceptive. For example, the booth is slaved to the body sway (upper left and lower right panels) to prevent changes in visual information. Similarly, information from the proprioceptive receptors in the ankles is cancelled by moving the foot platform (upper right and lower left panels). After spaceflight (Day 0) the astronauts posture is less stable in all conditions. (Adapted from Paloski et al. 1993)

When performed in conjunction with linear acceleration (such as "falls" simulated by bungee cords) the H-reflex amplitude is low in-flight, but very large postflight. Interestingly, sudden drops are perceived as falls or drops on Earth, and felt in-flight much as they did preflight. Later in-flight as well as postflight, drops were perceived as more sudden, fast, and hard. During those drops, the subjects did not have a falling sensation, but rather a feeling that "the floor came up to meet them".

Second, extensive dynamic postural testing with a moving platform was performed before and after space missions. Balance control performance has been systematically tested before and after the flight using a computerized dynamic posturography system widely employed for evaluation of balance disorders (Paloski et al. 1993). This system consists of a platform and a visual surround scene, both of which are motorized to simulate motion. Subjects complete multiple tests before and after the flight to establish stable individual performance levels and the time required recovering them. Two balance control performance tests are administered. The first test examines the subject's responses to sudden, balance-threatening movements of the platform. Computer-controlled platform motors produce sequences of rotations (toes-up and toes-down) and translations (backward and forward) to perturb the subject's balance. The second test examines the subject's ability to stay upright when visual or ankle muscle and joint information is modified mechanically (Figure 3-13).

Postflight measurements revealed significant deviations from the results obtained before flight. The strategy used by the individuals for balance on the moving platform is modified, and their behavior indicates a decrease in awareness of the direction and magnitude of the motion. On landing day, every subject exhibited a substantial decrease in postural stability (Figure 3-14). Some had clinically abnormal scores, being below the normative population 5th percentile. After flights ranging from 5-13 days, postflight re-adaptation took place in about 8 days and could be modeled as a double-exponential process, with an initial rapid phase lasting about 2.7 hours, and a secondary slower phase lasting about 100 hours. The effects of demographic factors like age, gender, and longer mission duration on these responses are currently evaluated.

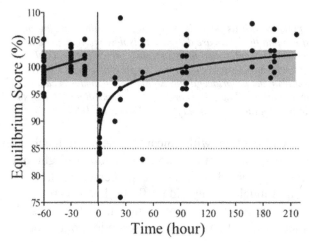

Figure 3-14. Sum of the equilibrium scores from the various sensory tests for 10 subjects before and after landing, compared with a large normative database. A few hours after landing, the average returning crew-member was below the limit of clinical normality (dashed line). Preflight stability levels were achieved by 8 days after landing, following a double exponential time course. (Adapted from Paloski et al. 1993)

Information obtained from these investigations is promising for ground-based clinical research. A relatively large number of individuals on Earth suffer from prolonged, frequently life-long, clinical balance disorders. Disorders like Ménière's disease and traumatic injuries to the inner ear can severely influence quality of life. Falls are the leading cause of injury-related deaths in the elderly and these numbers continue to grow. Inner ear disorders are thought to account for 10–50% of falls among senior citizens. Currently, human spaceflight is the only means available for studying the response to sustained loss and recovery of inner ear information. Comparison between data from astronaut-subjects and similar data from patients and elderly subjects demonstrates similarities between these balance disorders. One sensible difference is that the posture problems recover in a few days for the astronauts, whereas it can take weeks (or never recover) in the patients. It is hoped that a better understanding of the strategies used during the recovery process in the astronauts, and of the plasticity of this system in general, will help to improve rehabilitation treatments for patients with balance disorders on Earth.

3.3 Locomotion

The cautious gait of astronauts descending the stairs of the "white room" docked with the Space Shuttle and walking on the runway is an obvious example of changes in sensory-motor coordination.[*] Typically, locomotion in microgravity poses no problem and is quickly learned. However, adaptation continues for about a month. The astronauts who just visit the ISS note that the long-duration crewmembers move more gracefully, with no unnecessary motion. They can hover freely in front of a display when the new comers would be constantly touching something to hold their position.

When locomoting in space, the astronauts stop using the legs. Instead they use the arms or fingers to push or pull themselves. For clean one-directional movements, push must be applied through the center of gravity, i.e., just above the hips for a stretched-out body. When translating though, the natural place for the arms is overhead to grab onto and push off from things as they come whizzing by. This is the worst possible place from the physics of pushing and pulling for clean movements, for by exerting forces with arms overhead, some unwanted rotations will invariably occur, which have to be compensated with ever more pushes and pulls, giving an awkward look to the whole movement. "To cleanly translate, the best to keep the hands by the hips when exerting forces and boldly go headfirst. This way the pushing and pulling is directed through the body's center of gravity and gives nice controlled motions without unwanted rotations" (Pettit 2003).

Movement in a weightless environment obeys to the Newton's laws of motion. Friction forces are negligible and the angular momentum is always conserved unless acted on by an outside torque. Filmed sequences of astronauts performing a number of gymnastic moves in space were analyzed frame-by-frame. The principle of conservation of angular momentum was demonstrated as the astronauts tumbled, twisted and rotated in space. Throughout their motion and up until they entered in contact with the wall,

---

[*] The ritual of the crew walking on the runway and inspecting the vehicle immediately after landing is called a "walk-around" in NASA jargon. While the astronauts are "kicking the tires", the scientists are impatiently waiting to collect postflight data in the Flight Clinic. It is well known that re-adaptation to Earth gravity is very rapid and the possibility of testing this process at its earlier stage is fundamental for a full understanding of its mechanisms.

the angular momentum was constant at 35.7±1.2 kg. x m²/sec while rotating freely (Jones 1997).[*]

Since the legs are less used for locomotion, new sensory-motor strategies emerge in microgravity. Some of this newly developed sensory-motor program "carries over" to the postflight period, which leads to postural and gait instabilities upon return to Earth. Both U.S. astronauts and Russian cosmonauts have reported these instabilities even after short-duration (5-10 days) spaceflights. Subjects experienced a turning sensation while attempting to walk a straight path, encountered sudden loss of postural stability especially when rounding corners, perceived exaggerated pitch and rolling head movements while walking, and experienced sudden loss of orientation in unstructured visual environments. In addition, oscillopsia and disorienting illusions of self-motion and surround-motion occurred during head movement induced by locomotion.

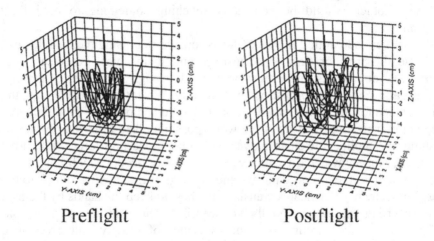

Preflight   Postflight

Figure 3-15. Head movements along the three axes (X, Y, and Z) in one subject during walking on a treadmill before spaceflight and on the day of landing. (Adapted from Bloomberg et al. 1999)

The beginning of the stance phase of locomotion, when initial foot-ground contact occur, is characterized by a rapid deceleration of the foot.

---

[*] Dan Barry, an astronaut of the STS-96 mission, got stranded in the middle of an ISS module, with the help of two fellow crewmembers. He then tried to kick himself over to the wall. He recalled later: "When I reached out an arm, my body moved back and my center remained in the middle of the room. I instinctively tried moving fast, then slow, then bicycled my legs. None of it helped. I just had to wait for the air currents to drift me to the wall. Sneezing and spitting didn't do much good either. On the other hand, throwing clothing as fast as I could produced enough reaction to send me to the opposite wall".

While the magnitude of this deceleration depends on walking speed and the rigidity of the support surface, it is common to observe decelerations in the order of 4-10 g. The forces created by the heel strike impact travel through the body and reach the head. The head-neck-eye complex then operates to minimize angular deviations in gaze during locomotion (Pozzo et al. 1990). After spaceflight, however, changes have been documented in both head-trunk and lower limb patterns of coordination. Bloomberg et al. (1997) reported changes in head pitch variability, a reduction of coherence between the trunk and compensatory pitch head movements, and self reports from crewmembers indicating an increased incidence of oscillopsia (the illusion of a visual surround motion) during postflight treadmill walking (Figure 3-15). A number of characteristics of walking also appear to be changed after spaceflight. For example, during the contact phase of walking, the foot "thrusts" onto the support surface with a greater force than that observed before flight.

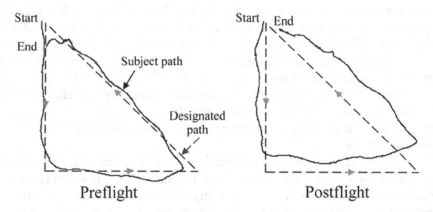

Preflight   Postflight

Figure 3-16. Examples of preflight and postflight (3 hr after landing) walking trajectories along a triangular path with the eyes closed. The length of the first two segments is about 3 m. (Adapted from Bloomberg et al. 1999)

The alterations in locomotion seen after spaceflight raise some concern about the crew capability for unaided egress from the Space Shuttle or the Soyuz in a case of emergency. As discussed earlier, many crewmembers experience marked vertigo when making head movements during entry, landing, and afterwards. This vertigo could be a major obstacle to successful egress if vision were impaired, as with a smoke-filled cabin. An interesting investigation was performed by Bloomberg et al. (1999), in which the ability for crewmembers to repeat a previously seen trajectory without vision was examined (Figure 3-16). When attempting to walk a triangular path after flight, blindfolded subjects showed both under- and over-estimations of the distances walked, but a correct estimation of the angle turned. These results suggest a difficulty for reconstructing motion cues from the otoliths, but not from the semicircular canals. However, the changes found could also be related to the lower walking velocity during postflight testing.

These results imply that mechanisms like computing self-displacement and updating spatial information (both of which being also called *navigation*) are disturbed by spaceflight and have to be reacquired after return to Earth.

3.4   Body Movement

On the Earth's surface, gravity significantly affects most of our motor behavior. It has been estimated that about 60% of our musculature is devoted to opposing gravity. For example, when making limb movements during static balance, anticipatory innervations of leg muscles compensate for the impending reaction torques and the changes in location and projection of the center of mass associated with these movements. Similar patterns of anticipatory compensations are seen in-flight, although they are functionally unnecessary. Also, rapidly bending the trunk forward and backward at the waist is accompanied on Earth by backward and forward displacements of hips and knees to maintain balance. The same compensatory movements of hips and knees are made in weightlessness. Since the effective gravity torques are absent during spaceflight, the innervations necessary to achieve these synergies in weightlessness are different from those needed on Earth. Consequently, these in-flight movements must reflect reorganized patterns of muscle activation.

During the first space experiments in which I participated in 1982, on board the Salyut-7 space station, we found that dorsi-flexor muscles (e.g., the *Tibialis anterior* leg muscle) assume a larger role in space than on Earth in regulating the orientation of the individual relative to his/her support. This is in contrast with the general use of muscle extensors on Earth, which are used to counteract gravity. This transfer of motor strategies from one muscle group to another explains the forward tilted posture of crewmembers placed in darkness when instructed to maintain a posture perpendicular to the foot support (see Figure 3-11).

Using a simple ball catching experiment in weightlessness, it has been elegantly shown that the central nervous system uses an internal estimate of gravity in the planning and execution of movements. During the act of catching a ball on Earth, the brain estimates the trajectory of the ball, accurately taking into account its downward acceleration due to gravity. In space, a seated astronaut had to catch a ball travelling at a constant velocity, in contrast to the constant acceleration that would occur on Earth (Figure 3-17). The ability to anticipate and predict is one of the nervous system's basic functions. When we catch a ball, the brain does not wait for it to touch the hand before stimulating arm flexor muscle contraction to compensate for the impact. About one third of a second before impact, the brain elicits just the right amount of contraction to counteract the force exerted, which itself depends on the weight of the object combined with the acceleration of its fall. The experiment led to the conclusion that the brain works by anticipating the effects of gravity on the ball rather than by making direct measurements of its acceleration. This anticipation ability remains even in conditions of weightlessness. Thanks to childhood experience, the brain possesses internal models of the gravity laws governing the behavior of a falling object, and perhaps more generally, Newton's law of mechanics. We see here the beginnings of an adaptation to new laws. A longer period in weightless flight would now be needed to assess how such an adaptation might develop (McIntyre et al. 2001).

Figure 3-17. Ball Catching experiment during the Neurolab STS-90 mission. A ball was thrown at the subject at a constant velocity. The trajectory of the subject's arm and the activity of his forearm muscles were recorded as he was trying to catch the ball. (Credit NASA)

Likewise, the analysis of astronauts' writing or drawing showed that such fine movements are not altered in microgravity. When cosmonauts were asked to draw "horizontal" ellipses in the air without the aid of vision, results indicated minimal changes as a function of microgravity, suggesting that the body (egocentric) reference system was not disturbed. The subjects were capable of maintaining a sense of verticality despite disappearance of the main factor contributing to verticality on Earth, i.e., the gravitational force (Gurfinkel et al. 1993). However, bending the head over the trunk causes the cosmonauts arm movement pattern to be more aligned with the head vertical axis, indicating that the head axis could also be used as a reference frame.

3.5   Eye Movement

Eye movement is probably the response of the vestibular system that has been the most studied during spaceflight. For several decades, the study of eye movements has been a source of valuable information to both basic scientists and clinicians. The singular value of studying eye movements stems from the fact that they are restricted to rotations in three planes and the eyeball offers very little inertia to the eye. This facilitates accurate measurement (for example using video eye recording in near infrared light), a prerequisite for quantitative analysis.

Eye movements must continuously compensate for head movements so that the image of the world is held fairly steady on the retina, and thus appears clear and stationary. During head movements, the vestibular apparatus measures head velocity and relays this information to those centers controlling eye position to generate compensatory eye movements; this reflex behavior ensures that vision is not blurred (Figure 3-18). When performed in darkness, this leads to a pattern of rhythmic eye movements known as *nystagmus*, consisting of slow phases in the direction opposite to the head and fast phases which bring the eye back when it reaches the extreme of its travel. The nystagmus response to a rapid head movement outlasts the changes in signals in the semicircular canals, through the activation of a velocity storage mechanism located in the brainstem.

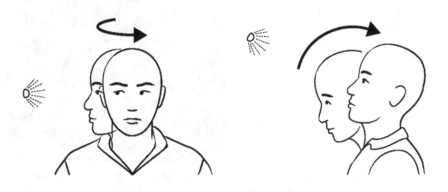

Figure 3-18. A compensatory eye movement, called the vestibulo-ocular reflex, is generated when subjects are fixating or imagining in darkness a wall-fixed target while moving their head in the yaw (left panel) or pitch (right panel) plane.

This so-called "vestibulo-ocular reflex" has been studied systematically in orbital flight, both during active (voluntary) and passive movements of the head (Clément 1998). With my co-investigators, we were the first to report that the amplitude of vertical eye movements was decreased during the first three days of weightlessness compared to normal value on Earth. In this experiment, the eye movements of an astronaut were recorded when he voluntary moved his head while either fixating a visual target or imagining that target in darkness. After four days in orbit, the vestibulo-ocular reflex gain returned to the pre-flight level, perhaps as a consequence of substituting neck receptor cues for vestibular receptor cues.

Several investigations have later reported that after short-duration spaceflight, the pattern of eye and head movements was significantly altered when subjects moved their heads and eyes to fixate a laterally displaced target (Figure 3-19).

Problems in hand-eye coordination and blurriness of the visual scene when reentering in normal gravity have also been reported after long-duration missions. Tracking of moving visual targets seems also to be altered, especially in the vertical direction (Figure 3-20). These deficits might pose a problem for piloting tasks during landing. The vestibular nuclei located in the brain stem are part of a system that allows one to fix the gaze on a stationary target during voluntary head motions as well as to track moving targets. This system appears to be disturbed during spaceflight, presumably as a consequence of altered vestibular receptor function due to the absence of gravity.

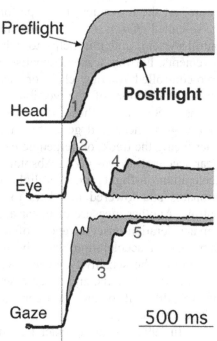

Figure 3-19. The ability to maintain visual fixation on targets while moving the head is diminished immediately after landing. Compared with the preflight response, the head movement is delayed and its amplitude is reduced postflight (1). As a consequence, the vestibulo-ocular reflex is initiated at inappropriate time (2), pulling gaze from target (3). Large eye saccades (4) are then required to direct gaze back on target (5).

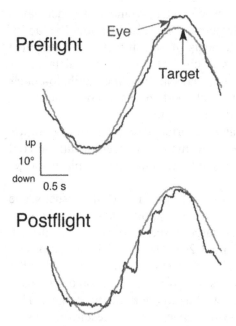

Figure 3-20. After landing, subjects have difficulties following a vertically moving visual dot. When the target moves up, the eyes try to catch-up the visual target with fast saccades rather than smooth pursuit.

One problem in studying eye movements by asking subjects to perform voluntary head movements is that the central nervous system is "aware" of the movement to be performed. A copy of the motor command (the so-called *efference copy*) is presumably sent to the eye-head coordination control system, and this helps to achieve the adequate, compensatory eye movements. For this reason, scientists also use passive rotation generated by servo-controlled rotating chair or sled in order to generate unpredictable inertial stimulation of the vestibular system, and to study the resulting responses. Several of these devices have flown on board the Spacelab. In 1985, a 4-m linear sled generated sinusoidal oscillations in subjects sitting either facing the track, or perpendicular to it, or lying on their back. The peak linear acceleration was 0.2 g. Absolute thresholds for the perception of linear acceleration in-flight and postflight were found to be elevated in some astronauts and lowered in others for some axes, relative to ground-based controls. Another measure of linear acceleration sensitivity, the time elapsed from acceleration onset to reports of self-motion (which varies inversely with magnitude of acceleration) have been more consistent. Results indicate an elevation of the sensitivity when linear accelerations are exerted along the body longitudinal axis, and a decrease in sensitivity for the other axes. It is, however, difficult to rule out a contribution of the somato-sensory sensation in these results.

In 1992, a rotating chair flew on board the Spacelab IML-1 mission, allowing the evaluation of the vestibulo-ocular reflex evoked by passive rotation of 4 crewmembers about the yaw, or pitch or roll axis, during the course of a 7-day spaceflight. Results showed that the responses generated by rotation in pitch and roll were the most affected in space.

More recently, in 1998, a human-rated centrifuge flew on the Neurolab mission, in which crewmembers were both exposed to angular and linear acceleration (see Figure 1-23). One objective of this experiment was to study the adaptation of the CNS by measuring the eye movements in response to the angular and linear acceleration in space. Eye rotations can compensate for both the rotational and the translational components of head motion. On the Earth's surface, two major sources of linear acceleration are normally encountered. One is related to the Earth's gravity: the gravitational force pulls the body toward the center of the Earth, and the body opposes this force to maintain an upright standing posture. The other sources of linear acceleration arise in the side-to-side, up-down, or front-back translations of the head, which commonly occur during walking or running, and from the centrifugal force sensed when turning or going around corners. The body responds by tending to align the longitudinal body axis with the resultant linear acceleration vector. Put in simple terms, we have to exert an upward force such as to balance gravity when standing upright and to tilt into the direction of the turn when in motion. As mentioned above, in microgravity, the otoliths are not stimulated by head tilt, and therefore the eye movements in response to head pitch or roll are likely to be altered during and after spaceflight. The results of the centrifuge experiment have not confirmed this hypothesis, though: the torsional (along the line of sight) eye movement elicited by the linear acceleration (known as *ocular counter-rotation*) was unchanged in-flight and postflight relative to preflight. More investigations are therefore necessary to fully understand the adaptation of the compensatory eye movements during spaceflight.

New tests of the otolith function are currently introduced in order to evaluate the re-adaptation of eye movements in response to body tilt after spaceflight. The eye movements and the perception of crewmembers exposed to body rotation about an axis tilted from Earth vertical (Figure 3-21) offer interesting capabilities. This *off-vertical axis rotation* (OVAR) causes, when rotation is in darkness at a constant low velocity, the perception of being successively tilted in all directions. Consequently, both a counter-rotation of the eyes and a perception of moving along the edge of an inverted cone, appear. At higher rotation velocity the illusion is that of being upright, but moving along the edges of a cylinder (hence more translational motion), and eye movements are predominantly horizontal. Another otolith test is achieved using a centrifuge where sitting subjects are displaced minimally from the rotation axis, so that one labyrinth becomes aligned on-axis, while the second labyrinth alone is exposed to the centripetal acceleration. This technique allows investigating subjective vertical and otolith-ocular responses during stimulation of the otolith on one side at a time. These tests should allow to

more accurately document a change in the reinterpretation of the otolith signals from tilt to perception in returning crewmembers, and validate or not the Otolith Tilt Translation Reinterpretation (OTTR) hypothesis.

Figure 3-21. A subject rotating at constant velocity about an axis tilted 20 deg relative to the vertical (off-vertical axis rotation) has the illusion of describing either a conical or a cylindrical motion when rotation is at low or high velocity, respectively. This stimulation allows distinguishing between tilt and translational response during stimulation of the otoliths.

Very recently, scientists have discovered that, on Earth, the eye movements also reflect an orientation to the resultant linear accelerations during turning. During either passive rotation, as in a centrifuge, or while walking or running around a curved path, the axis of eye rotation tends to align with the resultant axis of the summed linear accelerations. The same phenomenon occurs when viewing a visual scene that moves in the horizontal plane, but with the head tilted to the side. The eye movements (also called *optokinetic nystagmus*) are then oblique relative to the visual scene, as if they tried to align with the resultant of visual motion and gravity. Space experiments have showed that this gravity-oriented response was absent in microgravity, and that a return to the normal preflight response was observed two days after return to Earth.

On Earth, the eye movement responses to moving patterns tend to be asymmetric for upward and downward stimulation. It is generally easier to follow a visual scene moving upward than downward. The interpretation generally proposed for this phenomenon was the following: when we walk, there is an apparent downward motion of the floor. However, this motion would be ignored, and the downward eye movements suppressed in order to pay more attention to a further distance in case obstacles could occur. Space experiments showed that the vertical asymmetry tends to be eliminated in spaceflight, suggesting instead a role of gravity (presumably through a role of the otolith signals on the eye position) in this phenomenon (Clément 1998).

4 EFFECTS OF SPACEFLIGHT ON SPATIAL ORIENTATION

4.1 Visual Orientation

The visual system is addressed here principally in the context of its relationship to the vestibular system. Vision may compensate in large measure for modified otolith sensitivity. It helps in spatial orientation, and is essential to motor coordination.

Astronauts working in microgravity must rely much more on vision to maintain their spatial orientation, since otolith signals no longer signal the direction of "down". It has long been known that moving visual scenes can produce compelling illusions of self-motion ("seeing is believing"). These visually-induced illusions become even stronger in space, since visual cues are unhindered by constraints from the otoliths, which in microgravity do not confirm or deny body tilt. This has been confirmed with experiments where crewmembers observing a rotating visual field felt a larger sense of body rotation in space than on Earth (Lackner and DiZio 2000). It is interesting to note that frogs born in microgravity also showed stronger behavioral response to moving visual scenes when tested after their return on Earth than control animals born on Earth.

Crewmembers who remained seated in the relatively small Soyuz, Mercury, Gemini, and Apollo capsules rarely encountered orientation problems. However crews of the larger Skylab and Shuttle reported occasional disorientation, particularly when they left their seats, and worked in unpracticed, visually unfamiliar orientations. The problem occurred both inside the spacecraft, and also outside, as when performing an extra-vehicular activity (EVA) (Figure 3-22). For example, Bernard Harris, an astronaut of the STS-63 Shuttle mission reported: "As I was getting ready to step out of the spaceship, it felt like gravity was going to grab hold of me and pull me down toward Earth. Your natural response is to hesitate and grab on harder. I felt myself hanging on to the handrail and saying: "No, you're not going to fall toward the Earth, this is the same thing you've been seeing for the last five days."

Although episodes of visual disorientation are observed by many crewmembers, some seem more affected than other. In some individuals static visual cues become increasingly dominant in establishing spatial orientation in microgravity. Other subjects are more "body oriented" and align their exocentric vertical to be along their longitudinal body axis, and perceive the body axis relative to placement. Such individuals exhibit no problems in spatial orientation aloft even in the absence of visual cues for vertical orientation. Further, these individuals appear able to strengthen their perception of subjective verticality by using localized tactile cues, especially by pressure exerted on the soles of their feet.

Part of the difficulty of the people who predominantly rely on visual cues for spatial orientation is due to the natural tendency to assume that the surface seen beneath our feet is the floor. When working "upside down" in the spacecraft, the walls, ceiling, and floors then frequently exchange subjective identities. Also, when viewing another crewmember floating upside down in the spacecraft, they often suddenly feel upside down themselves, because of the subconscious assumption carried over from life on Earth that people are normally upright. Fluid shift and the absence of otolith cues also contribute, and make some crewmembers feel continuously inverted, regardless of their actual orientation in the spacecraft. The inversion illusion may be understood using a model that includes an internal (*ideotropic*) orientation vector. This vector may also explain the sensation of the "downs" (Mittelstaedt and Glasauer 1993).

Figure 3-22. Astronauts during EVA occasionally feel uncomfortable when working upside-down in the Shuttle payload bay when it faces the Earth, when there are no gravitational reference points and familiar cues, such as horizon. (Credit NASA)

There is also a natural tendency to perceive the Earth as "down". Consequently, when looking at the Earth out of a window "above" their head, some crewmembers may feel that they are just standing on their head (Figure 3-23). Astronauts often report that "if you lose something in weightlessness, you instinctively look down, which of course is not the solution" (Pettit 2003).

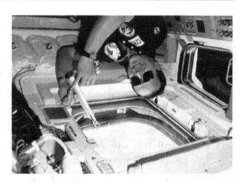

Figure 3-23. An astronaut looking at the Earth through a window located "above" his head might perceived himself inverted. (Credit NASA)

It was once thought that these inversion illusions could trigger attacks of space motion sickness during the first several days in weightlessness. Many crewmembers have reported to get sick when looking out the Shuttle middeck window and find the Earth at the top of the window frame instead of the bottom. However, though space sickness susceptibility eventually subsides, crewmembers on long-duration flights say that visual illusion episodes continue to occur. The observation that inversion illusions do not provoke space motion sickness as the flight progresses indicates a resolution of the factors that triggered the motion sickness early on. As a countermeasure for these visual illusions, it is thought that visual experience of working in unfamiliar orientations during preflight neutral buoyancy training (in a water tank) and virtual reality might help maintain spatial orientation while on orbit.

4.2   Cognition

The word "cognition" is often used in computer science-related fields to denote the level of activities that require "understanding" of what is going on, rather than merely signal-level reaction. We will review here the few cognitive functions that have been investigated during and after spaceflight.

4.2.1   Navigation

Vertebrate brains form and maintain multiple neural maps of the spatial environment that provide distinctive, topographical representations of different sensory and motor systems. For example, visual space is mapped onto the retina in a two-dimensional coordinate plan. This plan is then remapped to several locations in the central nervous system. Likewise, there is a map relating the localization of sounds in space and one that corresponds to oculomotor activity. An analogous multi-sensory space map has been demonstrated in the mammalian hippocampus, which has the important function of providing short-term memory for an animal's location in a specific spatial venue. This neural map is particularly focused on body position and makes use of proprioceptive as well as visual cues. It is used to resume the location at a previous site; a process called *navigation*.

This system of maps must have appropriate information regarding the location of the head in the gravitational field. So it follows that the vestibular system must play a key role in the organization of these maps. Only recently has this been demonstrated by experiments carried out in space. During an experiment performed on board Neurolab, rats ran a track called the Escher staircase, which guided the rats along a path such that they returned to their starting location after having made only three 90 deg right turns. On Earth, rats could not run this track. But in space, it provided a unique way to study the "place cells" in the hippocampus that encode a cognitive map of the environment. The rats had multi-electrode recording arrays chronically implanted next to their hippocampal place cells. Recordings in space indicated that the rats did not recognized that they were back where they started, after only three 90-deg right turns.

Such studies could help to explain the visual inversion illusions and the navigation difficulties experienced by some astronauts when they arrive in space. A weightless environment presents a true three-dimensional setting where Newton's laws of motion prevail over Earth-based intuition. We normally think in terms of two dimensions when we move from place to place. However, in orbit, one might decide the best way is to go across the ceiling and then sit on the wall. In addition, each module of the ISS provides a local visual frame of reference for those working inside. Once the ISS construction will be complete, the modules will eventually be connected at 90-deg angles, so not all the local frames of reference will be co-aligned. It might sometimes be difficult to remain oriented, particularly when changing modules. Even after living aboard for several months, it could be difficult to visualize the three-dimensional spatial relationships among the modules, and move though the modules instinctively without using memorized landmarks. Crewmembers will not only need to learn routes, but also develop three-dimensional "survey" knowledge of the station. Disorientation and navigation difficulties could be an operational concern in case an emergency evacuation is required in the event of a sudden depressurization or fire.

4.2.2   Mental Rotation

On Earth, gravity provides a convenient "down" cue. Large body rotations normally occur only in a horizontal plane. In space, the gravitational down cue is absent. When astronauts roll or pitch upside down, they must recognize where things are around them by a process of mental rotation that involves three dimensions, rather than just one.

It is well known that on Earth, a familiar visual environment, a face or a printed text cannot be recognized or analyzed when it is tilted by more than roughly 60 deg. In a very simple experiment, I once asked one crewmember to report the tilt angle of his body with respect to the inside of the spacecraft from which he had more difficulty in mentally rotating the visual features.

The reported angle was about 60 deg on the first day in-flight, 90 deg on the second day, but after three days in-flight his perception was independent of the respective orientations. One interpretation is that weightlessness, by providing a release of the gravity-dependent constraint on mental rotation, would facilitate the processing of visual images in any orientation with respect to the body axis.

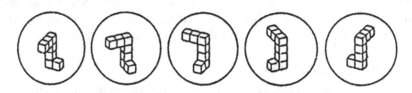

Figure 3-24. Examples of shapes used for a mental rotation test. When the shape on the extreme left is presented with, let's say, the shape on the extreme right (a 180-deg rotation), the time taken to decide that both shapes are the same is about 5 s. When the shape on the extreme left is presented with the shape in the middle (a 90-deg rotation), the response time is now only 2.5 s. Therefore the speed of mental rotation in this test is about 33 deg/sec.

In a series of subsequent mission, a mental rotation paradigm with pictures of three-dimensional objects was tested on several cosmonauts (Figure 3-24). Responses showed that the average rotation time per degree was shorter in-flight than on the ground. This difference seems to be particularly marked for stimuli calling for mental rotation about a roll or a pitch axis (an actual body rotation around both of these axes would induce different responses from the otolith organs in weightlessness compared to Earth). However, a later study in which the repertoire of objects was different between all experimental sessions to avoid a learning effect, showed no significant differences in rotation time in space versus ground data (Léone 1998). So, the results are inconclusive at this point and further studies are needed to investigate whether mental rotation is facilitated or not in microgravity. One concern is that a poorer ability to mentally rotate the visual environment could be a determinant factor for the apparition of space motion sickness. Another concern is the ability for the astronauts to recognize their fellow crewmembers when upside-down. However, preliminary tests suggest that after a few days in space it is less hard to identify an upside-down face (the so-called "inversion effect") in space than on Earth.[*]

Other experiments have investigated whether it was easier to detect the presence of a symmetry axis in absence of gravity. For example, it is well

---

[*] There was one instance on a Shuttle mission where a crewmember was "lost". Several of his crewmates looked for this individual but couldn't find him…yet all the while he was right in front of them. The lost crewmember was actually inverted relative to those looking for him (Millard Reschke, personal communication).

known that on Earth, a vertical axis of symmetry is faster to identify than a horizontal and an oblique axis of symmetry. A change in the position of the head relative to the trunk on Earth influences symmetry detection. One experiment performed in space on 5 astronauts indicated that both vertical and horizontal axes of symmetry were equally faster to identify (Léone 1998).

Interestingly enough, mental tasks that demand logical reasoning, decision-making, as well as memory retrieval functions, seem unimpaired during spaceflight . This result is in conflict with the frequent report by crewmembers of a difficulty in evaluating time periods while in space.

4.2.3   Mental Representation

An accurate representation of the visual environment is crucial for successful interaction with the objects in that environment. It is clear that humans have mental representations of their spatial environment and that these representations are useful, if not essential, in a wide variety of cognitive tasks such as identification of objects and landmarks, guiding actions and navigation, and in directing spatial awareness and attention.

In physics, a coordinate system, which can be used to define position, orientation, and motion is called a *reference frame*. It has been argued that the Earth's gravitational field is one of the most fundamental constraints for the choice of reference frames for the development and the use of cognitive representations of space. For example, a subject looking at a diamond-shaped figure (in retinal coordinates) perceives a square-shaped figure when he/she and the figure are both tilted relative to gravity. This result indicates that an object's form perception generally depends more on the orientation of this object in world (spatial) coordinates than on its orientation in retinal coordinates. In other words, gravity is critical for the extraction of an object's reference frame.

One problem with the ground-based studies is that tilting the observer's relative to gravity on Earth creates a conflict between perceived gravitational (extrinsic) vertical and retinal- or body-defined (intrinsic) vertical, but does not suppress the gravitational information. On the other hand, the loss of the gravitational reference in spaceflight provides a unique opportunity to differentiate the contribution of intrinsic and extrinsic factors to the spatial orientation system in astronauts.

Measuring the changes in the mental representation of an object throughout a space mission is a simple way to assess how the gravitational reference frame is taken into account for spatial orientation. Results of space studies by our group suggest that the absence of the gravitational reference system, which determines on Earth the vertical direction, influences the mental representation of the vertical dimension of objects and volumes. For example, I once asked one French astronaut to write with the eyes closed his name vertically and then horizontally on a notebook attached by Velcro to his knee.[*] The length of these words was compared between in-flight and preflight tests. Results showed that the length of the written words decreased in-flight for both vertical and horizontal directions, but the vertical direction was the most affected. In another astronaut, the reduction in the vertical length of words was observed during several days after returning from a 28-day space mission (Figure 3-25). It is interesting to note that in both experiments, the size of the letters did not change in-flight or postflight, but the vertical distance between them was decreased. These results suggest that adaptive changes in the mental representation of a vertical layout of letters take place when the gravitational frame of reference is removed (Clément et al. 1987).

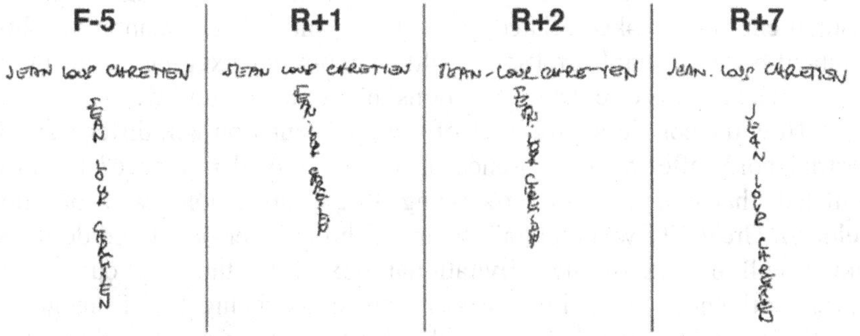

Figure 3-25. Horizontal and vertical writing test in an astronaut returning from a 28-day space mission. F-5: 5 days before flight, R+1: one day after landing, etc. Only the length of words in the vertical direction changed as a result of spaceflight.

During another test, two crewmembers had to draw the well-known Necker's cube. This figure is the simplest representation of a three-dimensional object in a two-axis coordinate system. Comparison between the length of line between the cubes drawn on the ground and the cubes drawn in space revealed a 9% decrease in size in the vertical dimension (i.e., the height) of the cubes drawn in weightlessness (Figure 3-26). Similar results have been found in another study involving two astronauts. The trajectory of hand-drawn ellipses in the frontal plane in the air with the eyes closed revealed a 10-13% decrease in the vertical length of the ellipses, whereas the horizontal length of the ellipses were basically unchanged. This result

---

[*] These tests are variants of tests traditionally used in oriental medicine (the Fukuda Writing test, the Square Drawing test) to diagnose patients with an impairment in motor function (when the size of all characters is irregular) from those with vestibular disorders (the writing or drawing is deviated to one side. Interestingly enough, the astronauts responses are close to those of patients with otolithic disorders on Earth.

supports our hypothesis that the mental representation of the vertical dimension of objects or volumes is altered during exposure to weightlessness.

The results of these studies may have important consequences for human performance during spaceflight. For example, if an astronaut cannot accurately visualize the station, navigation of the station may cause delays and frustration. There may also be consequences for space habitat design if squared volumes do not look square to astronauts. Virtual reality training may be a way to train the astronauts to compensate for such altered spatial representation.

Further investigations carried out in space will perhaps reveal that other higher cortical functions are impaired in weightless conditions. The combination of virtual reality with multi-EEG recordings (for the measurement of evoked-related potentials and brain mapping), both equipment being available on ISS, should soon provide exciting results on the adaptive mechanisms of cerebral functions in absence of gravity.

How the cognitive processes of spatial orientation will differ from the terrestrial norm after a long absence of a gravitational reference? It can be speculated that the way of processing three dimensions will be more developed. Creativity will certainly be more three-dimensional and definitely thinking will be out of the gravitational box. Like the way culture and language influences our ability to creatively think, being free from gravity will entice thoughts never before possible for the human mind, and thus give opportunities for new art and scientific discoveries (Pettit 2003).

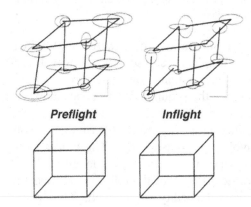

Preflight   **Inflight**

Figure 3-26. Top: Mean ± SD of each point of cubes drawn preflight and in-flight. Bottom: Averaged mean preflight and in-flight cubes. Horizontal and oblique lines were unchanged, but the vertical lines were shorter in microgravity. (Adapted from Lathan et al. 2000)

5   WHAT DO WE KNOW?

5.1   Space Motion Sickness (SMS) Experience

The severity of SMS is categorized depending on its impact upon crew performance (Table 3-01). For example, "Mild" SMS has no operational impact, because the crewmember can still perform all the required activities. "Moderate" or "Severe" SMS are operational concerns since the workload must be redistributed among the remaining, unaffected crew.

In the first 36 missions of the Space Shuttle program, about 71% of the 109 crewmembers making their first flight reported symptoms of SMS. About 33% reported mild symptoms, 27% moderate symptoms, and 11% severe symptoms. Most recovered by the end of the third day in space. In a few cases in the Russian and U.S. missions, however, crewmembers were ill for 7-14 days.

The severity of SMS among those making a second flight remained unchanged in 56% of crewmembers, whereas a slight improvement was observed in 35%, but even more symptoms were noted in 9%. This indicates that symptoms are not significantly reduced on a following flight.

In addition to feelings of vertigo and nausea, SMS can cause Sopite Syndrome, which includes lack of motivation to work or interact with others, drowsiness, fatigue, and the inability to concentrate. Sopite syndrome is often a byproduct of dizziness experienced by astronauts during space travel.

Space motion sickness is self-limited. Complete recovery from major symptoms (i.e., adaptation to the spaceflight environment) occurs within two to four days. After complete adaptation occurs, crewmembers appear to be immune to the development of further symptoms. This development of immunity to further SMS symptoms was eloquently demonstrated by rotating chair tests, designed to provoke an SMS response, that were conducted in-flight during Skylab missions.

None	*No signs or symptoms reported*
Mild	*One to several transient symptoms* *No operational impact* *All symptoms resolved in 36-48 hours.*
Moderate	*Several symptoms of a persistent nature* *Minimal operational impact* *All symptoms resolved in 72 hours.*
Severe	*Several symptoms of a persistent nature* *Significant performance decrement* *Symptoms may persist beyond 72 hours*

Table 3-01. NASA categorization of Space Motion Sickness according to the severity of symptoms.

5.2   Theories for Space Motion Sickness

Two major theories advanced to account for SMS are the *fluid shift* theory and the *sensory conflict* (also known as the neural mismatch, sensory mismatch, or sensory rearrangement) theory (Crampton 1990). Although both theoretical positions have some merit and neither is ideal, the fluid shift theory may have the most difficulty in dealing with the development of motion sickness during spaceflight (we don't get sick when lying in a bed). While the fluid shift theory of SMS could be associated with sensory conflict, there are mechanisms whereby the headward fluid shift accompanying microgravity could bypass the classic vestibular inputs to induce vomiting.

Briefly, the sensory conflict theory of motion sickness assumes that human orientation in three-dimensional space, under normal gravitational conditions, is based on at least four sensory inputs to the central nervous system. The otolith organs provide information about linear accelerations and tilt relative to the gravity vector; angular acceleration information is provided by the semicircular canals; the visual system provides information concerning body orientation with respect to the visual scene or surround; and touch, pressure, and somato-sensory (or kinesthetic) systems supply information about limb and body position. In normal environments, information from these systems is compatible and complementary, and matches that expected on the basis of previous experience. When the environment is altered in such a way that information from the sensory systems is not compatible and does not match previously stored neural patterns, motion sickness may result.

The sensory conflict theory postulates that motion sickness occurs when patterns of sensory inputs to the brain are markedly rearranged, at variance with each other, or differ substantially from expectations of the stimulus relationships in a given environment. In microgravity, sensory conflict can occur in several ways. First, there can be conflicting information (i.e., regarding tilt) transmitted by the otoliths and the semicircular canals. Sensory conflict may also exist between the visual and vestibular systems during motion in space; the eyes transmit information to the brain indicating body movement, but no corroborating impulses are received from the otoliths (such as during car sickness). A third type of conflict may exist in space because of differences in perceptual habits and expectations. On Earth, we develop a neural store of information regarding the appearance of the environment and certain expectations about functional relationships (e.g., the concepts of "up" and "down"). In space, these perceptual expectations are at variance, especially during the inversion illusions described above.

It is important to note that no single course of sensory conflict appears to entire account for the symptoms of space sickness. Rather, it is the combination of these conflicts which somehow produces sickness, although the exact physiological mechanisms remain unknown. Thus, sensory conflict explains everything in general, but little in the specific. Shortcomings of the sensory conflict theory include: a) its lack of predictive power; b) the inability to explain those situations where there is conflict but no sickness; c) the inability to explain specific mechanisms by which conflict actually gives rise to vomiting;[*] and d) the failure to address the observation that without conflict, there can be no adaptation. The hypotheses outlined below may be helpful in overcoming some of the weaknesses associated with the construct of this theory.

Some investigators have proposed a mechanism complementary to the sensory conflict theory to explain individual differences in SMS susceptibility. They suggest that some individuals possess slight functional imbalances (for example, weight differences) between the right and left otolith receptors that are compensated for by the central nervous system in 1 g. A weight imbalance between the left and right otoconia is reasonable since there is a continual turn-over of otoconia, and it is unlikely that the two otoliths would ever weigh exactly the same. This compensation is inappropriate in 0 g, however, since the weight differential is nullified and the compensatory response (either central or peripheral) is no longer correct for the new inertial environment. The result would be a temporary asymmetry producing rotary vertigo, inappropriate eye movements, and postural changes until the imbalance is compensated or adjusted to the new situation. A similar imbalance would be produced upon return to 1 g, resulting in postflight vestibular disturbances. Individuals with a greater degree of asymmetry in otolith morphology would thus be more susceptible to SMS.

A sensory compensation hypothesis has also been proposed. Sensory compensation occurs when the input from one sensory system is attenuated and signals from others are augmented. In the absence of an appropriate graviceptor signal (or perhaps the presence of atypical signals) in microgravity, information from other spatial orientation receptors such as the eyes, the semicircular canals, and the neck position receptors would be used to maintain spatial orientation and movement control. The increase in reliance on visual cues for spatial orientation could be explained by this mechanism. Closely related to this sensory compensation hypothesis is the OTTR hypothesis mentioned above (see Figure 3-06).

5.3   Countermeasures

The disruptive nature of SMS, occurring as it does during the early, critical stages of a mission, has led to a variety of approaches for the prevention or control of this medical problem.

Prediction of susceptibility has been a major objective of the SMS research. Various approaches ranging from the use of questionnaires, history,

---

[*] It has been proposed that motion sickness results from the activation of a vestibular mechanism whose physiological function is the removal of poisons from the stomach. Nausea and vomiting would also tend to keep a disoriented or dizzy individual from moving about the environment in search of food when he would be at risk doing so (Money 1990).

experience or personality traits, vestibular function tests, physiological correlates, and tests in specific nauseogenic environments have been directed toward the question of SMS susceptibility. However, striking differences were found in the pattern of symptoms generated during flight compared to the pattern generated during the ground-based tests. Further, the specific nature and time course of in-flight symptomatology were highly variable. The preflight questionnaire results did not correlate with the reported incidence of SMS. Differences between susceptible and non-susceptible crewmembers in results for each of the preflight tests were not significant, nor were correlation between susceptibility to motion sickness in the ground-based tests and susceptibility to SMS. Individual variations in preflight experience, medications, in-flight tasks (i.e., mobility), and personal strategies for symptom management have further compounded the problem. Consequently, the use of a single ground-based parameter or test procedure is inadequate for predicting SMS susceptibility. Despite an inability to identify ground-based predictors of SMS susceptibility, one reasonably accurate predictor was identified, and that is spaceflight itself. Of 16 crewmembers who had flown two or more space missions, the response pattern of only 3 changed from one flight to the next.

While research on predictors of SMS has been inconclusive, some progress has been made in the development of countermeasures. Current areas of investigation include preventive training techniques, in-flight techniques for minimizing head and body movement, and use of anti-motion sickness drugs.

Attempts by the Russian program to prevent SMS by pre-selection of individuals with a high tolerance to motion sickness during complex vestibular stimulation have not met with success. Vestibular testing was once used in the U.S. space program for the early selection of astronauts, but it is no longer used for Shuttle and ISS crewmembers. Vestibular training prior to spaceflight in the Russian space program has primarily involved Coriolis and cross-coupled angular accelerations. However, this training is rather demanding for the crewmembers and its efficacy against SMS has never been proven.

One preventive technique, developed at the NASA Ames Research Center, is a combined application of biofeedback and autogenic therapy (a learned self-regulation technique). This technique proved quite successful in controlling some symptoms of SMS associated with the autonomic nervous system, such as nausea and vomiting. In some individuals, autogenic feedback has produced improvement in motion tolerance with as little as six hours of training. However, it does not work with all individuals.

Training procedures that pre-adapt astronauts to the sensory stimulus rearrangements of microgravity gave promising results. The NASA Pre-flight Adaptation Trainer (PAT) provides astronauts with demonstrations of and experience with altered sensory stimulus rearrangements that produce perceptual illusions of various combinations of linear and angular self- or surround-motion (Figure 3-27). Crewmembers who were exposed to this training before flight had a significant reduction (19%-54% depending on the symptoms) in the severity of SMS symptoms by comparison with those who were not exposed to it.

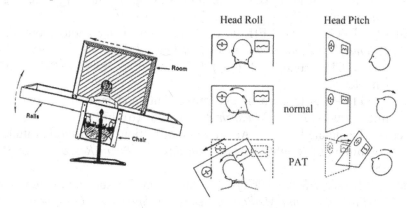

Figure 3-27. Preflight Adaptation Training. Before the flight, crewmembers are passively tilted in roll or in pitch while exposed to a lateral translation of the visual scene (such as on the left panel) or to a fore/backward translation, respectively, in order to induce a reinterpretation of their otolith signals by the visual system. (Credit NASA)

Because crewmembers have reported that rapid head movements worsen the nausea and spatial disorientation associated with SMS, head and neck restraints that restrict such movements have been used, but with limited success.

Drugs that diminish the SMS symptoms are being used and studied. During the Apollo, Skylab and the first Shuttle missions, scopolamine and a combination of scopolamine and dextroamphetamine, given orally, were used to treat SMS, with limited success. At present, the drug promethazine is available to crewmembers. The use of this medication has been quite successful, decreasing the symptoms of space motion sickness in most, but not all, crewmembers. Recent research indicates, however, that this medication can cause deleterious side effects that further degrade human performance (e.g., reaction time, grammatical reasoning ability, pattern recognition) and negatively impact mood and sleep.

Although past research has yielded a great deal of information applicable to SMS, a definitive solution to this vexing problem is urgent. Among the objectives of current SMS research is the development of: a) more precise predictive indices; b) more effective drug treatments; c) more efficient preflight adaptation procedures; d) methods to evaluate performance impairment induced by SMS and anti-motion sickness drugs; and e) the early detection of incipient symptoms.

6   REFERENCES

Bloomberg JJ, Peters BT, Smith S.L, Huebner WP, Reschke MF (1997) Locomotor head-trunk coordination strategies following spaceflight. *Journal of Vestibular Research* 7: 161-177

Bloomberg JJ et al. (1999) Effects of spaceflight on locomotor control. In *Extended Duration Orbiter Medical Project. Final Report 1989-1995.* Sawin CF, Taylor GR (eds) Houston, TX: NASA SP-1999-534, Chapter 5.5

Clément G (1998) Alteration of eye movements and motion perception in microgravity. *Brain Research Reviews* 28: 161-172

Clément G, Lestienne F (1988) Adaptive modifications of postural attitude in conditions of weightlessness. *Experimental Brain Research* 72: 381-389

Clément G, Reschke MF (1996) Neurosensory and sensory-motor functions. In: *Biological and Medical Research in Space: An Overview of Life Sciences Research in Microgravity.* Moore D, Bie P, Oser H (eds), Heidelberg: Springer-Verlag, Chapter 4, pp 178-258

Clément G, Moore S, Raphan T, Cohen B (2001) Perception of tilt (somatogravic illusion) in response to sustained linear acceleration during spaceflight. *Experimental Brain Research* 138: 410-418

Clément G et al. (2003) Perception of the Spatial Vertical during Centrifugation and Static Tilt. In: *The Neurolab Mission: Neuroscience Research in Space.* Buckey JC, Homick JL (eds) Johnson Space Center, Houston TX: U.S. Government Printing Office, NASA SP-2003-535, pp 5-10

Crampton GH (ed) *Motion and Space Sickness.* Boca Raton: CRC Press, 1990

Davis JR, Vanderploeg JM, Santy PA, Jennings RT, Stewart DF (1988) Space motion sickness during 24 flights of the Space Shuttle. *Aviation Space Environmental Medicine* 59: 1185-1189

Godwin R (1999) *Apollo 12 NASA Mission Reports.* Burlington, Canada: Apogee Books, CG Publishing Inc.

Gorgiladze GI, Bryanov II (1989) Space motion sickness. *Space Biology and Aerospatial Medicine (in Russian)* 23: 4-14

Gurfinkel VS, Lestienne F, Levik YS, Popov KE, Lefort L (1993) Egocentric references and human spatial orientation in microgravity. II. Body-centered coordinates in the task of drawing ellipses with prescribed orientation. *Experimental Brain Research* 95: 343-348

Homick JL, Miller EF II (1975) Apollo flight crew vestibular assessment. In: *Biomedical Results of Apollo.* Johnston RS, Dietlein LF, Berry CA (eds) Washington DC: U.S. Government Printing Office, NASA SP-368, pp 323-340

Jones JA (1997) Angular momentum conservation: Astronauts at play. The College of Wooster, Junior Thesis, Physics Department. Available at: http://www.wooster.edu/physics/JrIS/Files

Lackner J, DiZio P (2000) Human orientation and movement control in weightless and artificial gravity environments. *Experimental Brain Research* 130: 2-6

Lathan C, Wang Z, Clément G (2000) Changes in the vertical size of a three-dimensional object drawn in weightlessness by astronauts. *Neuroscience Letters* 295: 37-40

Léone G (1998) The effect of gravity on human recognition of disoriented objects. *Brain Research Reviews* 28: 203-214

McIntyre J, Zago M, Berthoz A, Lacquaniti F (2001) Does the brain model Newton's laws? *Nature Neuroscience* 4: 693-694

Mittelstaedt H, Glasauer S (1993) Crucial effects of weightlessness on human orientation. *Journal of Vestibular Research* 3: 307-314

Money KE (1990) Motion sickness and evolution. In: Crampton G (ed) *Motion and Space Sickness*. Boca Raton, FL: CRC Press

Oman CM, Lichtenberg BK, Money KE (1990) Space motion sickness monitoring experiment: Spacelab 1. In: *Motion and Space Sickness*. Crampton GH (ed) Boca Raton, FL: CRC Press, pp 217-246

Paloski WH, Black FO, Reschke MF, Calkins DS, Shupert C (1993) Vestibular ataxia following shuttle flights: effects of microgravity on otolith-mediated sensorimotor control of posture. *American Journal of Otology* 14: 9-17

Parker DE, Reschke MF, Arrott AP, Homick JL, Lichtenberg BK (1985) Otolith tilt-translation reinterpretation following prolonged weightlessness: implications for pre-flight training. *Aviation Space Environmental Medicine* 56: 601-606

Pettit D (2003) Expedition Six Space Chronicles. Available at http://spaceflight.nasa.gov/station/crew/exp6/spacechronicles.html

Pozzo T, Berthoz A, Lefort L (1990) Head stabilization during various locomotor tasks in humans. 1. Normal subjects. *Experimental Brain Research* 82: 97-106

Ross HE, Brodie EE, Benson AJ (1986) Mass-discrimination in weightlessness and readaptation to earth's gravity. *Experimental Brain Research* 64: 358-366

Ross MD, Tomko DL (1998) Effects of gravity on vestibular neural development. *Brain Research Reviews* 28: 44-51

Thornton WE, Biggers WP, Thomas WG, Pool SL, Thagard NE (1985) Electronystagmography and audio potentials in spaceflight. *Laryngoscope* 95: 924-932

Watt DGD (1997) Pointing at memorized targets during prolonged microgravity. *Aviation, Space and Environmental Medicine* 68: 99-103

Watt DGD (2001) Background Material On H-REFLEX Experiment. Available at: http:/space.gc.ca/asc/eng/csa_sectors/space_science/life_sciences/h-reflex.asp

Young LR, Oman CM, Watt DGD, Money KE, Lichtenberg BK, Kenyon RV, Arrott AP (1986) MIT/Canadian vestibular experiments on the Spacelab-1 mission: 1. Sensory adaptation to weightlessness and readaptation to one-g: an overview. *Experimental Brain Research* 64: 291-298

Additional Documentation:

Clément G, Droulez J (1983) *Microgravity as an Additional Tool for Research in Human Physiology with Emphasis on Sensori-Motor Systems.* Noordwijk, NL: European Space Agency Publication Division, ESA BR-15

Clément G (2001) The human sensory and balance system. In: *A World Without Gravity.* Fitton B, Battrick B (eds) Noordwijk, NL: European Space Agency Publication Division, ESA SP-1251, pp 93-111

Crampton G (ed) (1990) *Motion and Space Sickness.* Boca Raton, FL: CRC Press

International Workshop on Human Factors in Space (2000) *Aviation Space Environmental Medicine* 71, No. 9, Section II, Supplement

Man-Systems Integration Standards (1995) Revision B, Volume 1, NASA-STD-3000. Houston, TX: National Aeronautics and Space Administration

Space Neuroscience Research (1998) *Brain Research Reviews*, Volume 28, Numbers 1 and 2, Special Issue

The Neurolab Mission: Neuroscience Research in Space (2003) Buckey JC, Homick JL (eds) Johnson Space Center, Houston TX: U.S. Government Printing Office, NASA SP-2003-535

Chapter 4

THE CARDIO-VASCULAR SYSTEM IN SPACE

One of the major concerns for both short- and long-term spaceflight is the phenomenon of cardio-vascular deconditioning. This chapter introduces the principles of cardio-vascular fluid and electrolyte control, in order to understand better the symptoms typically reported by astronauts during and after spaceflight. Data from flight experiments are discussed as well as the value of ground-based models such as bed rest studies. The value of exercise, inflatable suits, saline loading, and artificial gravity is also discussed.

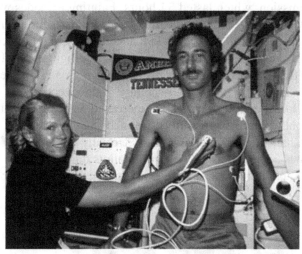

Figure 4-01. Two astronauts of the Space Shuttle are performing measurements of the cardio-vascular function using an onboard echo-cardiograph. (Credit NASA)

1   THE PROBLEM: POSTFLIGHT ORTHOSTATIC INTOLERANCE

The cardio-vascular system has the primary function of circulating blood through the body, and is composed of the heart, the circulatory system, the lungs and the kidneys. Blood supplies nutrients to and collects wastes from cells, and maintains the body's internal environment by regulating the acid/base balance, fluid content and temperature of the body, a process called *homeostasis*. Clearly, this process is crucial to health. Although the responses of the cardio-vascular system to microgravity seem to have been relatively free of major threats to well being and performance during flight, problems such as orthostatic hypotension and diminished exercise capacity are commonly observed after return to Earth.

Orthostatic intolerance is characterized by a variety of symptoms that follow standing: lightheadedness, increase in heart rate (tachycardia), altered blood pressure, and pre-syncope or syncope (fainting). Diminished exercise capacity is the observed decrement in ability to perform given amounts of work and is usually measured by duration of treadmill or stationary bicycle exercise up to a maximum level of oxygen consumption (VO_2 max). Both orthostatic intolerance and diminished exercise capacity become more severe with longer exposure to microgravity and require more lengthy recovery times after returning to Earth.

Orthostatic hypotension has been noticed since the earliest human spaceflights. A modest increase in heart rate was observed after the Mercury-8 mission, which lasted only 9 hours. More significant increase in heart rate (132 beats/min supine; 188 standing) was measured after the Mercury-9 mission, which lasted 34 hours. Fainting episodes were later observed during Gemini missions and heart rhythm disturbances during Apollo missions.

It can be estimated that orthostatic intolerance affects about two-thirds of the astronauts returning from spaceflight even of relatively short duration (Buckey et al. 1996a). This is an even greater problem for Shuttle pilots, who must perform complex reentry maneuvers in an upright, seated position. Operational concerns have increased since astronauts began using the heavy, bulky partial-pressurized launch and entry suit, required for all post-Challenger flights. However, virtually no testing of cardio-vascular function has been performed on Shuttle pilots during reentry. Another threat is whether a debilitated crew can respond to an emergency upon landing.

Figure 4-02. After a long-duration stay in orbit, cosmonauts are so severely debilitated that they could not egress the Soyuz capsule without assistance from ground personnel. (Credit Intercosmos)

The extent of orthostatic intolerance postflight is variable and depends on the duration of the flight, inter-individual differences in cardio-vascular function among the astronauts, and the time and method of postflight testing. Recovery to preflight level of orthostatic tolerance occurs within a day or so following flights of less than one-month duration, but longer recovery is associated with longer-duration flights (Watenpaughm and Hargens 1995).

Recovery of exercise capacity is also relatively rapid but takes about one week following a short-duration spaceflight. Following long-duration spaceflights on board the Russian space stations Salyut and Mir, many returning cosmonauts were incapacitated and were unable to egress the capsule without assistance from ground personnel (Figure 4-02). As a standard routine, crews returning from a 6-month flight on board Mir underwent many weeks of rehabilitation, with graduated exercises, guided movements in a warm swimming pool, and massage. Even with this rehabilitation program, after a few months, some reported that they couldn't jog without becoming short of breath.

The effects of exposure to microgravity beyond 9 months on the cardio-vascular system are entirely unknown. This is of great concern, because such effects may involve not only amplification of reversible changes already known, but also the emergence of unrecognized and irreversible alterations in cardio-pulmonary function. For example, some observers have speculated that there is a loss of cardiac mass during prolonged microgravity exposure. Will lengthy missions render space travelers unfit for return to a 1-g environment?

Figure 4-03. During acrobatic flying, pilots often fly upside-down and are exposed to –1 g with no problems.

Jet pilots often fly upside down at –1 g with no problem (Figure 4-03). Why then would flying in 0-g pose a problem? How can these space travelers perform nominally on orbit and then be so debilitated when they land? Symptoms of orthostatic hypotension are seen on Earth with patients who have certain types of cardio-vascular disease. Did the space agencies pick individuals who were susceptible to these problems? In fact, the astronauts selected by NASA are initially screened for significant cardio-vascular diseases (see Chapter 7, Section 2.2). They are in excellent physical shape and many have increased heart muscle mass compared to the terrestrial "normal" population.

So, what is the risk? The risk of cardio-vascular medical events happening during or after a mission can be divided into two categories: a) those medical events which occur as a consequence of pre-existing cardio-vascular disease (not detected during the selection medical examination) which are aggravated by spaceflight; and b) those medical events which occur as a consequence of the expected cardio-vascular physiological changes induced by spaceflight. A recent study based on experience with air pilots has determined that the risk of a "mission loss" due to serious cardio-vascular event (e.g., heart attack, sustained rhythm disturbance) during a 16-day spaceflight with a six-person crew is 0.3% per flight. The worst-case cardio-vascular risks of an incapacitating event for ISS crews over a 1-year period is 1% per person per year if all other space-related factors are ignored (Hamilton 2003).

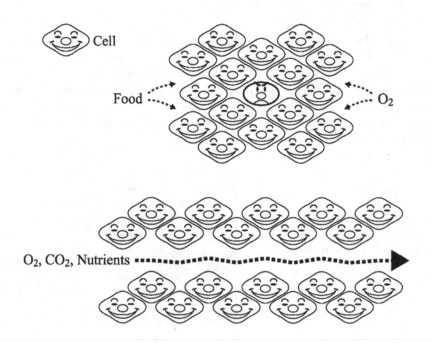

Figure 4-04. Why do we have a cardio-vascular system? When single-celled organisms, living in an aqueous environment, evolved to larger and larger multi-cellular organisms, the metabolic needs of the cells in the interior of the enlarging structure could no longer be satisfied by simple diffusion with the surrounding liquid environment. A vascular tree was eventually required for bringing-in the necessary materials to the individual cells and for carrying-out the undesired by-products of their metabolism. Thus, the primary function of the cardio-vascular system is to deliver a flow of blood to local tissues so that individual cells in those tissues can be maintained in optimal condition. (Churchill 1999)

2 PHYSIOLOGY OF THE CARDIO-VASCULAR SYSTEM

2.1 Basics

The major functions of the cardio-vascular system include: a) delivery of O_2 and nutrients to all areas of the body, and b) removal of CO_2 and cell metabolic wastes to specific organs, such as the lungs or kidneys. Other functions of the cardio-vascular system as it circulates blood include the transport of hormones, transport of immune system components such as white blood cells or antibodies, and heat regulation.

These functions were originally localized within the unique cell of the early, unicellular organisms. However, as organisms increased in size and in number of cells, isolated individual cells also needed to ingest nutrients and excrete waste (Figure 4-04). The development of a vascular tree, able to reach all individual cells, was therefore required. Such system includes both the heart and the vessels that pass to and from the heart to the tissues of the body. The vessels include the large arteries that receive blood from the heart, branching into smaller arterioles that branch further into capillaries. From the capillaries, blood flows into small collecting venules, then into larger and larger veins for return to the heart.

The overall organization of the cardio-vascular system consists of a driving pump, the heart, and two key circulatory systems that it powers: the *pulmonary* circulation (the lungs) and the *systemic* circulation (the rest of the body). In the systemic circulation, the arteries (red blood cells) transport oxygenated blood under relatively high pressures to the body at approximately 80 to 90 mmHg pressure, and the veins return deoxygenated blood to the heart at lower pressures of 5 to 15 mmHg. The capacity of the venous system is large and at least 70% of the blood volume in humans is found in the veins. The mean arterial pressure represents the average pressure tending to push blood through the capillaries and other vessels of the systemic circulation—it gives insight into average blood flow to the tissues. On Earth, there is a large pressure gradient from the head to the feet, with the mean arterial blood pressure being about 70 mmHg at the head level, 100 mmHg at the heart level, and 200 mmHg at the feet level. This is because the vascular system is essentially a set of vertical "columns" of blood, and pressure in these columns increases with depth, just as pressure increases with depth in the ocean (Figure 4-05).

In the pulmonary circulation, the deoxygenated blood is carried by perfusion from the right ventricle of the heart to the lungs via the pulmonary artery. This is a low-pressure system, typically 10-20 mmHg, with low resistance to blood flow. In an individual standing upright, this pressure may be insufficient to overcome hydrostatic gradients, and so very little flow reaches the upper regions of the lungs, but a relatively large portion of pulmonary blood flow perfuses the dependent portions of the lungs. Air in the alveoli flows somewhat preferentially into the middle and upper regions of the lungs. These regional differences create a mismatch of air ventilation and blood perfusion and are the basis for the system of classification of lung zones (West 1968). As the blood flows through the capillaries in the lungs, waste gases (i.e., CO_2) are released and oxygen is absorbed by simple diffusion. Blood then passes to the left atrium of the heart via the pulmonary veins. From there, the oxygenated blood passes to the left ventricle to be pumped out of the heart again.

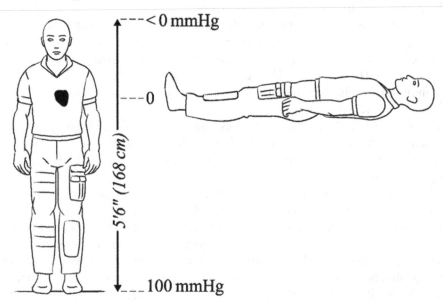

Figure 4-05. On Earth, in the upright posture, there is a gradient established in hydrostatic pressure from head-to-foot. The veins of our ankles sustain a pressure of approximately 100 mmHg when we are standing. An immediate effect of transition to microgravity is loss of hydrostatic gradient in the venous vascular system, similar to being supine on Earth.

The contraction of the heart, first atria and then ventricles is termed *systole*. Between beats, the heart pauses briefly for the atria to refill with blood for the next contraction. This period is termed *diastole*. Blood pressure in the arteries fluctuates during these phases of the full cardiac cycle. Arterial pressure rises sharply during ventricular systole and drops during ventricular diastole. Flow through a blood vessel is determined by both the force that pushes the blood through that vessel and the resistance of the vessel.

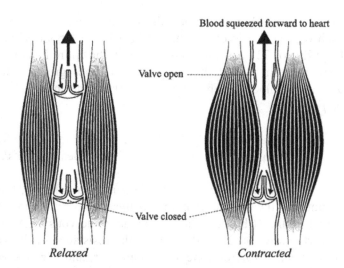

Blood squeezed forward to heart

Valve open

Valve closed

Figure 4-06. The contraction of skeletal muscles in the legs helps to pump blood toward the heart, but is prevented from pushing blood away from the heart by closure of the venous valves. (Adapted from Lujan and White 1994)

Relaxed   *Contracted*

It is also important to note that the flow of blood in the body is directly influenced by gravity. When a person is standing, gravity causes blood to pool in the relatively compliant leg veins. The force of gravity also makes it more difficult for the blood to flow upward to return to the heart and lungs for more oxygen. Since the veins are a low-pressure system with very distensible walls, in many cases, blood would tend to pool in them unless something aided flow. Most veins have valves in them, which prevent any "back-flow". In addition, a periodic pushing of blood forward every time skeletal muscles in the legs contract assists venous return (Figure 4-06). This mechanism effectively counteracts the force of gravity.

2.2   Control Mechanisms

2.2.1   Control of Blood Pressure

The cardio-vascular system is well known as the major fluid transportation system of the human body. Its regulated transport to and from the capillaries is central to life from the cellular to whole-person level. Yet the plasma of the cardio-vascular system contains only about 3 liters of total body water of 42 liters (about 7%) for a 70-kg "average" male. Another 11 liters of water are located in the extracellular interstitial spaces (between cells) and in the lymphatic vessels that drain those spaces. The remaining 28 liters of water is the intracellular fluid inside cells of the body.

Except in the case of bleeding, changes in blood volume usually occur because of changes in the water content of the blood plasma. Increasing total blood volume ultimately increases the "filling pressure" of the vascular system and the amount of blood to be ejected by the heart with each stroke.

The volume of blood discharged from the heart left ventricle with each contraction is called the *stroke volume* (about 70 mL at rest). With a heart rate of 72 beats/min, blood flow in the entire human circulation is about 5040 mL/min at rest, but may be 5-6 times greater during exercise. The amount of blood pumped by the heart in one minute is called the *cardiac output*:

Cardiac Output   =   Stroke Volume   x   Heart Rate
(mL of blood/min)   (mL of blood/beat) (beats/min)

Both heart rate and stroke volume can change, thus varying cardiac output and the supply of blood to the entire circulation. Controlling blood vessel radius (through the sympathetic nervous system) is also a very powerful way for the body to vary the resistance of the vessel to the blood flow, divert flow from one area to another, and vary blood pressure overall. Blood pressure will also vary as a function of the viscosity of the blood, such as the quantity of blood cells within the plasma (Table 4-01).

Factors	*Why blood pressure increases*
Increase in blood volume	*Increased total "filling pressure" in the semi-flexible cardio-vascular system; increased venous return to the heart, leading to higher stroke volume*
Increase in heart hate	*Increased cardiac output which, without a countering change in peripheral resistance, increases pressure*
Increase in stroke volume	*Same as increased heart rate*
Increase in peripheral resistance	*Normally varied by changing vessel diameter, particularly in the arterioles, increased constrictive resistance increase pressure in the vessels leading up to it*
Increase in blood viscosity	*Increased resistance, as thicker blood does not flow as easily*

Table 4-01. Some factors that influence arterial blood pressure and the associated mechanisms (Adapted from Reed 1999).

Both heart rate and blood pressure are controlled by the autonomic (unconscious) nervous system, which consists of two parts: the parasympathetic and the sympathetic nervous systems. These two systems have opposing roles and are activated according to the different needs of the individual. The parasympathetic nervous system is activated during rest and assists in energy restoration by means of the digestion and absorption of food. This system also acts to decrease heart rate. The sympathetic nervous system, on the other hand, prepares the body for an emergency and counteracts the parasympathetic nervous system in order to maintain the required energy supply. During any emotional or physical stress, adrenaline is released by the sympathetic nervous system, which acts to increase heart rate and blood pressure.

Accordingly, the balance between parasympathetic and sympathetic nervous systems activity controls heart rate. On a beat-to-beat basis, however, it has been observed that heart rate is not constant and there are periodical fluctuations indicative of the relative contributions of each of these two components of the autonomic nervous system. There have been various methods employed in an attempt to quantify the relative contributions of each of these systems. One of the most commonly used methods is the frequency domain analysis of heart rate variability. This method uses highly sophisticated techniques to determine different frequencies of heart rate, and from this analysis can identify which of the two systems is predominantly active during both rest and exercise.

2.2.2 Baroreceptor Reflexes

Baroreceptors ("pressure receptors") are specialized nerve endings located in both the arterial and venous systems, which are stimulated when the blood vessels are stretched by increased pressure. The baroreceptors in the arterial system are located in the neck as the carotid artery ascends to the brain, and in the aortic arch, immediately as blood leaves the heart in the aorta. When blood pressure increases, the corrective response via the stimulation of the baroreceptors and sympathetic nervous system includes a decrease in heart rate and stroke volume and a vasodilatation of the arterioles in order to decrease vascular peripheral resistance. In addition, secondary effects act on the kidney to allow increased urine production. The reverse effects take place if blood pressure is decreased.

Baroreceptors in the venous system are rather diffusely located and less understood classically. In general, these receptors are scattered in the major veins entering the heart, the atria and the pulmonary vessels (also called the vena cava). Because the large veins are very compliant, changing greatly in volume with small pressure changes, the venous baroreceptors are actually monitoring rather significant changes in venous blood volume in the upper body. Although less well characterized than their arterial counterparts, it is possible that these are the first baroreceptors activated by fluid shifts seen in spaceflight.[*]

---

[*] Earth's gravity may determine the location and size of internal organs such as the heart. For example, Lillywhite et al. (1997) noticed that the heart of the tree snake, who is crawling up and down trees and therefore must cope with gravity, was closer to the brain than the land or sea snakes, who spend most of their life in a horizontal position or are neutrally buoyant. The tree snake was the most tolerant to centrifugation, suggesting that it would be more gravity tolerant than the other snakes

There is also more and more evidence that the vestibular system plays a significant role in the regulation of blood pressure. During the transition from various postures, the stimulation of the otolith organs by the changes in head orientation relative to gravity could be used as a signal for triggering fast adjustment in blood pressure (Yates et al. 1996). This response would, of course, be altered in microgravity.

Overall, although the baroreceptors and vestibular receptors respond rapidly to pressure and acceleration changes, respectively, the response is not immediate. Anyone who has stood very quickly though from a reclining or kneeling position, knows that there can be a few seconds of dizziness during this process. This dizziness is the period when the brain is receiving too little blood flow while the body's reflexes work to correct blood pressure in the upper body. Jet pilots flying high-g turn could experience the same situation. There is a "lag" period before cardio-vascular responses begin to accommodate to the g-induced movement of blood downward from their heads to their feet. During, and even before, this critical time pilots must be especially vigilant to maintain their blood pressure by special straining maneuvers that increase blood pressure (further aided by a rapid-reacting anti-g suit that inflates on their legs and abdomen to push blood back towards the upper body). Such lag times in body reflexes are not so important in microgravity, because the headward fluid shift to which the baroreceptors are responding is a much slower and chronic condition, beginning as the astronauts recline in their seats for launch and continuing on-orbit.

2.2.3 Fluid Volume Regulation

Long-term regulation of blood pressure and related blood volume primarily involves coordination with the kidneys. The main blood factor controlled is blood plasma volume, the liquid portion of the blood. Ultimately, the cellular components of the blood can also be controlled (i.e., by producing or destroying red and white blood cells, and platelets); however, this is a secondary and longer-term phenomenon.

The kidneys play a large part in the regulation of fluid volume in the body and aid in the control of red blood cell production and blood pressure. The kidneys also help to maintain the normal concentrations of water and electrolytes within body fluids. This control is dependent upon hormones regulating salt and water balance. For example, the anti-diuretic hormone (ADH) is a polypeptide hormone released through the posterior pituitary gland of the brain when cells sense an increase in plasma osmolarity (relative salt concentration). The ADH hormone acts directly on the kidneys to cause them to retain more water. This water then dilutes the blood plasma and

---

as it did not have to carry blood over as great a distance from the heart to the brain (Morrey-Holton 1999).

increases plasma volume. Alternately, a reduction in ADH levels will cause the elimination of more water from the body.

Certain kidney cells are also sensitive to arterial pressure in a baroreceptor fashion, sympathetic nervous system activity, or circulating epinephrine from the sympathetic nervous system. In response to low blood pressure or increased sympathetic nervous activity, these cells initiate a rather complex, multi-stage process, that serves to retain or eliminate water and electrolytes.

In summary, I would like to cite the analogy proposed by Levine (1999): "It is helpful to think about the system as made up of "the plumbing", comprised of the "pump" (the heart) and "pipes" (the blood vessels), and the "control system" (the autonomic nervous system, hormones regulating salt and water balance, and local endothelial derived mediators of microvascular flow). For acute demands, such as during exercise or rapid changes in posture, higher order centers in the brain initiate an increase in the heart rate, termed "central command". Sensors located in skeletal muscle respond to changes in both metabolic and mechanical state and send back signals to the brain reflective of the intensity of effort. The heart itself is a sensory organ and detects the adequacy of hydration and cardiac filling through "mechanoreceptors". Pressure sensors or "baroreceptors" in the walls of the large blood vessels detect the pressure within the vasculature. These signals are integrated in special centers in the brain, which respond by regulating both the strength and frequency of the heart's contraction and the resistance of the blood vessels, primarily by neural mechanisms."

3   EFFECTS OF SPACEFLIGHT ON THE CARDIO-VASCULAR SYSTEM

Rapid transition between upright, sitting, and lying down postures requires that the heart and blood vessels adjust very quickly. On Earth, this is achieved by very sophisticated control centers. These control centers are challenged during spaceflight. When hydrostatic gradients are removed, such as changing from the upright to the supine position or exposure to microgravity, blood is shifted from the lower part of the body towards the chest virtually doubling the amount of blood within the heart. The heart responds to this volume load by increasing the amount of blood it pumps, and by initiating both a redistribution and elimination of plasma.

Research studies have focused on understanding the effects of spaceflight on the cardio-vascular system by studying cardiac output, heart rate, blood vessel behavior, blood pressure, and blood volume during spaceflight and upon return to Earth. One aim of these studies is to determine precisely when fluid shifts occur, because they are believed to be the precursor of other physiologic changes that occur in microgravity.

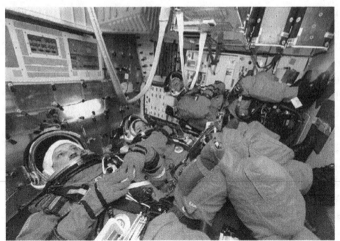

Figure 4-07. During launch, when the largest acceleration affects the Space Shuttle, the astronauts are reclined on their backs with their legs higher than their heads. This prevents blood from pooling in the legs during ascent and assists the heart in pumping blood to the rest of the body. (Credit NASA)

3.1   Launch Position

It is important to realize that the astronauts are oriented in almost a horizontal position while waiting for a launch in the Space Shuttle or Soyuz (Figure 4-07). The crew is placed in this position approximately 2.5 hours prior to the expected launch time, and it can stay there for as long as 4 hours before Mission Control considers a launch scrub. This supine position with a 90-deg hip and knee flexion is chosen in order to limit the launch acceleration +Gx, horizontal (back-to-chest) direction, for which the tolerance is greater (Table 4-02) (see also Figure 1-18).

The effect of this specific supine position is that significant blood volume is placed above the heart, thereby increasing pre-load to the heart (central venous pressure) and cardiac output. During the early portion of this orientation, a subject's stroke volume increases from about 75 mL/beat to about 90 mL/beat. This is entirely expected, because there is a rush of fluids to the upper part of the body and the heart has then more blood to force out during each beat. The body compensates in part for this situation by reducing blood volume through urination and reduced thirst. Shuttle astronauts wear undergarments with a fluid-absorbent material that permits them to urinate inside the launch suits, if necessary. However, some astronauts sometime prefer to restrict their fluid intake from 12 to 24 hours before launch and "fly dry" to prevent using the absorption garment. This may work in the short term; however, it puts them into orbit in a fluid-depleted state. In addition, such reduction in blood volume on the launch pad may impair their ability to emergency egress and provoke syncope upon fast standing.

Shuttle emergency egress plans during launch or landing call for the crew to escape through the side hatch or flight deck windows. During these phases, the crew is wearing a 41-kg space suit, which includes a life support system and a parachute. It is pertinent to ask whether, after several hours in the launch supine position, every crewmember would be able to use the escape system, and egress rapidly the launch or landing site without any assistance.

Vertical acceleration (g up is positive)	Event or Symptom
16	limit of human tolerance, centrifuge* ejection seat acrobatic
12-14	airplane loss of consciousness complete loss of vision (black-
11.4	out) partial loss of vision (gray-out) roller-coaster, maximum
4.5-6.3	at bottom of first dip congestion of blood in head severe blood
3.9-5.5	congestion, reddening of vision (red-out) limit of sustained
3.4-4.8	human tolerance
4.5	
-1	
-2	
-5	

Horizontal acceleration (g magnitude only)	Event or Symptom
0.4	"pedal to the metal" in a typical car "pedal to the metal" in a
0.8	high performance sports car Extreme Launch™ roller-coaster
2	at start
3	Space Shuttle, maximum at takeoff**
	jet fighter landing on aircraft carrier
8	limit of sustained human tolerance
21-35	limit of human tolerance, centrifuge*, 5 s duration
40-80	USAF chimpanzee, centrifuge*, 60 s duration
60	chest acceleration limit during car crash at 48 km/h with airbag
70-100	car crash that killed Diana, Princess of Wales, 1997
83	human subject, rocket powered impact sled, 0.04 s duration
247	USAF chimpanzee, rocket powered impact sled, 0.001 s
3400	impact acceleration limit for crash-survivable flight recorder

Table 4-02. Tolerance to vertical (Gz) and horizontal (Gx) acceleration (Source Elert 2002).
* The passenger capsule of a human centrifuge pivots so that a test subject in a seat would experience a vertical acceleration while a test subject lying down would experience a horizontal acceleration.
** During lift off, the Space Shuttle (which is pointing more or less upward) is accelerated in the direction of its vertical axis, but the passengers (who are lying on their backs) are accelerated in the direction of their horizontal axes.

3.2 Early On-Orbit

3.2.1 Fluid Shift

Once launch is over, the headward fluid shift continues in microgravity relative to normal Earth conditions. This shift is thought to occur because the mechanisms that normally act to counter the pooling of blood in the lower extremities continue to act even in the presence of gravity. This headward fluid shift actually creates a more even distribution of fluids and a more even distribution of blood pressures than is seen on Earth (Figure 4-08B). This initial fluid shift occurs rapidly and is virtually complete within the first 6-10 hours of flight. The effects of this headward fluid shift tend to last during the entire duration of the flight. The most obvious effect is a visible distension of veins in the head and neck region, as well as puffiness around the eyes. The astronauts sense this fluid shift and describe it as a "fullness in the head" or a nasal stuffiness similar to chronic sinus congestion. The senses of smell and taste may be altered, as happens on Earth when one has a cold. Some astronauts also report increased pressures inside the eye for a few days and pain in the eyes when they execute large ocular saccades. Occasional headaches have been reported, and intracranial pressure is still being studied to see how and if it changes in microgravity as a possible correlate to this and other effects (such as nausea).

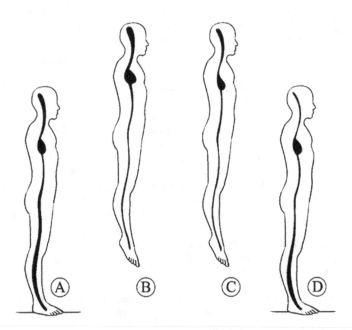

Figure 4-08. Fluid shift during spaceflight. A. On Earth, because of the downward pull of gravity, the body easily supplies blood to the lower limbs. B. Early on-orbit, blood volume shifts toward the chest and head, resulting in more blood than usual in the upper portion of the body. C. This increase triggers the receptors, which then cause the body to reduce the volume of fluid. The body functions with less fluid and the heart become smaller. D. When astronauts return to Earth, the bulk of the blood goes back down to the legs, and because the total amount of blood has decreased, there isn't enough to fill up the whole system of blood vessels. This contributes to the occurrence of orthostatic hypotension.

In contrast to the upper body, the legs experience a net loss of fluids as general capillary pressures decrease there. This leads to a so-called "Chicken Leg Syndrome" as leg volume decreases with time in microgravity. Studies have shown that leg circumference may decrease 10-30%, mostly in the fleshier thighs, as up to 2 liters of fluid shift headward. Fluid shifts clearly account for the first phase of this decrease (Moore and Thornton 1987).

Thirst is generally decreased early in-flight, and astronaut fluid intake is down. In part, this may be due to headward fluid shifts and suppression of normal thirst reflexes. However, there are no indications that urine output is increased in space (Norsk 2001). The reduced fluid intake may also be due to the effects of space motion sickness in many crewmembers for the first 1-3 days of spaceflight. Use of anti-motion sickness drugs, mission activities, and many other factors can also impact hydration and urine volume in space.

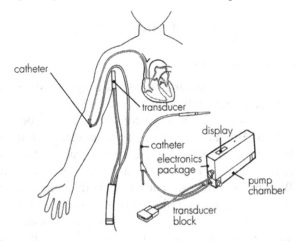

Figure 4-09. One Spacelab experiment used a catheter inserted preflight into an arm vein of an astronaut and later moved nearer to the heart. This catheter had a sensor attached, which measured the blood pressure closest to the heart. The experiment showed that the astronauts experience a much more rapid fall in central venous blood pressure than what was predicted by the model. (Adapted from Lujan and White 1994)

3.2.2 Blood Pressure

One measurement of special interest in cardio-vascular physiology is the central venous pressure (CVP). The CVP is the pressure in the vena cava, which are the large veins returning blood to the heart's right atrium from the systemic circulation. This pressure represents the blood "available" to be picked up by the right atrium before each contraction cycle. This measurement should therefore establish the amount of fluids that redistribute or shift to the upper part of the body and how rapidly that shift of fluids occurs. The first direct measurements of CVP in humans occurred during a Spacelab mission in 1993, when one astronaut was launched with a catheter extending into the inferior vena cava near the heart (Figure 4-09). Results were confirmed in later Spacelab missions. Data have indicated an increase in CVP before launch, when the astronauts are in the knees-up seated position, and a further increase during launch and ascent. One minute after reaching microgravity though, CVP decreased below pre-launch levels and stayed lower than normal (Foldager et al. 1996, Buckey et al. 1996b). The increase in CVP before and during the actual launch phase were expected, because of the rush of fluids to the right atrium due to the supine position and the g forces compressing the chest area, respectively. However, the decrease in CVP once the astronauts arrived in space, even though the body fluids continued to shift upward, and the fact that CVP fell to a level below normal within one minute upon entering the space environment, were totally unexpected.

In microgravity, the decreased CVP may indicate that venous compliance of thoracic blood vessels increases rapidly to hold the increased fluids at a lower pressure. This increased space for fluid build-up around the heart could be enhanced, for example, by removal of the weight of the lungs on veins that surround the heart. Other short-term changes include an increase in heart rate and an increase in heart size due to excessive fluid volume.

3.3   Later On-Orbit

3.3.1   Fluid Shift

The headward fluid shift triggers the baroreceptors, which inform the control centers, which in turn eliminate the excess of fluid in the upper body (Figure 4-08C). Over several days, the blood volume decreases, due to a decrease in thirst and an increase in water output by the kidneys. Total loss of fluid from the vascular and tissue spaces of the lower extremities has been found to be 1-2 liters (about a 10-15% volume change compared to preflight). Within 3 to 5 days in space, total body water stabilizes at about 2 to 4% below the normal level and plasma volume decreases by about 22%.

However, surprisingly, total body water (measured using an isotope-dilution technique) is unchanged, although the extracellular fluid and plasma volume are decreased. These results imply that the two liters lost in the vascular and interstitial compartment of the lower extremities are partially relocated in the intracellular space. After a reduction in blood volume, secondary mostly to plasma volume and red cell mass loss, an astronaut will reach a new state of intravascular hydration that, while adapted for microgravity, is profoundly hypovolemic for 1-g gravity.

It was also found that exposure to microgravity impairs the efficiency of the baroreflex loop. A closely fitting neck collar (similar to a whiplash collar) was used on astronauts during the Spacelab SLS-1 mission to test and record two blood pressure sensing areas located in the neck. By the eighth day of flight, astronauts had significantly faster resting heart rates, less maximum change of heart rate per unit of neck pressure change, and a smaller range of heart rate responses. The changes that developed were large, statistically significant, and occurred in all astronauts studied.

Measurements on humans before, during, and after several space flights have also provided echo-cardiographic data on cardiac dimensions and function. Ultrasound imaging revealed that heart volume increases dramatically when the astronauts first arrive in space, probably because of the increased volume of blood flowing into the heart (Figure 4-10). The heart volume then slowly decreases as the astronaut's body adapts to the space environment, ending up smaller compared with its size on Earth. This decrease in size can be explained by the fact that the excess of blood and fluids has now been eliminated, and the heart does not have to pump the blood against gravity. In addition, physical work requirements are generally less in space. The blood vessels also appear to get slightly smaller and stiffer.

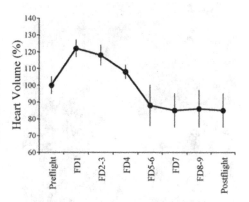

Figure 4-10. Changes in the volume of the left ventricle just before contraction (left–ventricular end-diastolic volume) during spaceflight (FD = flight day). Ventricular filling increases early in-flight and decreases after a few days relative to preflight levels (Adapted from Lujan and White 1994).

3.3.2 Maximal Exercise Capability

Since the amount of blood that passes through the lungs is equal to the amount of blood that flows out of the heart (cardiac output) and proportional to the amount of gas absorbed by the lungs, measuring gas exchanges can be used to determine cardiac output (Figure 4-11).

When the astronauts are required to exercise at their maximal capability and their consumption of oxygen (VO_2 max) is measured in orbit, the measurements are not different from preflight. This indicates that the maximal capacity of the cardio-vascular system, as reflected by maximal exercise capability, is well maintained during short-term spaceflight. Consequently, the loss of fluid volume and the heart changes seem to reflect a normal adaptation of the human body to an extreme environment change. This adaptation is a physiological, rather than a pathological response and does not appear to be associated with impaired function.

Exercise has also proven to be the single most effective method for reducing re-adaptation effects (such as the postflight orthostatic intolerance) and maintaining a healthy cardio-vascular system in space, as on Earth. For example, just a few exercise sessions were scheduled during the first two

Skylab missions. In the later missions, more exercise devices were used and the number of sessions increased. During the 84-day Skylab mission, some astronauts actually improved their cardio-vascular fitness, probably because the rigorous exercise requirements of the mission exceeded their preflight training practices.

On the contrary, VO_2 max is generally reduced immediately postflight. It is interesting to note that fit subjects demonstrated larger (16%) reduction in VO_2 max and plasma volume than unfit subjects (only 6% reduction) did. Maximal exercise capacity is restored to preflight levels within a week, at least after short-term spaceflight. The data for long-term spaceflight are scarcer.

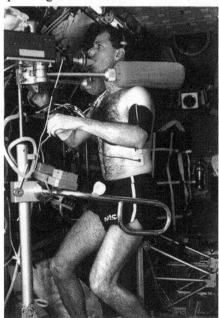

Figure 4-11. The amount of gas absorbed by the lungs and heart rate is measured during exercise on a cycle ergometer to determine cardiac output and maximal oxygen consumption. (Credit NASA)

3.3.3   Extra Vehicular Activity

During activities outside the vehicle (EVA) astronauts may be working hard for up to 6 hours in a space suit with mean metabolic rates of 800 kcal/hr and 5-10 minute peaks of over 1500 kcal/hr. Astronauts lose 0.7-2.2 kg (mostly fluids) during a typical EVA. The space suit is equipped with a liquid cooling garment, which accommodates the heat produced by these high workloads requirements. However, overcooling of the extremities is frequently observed, presumably due to the relative decrease in blood flow in microgravity. The astronaut can drink, using a straw, from small containers of liquids within the suit. Failure to do so would rapidly cause dehydration. This is a concern since the acute dehydration caused by EVA might aggravate an already deconditioned state caused by exposure to microgravity.

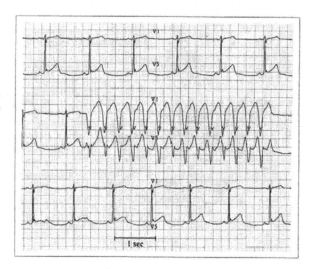

Figure 4-12. During the U.S.-Russian Shuttle-Mir Program, ambulatory ECG recordings were obtained on several crews. A recording in one cosmonaut revealed a non-sustained 14-beat run of ventricular tachycardia. (Adapted from Fritsch-Yelle et al. 1998)

3.3.4   Heart Rhythm

In the early phase of the U.S. space program, the presence of irregular heart rhythm (dysrhythmia) was taken as presumptive evidence of cardiac pathology. As a matter of fact, the first cardiac grounding of an astronaut occurred because of a heart rhythm disturbance (Johnson and Dietlein 1977). The presence of rhythm abnormalities during actual spaceflight can only be assessed when a crew is physiologically monitored, e.g., during extra-vehicular activity, exercise, and application of lower body negative pressure. Consequently, these heart rate irregularities seem more prevalent during these activities.

Serious heart rhythm disturbances were noted during the Apollo program, both during EVA on the Moon and after return to Earth (Berry 1974). Some of these abnormalities have been attributed to electrolyte balance or stress. For example, the astronauts were given excessive workloads on the Moon and had low blood potassium levels upon return to Earth.

Significant dysrhythmias were also observed during Skylab, Mir, and Shuttle flights. A crewmember during the Mir-2 (1987) mission developed a persistent dysrhythmia during EVA (Figure 4-12). This resulted in the mission duration being shortened. Earlier on, a Soviet cosmonaut was returned earlier from the Salyut-7 space station due to an intermittent cardiac dysrhythmia which originated during the course of a minor mishap during an EVA. The Russians have also reported a cosmonaut suffering from a massive myocardial infarction at the age of 49 years, just 2 years after his third short-duration flight. Moreover, the Russian medical community has reported to NASA that over the last 10 years of Mir operations, they observed approximately 31 abnormal electrocardiograms and 75 dysrhythmias (Hamilton 2003).

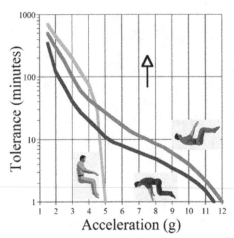

Figure 4-13. Tolerance to acceleration, measured as the time before fainting during a run in a centrifuge in various body positions. The arrow indicates the direction of the acceleration relative to the subject's body position. Tolerance is stronger in the supine or prone than in the upright position due to the effect of acceleration on the column of blood.

3.4   Postflight

As mentioned above, the reentry forces in the Soyuz capsules can go up to 3.8 g exerted along the Gx axis (chest to back).[*] The position the most effective for g-tolerance, lying-on-back with the legs flexed, was chosen because the need for the cosmonaut to "fly" the vehicle is minimal (Figure 4-13). By contrast, the astronauts in the Space Shuttle are exposed to reentry forces along the Gz axis (head-to-toe), due to the gliding attitude of the vehicle. These forces do not exceed 1.3-1.5 g for approximately 20 minutes. However, such small forces during reentry after 16 days of cardio-vascular deconditioning in microgravity may be as provocative as forces of 5-6 g in a fighter aircraft with a fit pilot. In the worst case, loss of consciousness (syncope) may result from a decrease in blood flow to the brain (cerebral hypoperfusion) (see Table 4-02).

Both the heart rate and arterial pressure increase during reentry and just after landing (Figure 4-14). After landing, all first-time astronauts are tested for the degree of their intolerance either by laying them on a "tilt table" that moves rapidly from the horizontal to vertical position, or by comparing their heart rates and blood pressure between supine rest and upright standing. After short-duration spaceflight about 27% of crewmembers are unable to complete a 10-minute stand test on landing day, and are forced to sit down to prevent syncope. The causes for this orthostatic intolerance are diverse. In some subjects, the syncope is due to a decreased blood flow (less than 30

---

[*] During the ISS Expedition-6 crew return in May 2003, the Soyuz headed down at a steeper angle, thus decelerating faster than planned. As a result, the crew was subjected to 8-10 g for several minutes.

mL/min per 100 g of brain tissue); in others it is due to a decreased arterial pressure (mean arterial pressure less than 40 mmHg).

A very evident cardiac response postflight is a markedly elevated heart rate (Figure 4-14). This is probably a reflexive action to increase cardiac output and, thus, blood pressure. This increased heart rate seems especially important to cardio-vascular function, since tests have shown that other mechanisms to increase short-term blood pressure (i.e., increasing peripheral resistance or increasing stroke volume through increased heart contractility) via the sympathetic nervous system seem less efficient after spaceflight. Decreased baroreceptor sensitivity acquired in microgravity can also slow the total response.

Recent studies performed on six astronauts before and after long-duration (129–190 days) spaceflights revealed that orthostatic intolerance is even more severe after long-duration than after short-duration flight. Five of the six astronauts studied became pre-syncopal during tilt testing after long-duration flights, whereas only one had become pre-syncopal during stand testing after short-duration flights (Meck et al. 2001).

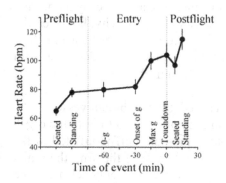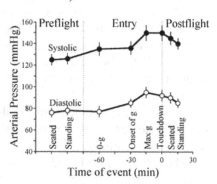

Figure 4-14. Changes in heart rate and arterial pressure when going from the seated to a standing position preflight and postflight, and during reentry and landing (0-g to Touchdown). When standing postflight, despite an increase in heart rate, there is a decrease in arterial pressure, which can lead to pre-syncope or syncope. (Adapted from Sawin et al. 1998)

4   WHAT DO WE KNOW?

4.1   Orthostatic Intolerance

Postflight orthostatic intolerance is due to more than just loss of fluid. Orthostatic intolerance is presumably caused by three factors that are related to each other: the volume of blood in the blood vessels, the ability of blood vessels to expand or constrict to maintain blood pressure, and the functioning of the heart itself.

We have already seen that upon return to Earth, when gravity pulls the fluid downward and there is not enough fluid, the system cannot function normally and this contributes to the occurrence of orthostatic hypotension (Figure 4-08).

Another contributor to this problem is the autonomic nervous system, which helps control blood pressure, among other things. Normally, this system is responsible for making minute and immediate adjustments to the cardio-vascular system in order to maintain the blood flow and pressure during changes in posture. The system does this by releasing a neurotransmitter called norepinephrine that causes the blood vessels to constrict to keep the pressure at the appropriate level to supply an adequate amount of blood to the body's organs. In space, when hydrostatic gradients are removed, such as changing from the upright to the supine position, perhaps those mechanisms "forget" their function.

Work on animals has underlined the importance of the baroreceptors in the regulation of blood pressure. Dogs whose carotid sinus baroreceptors have been excised have a variation from 40 to 200 in their mean blood pressure. Manual compression of carotid sinuses can cause syncope and bradycardia (low heart rate in a healthy person). In microgravity, however, the carotid sinus would fire nerve impulses at a more or less constant rate. The extent to which the sensitivity of the baroreceptors may decrease with prolonged spaceflight, and their ability to regain the lost sensitivity, is unknown.

Orthostatic hypotension is also caused by the blood vessels. The vessels themselves can try to control the amount of blood getting to the organ they are serving. For example, when a blood pressure cuff is inflated to the point where there is no blood flow through the vessels of the arm, the arm's vessels dilate to try to get flow going. If the cuff is suddenly released, the dilated vessels allow the blood to rush back into the arm. This increases shear stress, which actually stimulates the lining of the blood vessels to release additional vasodilators, ensuring that the arm will get better flow. This mechanism is called reactive hyperemia. Similarly, when an astronaut returns to Earth and blood rushes to his/her legs, the vessels might respond not by constricting, to force the blood back up, but by dilating further, which permits more flow downward and less pressure, resulting in less blood in the astronaut's upper body and head.

Another trend observed in recent studies is that there is a difference between women and men in their bodies' abilities to maintain blood pressure after spaceflight. Women generally have a higher heart rate and a lower vascular resistance than men do. Thus, when female astronauts return from space, their vascular resistance, already low, is insufficient to combat the lower blood volume. In a recent postflight analysis after a short-duration spaceflight, all female astronauts became syncopal versus only 20% of male astronauts.

Aging might also be a factor. The differences that age had made in his body better equipped John Glenn, at age 77, to handle the cardio-vascular adaptations to a microgravity environment. When he returned from orbit, far from feeling faint or suffering from cardio-vascular stress, he was calm and stood upright with no problem. An older person has a different strategy than a young person in maintaining blood pressure. Glenn had a high release of norepinephrine, which helps maintain pressure, both before and after spaceflight. This response is typical of the elderly and for Glenn, this resulted in a normal level of vascular resistance. Glenn also had a higher cardiac output than the other male astronauts, possibly due to a greater venous return. This higher output coupled with a normal vascular resistance, presumably enabled Glenn to maintain an adequate blood pressure on landing day (Rossum et al. 2001).

Figure 4-15. The ultrasound imaging system on board the ISS provides three-dimensional image view of the heart and other organs, muscles, and blood vessels. (Credit NASA)

Finally, although there is evidence that both the autonomic nervous system and hormone secretion are altered, their effects on the kidneys, blood vessels, and heart have yet to be fully understood and must be studied over varying duration of exposure to weightlessness. Elucidation of the mechanisms of these effects promises to shed light on some clinical, non-spaceflight problems such as high blood pressure and heart failure.

In summary, hypovolemia, cardiac atrophy, and autonomic dysfunction have each been hypothesized to contribute to postflight orthostatic intolerance, but their relative importance is unclear. Furthermore, it is unknown whether actual abnormalities in the myocardium itself develop with long-term spaceflight. Therefore, reliable portable noninvasive methods are needed in order to detect and quantify these changes. The imaging modalities of radiography, magnetic resonance imaging and computerized tomography would be state-of-the-art techniques. A specially modified commercial ultrasound echocardiograph instrument is being used in the Human Research Facility of the ISS for medical diagnosis and physiology research (Figure 4-15). To date, echocardiography has the most versatile ability to characterize cardio-vascular anatomy and physiology in ground-based models, pre- and postflight, and most importantly during flight (Martin et al. 2003).

4.2 Pulmonary Function

The pulmonary system works in tandem with the cardio-vascular system to supply the body with the oxygen needed for life. Unlike the cardio-vascular system, no pulmonary system problems have been associated with weightlessness per se, and researchers have devoted less attention to its physiology in microgravity. In fact, an increased lung blood flow and more uniform flow distribution were observed in microgravity, suggesting that overall lung function was actually improved in space (West et al. 1997). However, in the long term, lung function can be altered by changes in vascular pressure and volume. Also, it is possible that lengthy alterations in the relative flow distribution of blood and air in different lung regions might permanently affect right heart function (Linnarsson 2001). Dysbarism, the condition that results from exposure to decreased or changing barometric pressure, is also a problem of increasing magnitude during extra-vehicular activity (see Chapter 7, Section 3.2).

Along with alterations due to changes in vascular pressures and volumes, inhaled gases, vapors, and aerosols can damage the lungs. Integrity of the pulmonary system cannot be assumed simply because of lack of symptoms or overt clinical signs. The factors affecting the selection of cabin atmospheres and pressures for spaceflight, as well as the problems of cabin atmosphere maintenance and contamination in open or closed environmental systems will be discussed later (see Chapter 7, Section 4.1).

The kidneys are central to the above-mentioned physiologic questions. Renal problems may occur in the space environment. As we will see in the next chapter (see Chapter 5, Section 1.2), weightlessness causes a monthly 0.4% resorption of bone calcium, which is excreted in the urine. With increased concentration of urinary calcium and some other changes induced by weightlessness (such as urine alkalinity and possible reduction in urine volume), kidney stones may form more easily. In addition to debilitating pain, kidney stones might obstruct the urinary tract and precipitate infection, which is potentially quite dangerous in space. Thus, kidney function must be understood better with regard to calcium metabolism as well as its relation to cardio-vascular phenomena.

We also still need to understand more completely the actions of drugs that affect cardio-pulmonary and renal systems in space. This will be essential for adequate health maintenance. Ordered in descending priority, the following classes of agents must be investigated: anti-arrhythmics, bronchodilators, anti-allergy and anti-anaphylactic drugs, analgesics (including narcotics), hypnotics and psychotropics, diuretics, and anti-coagulants.

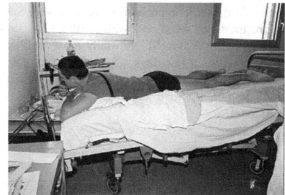

Figure 4-16. During bed rest studies, volunteers carry out all their daily tasks, including eating, in bed. They also shower in the head-down tilt position. (Credit ESA/CNES/MEDES)

4.3   Bed Rest

Since the beginning of the space program, human bed rest has been commonly used as a ground-based model to test the effects of weightlessness and proposed countermeasures upon the cardio-vascular system. In this model research subjects are required to remain in bed tilted at 6-deg head-down for lengths of time from weeks to several months (Figure 4-16).

In fact, studies had been conducted on bed rested patients and on normal, healthy subjects starting as early as 1855. Physicians have utilized prolonged rest in bed to immobilize and confine patients for rehabilitation and restoration of health even before that time. The rationale is that the horizontal position relieves the strain of the upright posture on the cardio-vascular system, bone fractures, muscle injuries or fatigue. Consequently, there is an almost complete loss of hydrostatic pressure (see Figure 4-05), virtual elimination of longitudinal compression of the spine and long bones of the lower extremity, and reduced muscular force. Patients on prolonged bed rest experience headward fluid shift, decrease in blood pressure and blood volume, and reduced physical activity similar to those that occur in astronauts in space (Gharib and Custaud 2002).

Bed rest studies were the first to indicate that baroreceptor reflexes may be impaired with time. In effect, when continuously exposed to increased pressure, the baroreceptors apparently become less sensitive and responsive

(Eckberg and Fritsch 1992). Similar findings have later been found during spaceflight. These changes in sensitivity and responsiveness may take days to occur and, similarly, may take several days to readjust upon return to Earth. Ongoing studies are being conducted to verify these results.

Following a bed rest, volunteers are monitored while placed supine on a tilt table and suddenly brought upright (Figure 4-17). Getting upright again after three months in the horizontal position evidently has its cost! Even after a few days in bed most people experience problems with balance and dizziness. After three months in bed the same and more expressed problems, such as orthostatic hypotension are observed. Subjects need considerable time before they can stand and walk without assistance. Rehabilitation activities are then being evaluated. Normally most problems with standing and walking are over within a few days, after which the rebuilding of muscle strength can begin. The full recovery of muscle and in particular bone tissue may take much longer, up to 6 months, although not necessarily being felt by the individual.

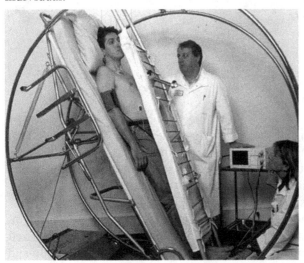

Figure 4-17. At the end of a bed rest study, the volunteers are placed on a motorized tilt table. Their cardio-vascular responses are measured when they are passively moved from a supine to an upright posture. (Credit MEDES)

5   COUNTERMEASURES

Researchers remain concerned with devising and refining countermeasures to prevent or avoid cardio-vascular problems associated with the return from microgravity to Earth gravity. As discussed above, the cardio-vascular problems associated with spaceflight are multifactorial. Four or five different countermeasures in some combination will probably be needed to solve the problem completely. Current countermeasures include preflight and in-flight exercise, application of lower body negative pressure (LBNP) in-flight, fluid loading prior to reentry, and rehabilitation after return to Earth.

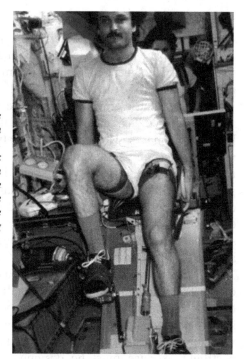

Figure 4-18. A Russian cosmonaut is exercising on the cycle ergometer with thigh cuffs to minimize the headward shift of blood. A recent experiment performed on six cosmonauts showed that the use of these thigh cuffs compensated only partially for the cardiovascular changes induced by exposure to microgravity: orthostatic intolerance occurred in all of the six cosmonauts.during postflight stand test (Herault et al 2000). (Credit Intercosmos)

5.1   In-flight

As a rule, crewmembers are encouraged to drink adequate amounts of fluids and to maintain a regular exercise schedule.

5.1.1   Exercise

Exercise has a protective effect on the increase in heart rate and fall in blood pressure during standing after flight. One way to prevent harmful effects of cardio-vascular deconditioning is to start the spaceflight at a higher level of conditioning by athletic training before the flight. Many of the astronauts run, jog, or participate in aerobic exercise as part of their daily routine training.

An aggressive in-flight aerobic exercise program seems to be partially effective in maintaining postflight aerobic capacity, but its effects on orthostatic tolerance are largely unknown. It is important to note that the discussion of exercise as a countermeasure here is aimed at cardio-vascular conditioning. Other types of exercise (i.e., resistance training) can also be important as countermeasures for skeletal and bone loss seen in microgravity (see Chapter 5, Section 6.1).

Special restraint systems (e.g., bungee cords) are required to hold the astronauts in place during exercise sessions. Another practical issue regarding exercise in space is that it generates vibrations that are transmitted through the structure of the spacecraft and might impact experiments requiring very low gravity levels. Also, there is no shower!

The Shuttle crews exercise once every second day after being on orbit more than three days. More stringent daily physical exercise is scheduled for ISS crewmembers. However, the duration of required exercise to balance mission needs and achievable cardio-vascular conditioning remains debated. Following the Mir experience, exercise is generally recommended for 2 hours per day. While exercising, the Russian cosmonauts sometimes use thigh constriction cuffs to decrease headward fluid shift (Figure 4-18). Data on the effectiveness of these cuffs are lacking, but many cosmonauts report significant relief from the head congestion and facial edema otherwise associated with microgravity.

It is perhaps surprising that, to date, we have very little understanding of the exact physiologic effects (beneficial or harmful) of various types of exercise on the phenomenon of cardio-vascular deconditioning. For example, some evidence suggests that the aerobically trained individual may be more vulnerable to orthostatic intolerance. Protocols for preflight, in-flight, and postflight exercise must be designed and tested in a rigorous manner to determine what, if any, types of exercise may be the best countermeasures to deconditioning. Integrated into the problem of understanding the effects of exercise on cardio-vascular deconditioning is also understanding the responses of blood gases, electrolytes, glucose, insulin, growth hormone, glucagon, and cortisol.

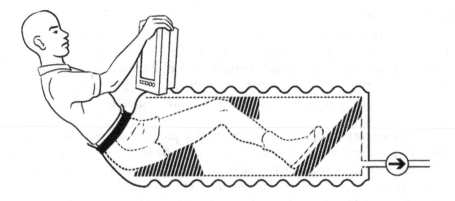

Figure 4-19. A Lower Body Negative Pressure (LBNP) device causes the intravascular volume to shift towards the lower extremities in microgravity, in a manner similar to the orthostatic load caused by assuming an upright posture in Earth gravity.

5.1.2   Lower Body Negative Pressure

A lower body negative pressure (LBNP or *Chibis*) is a device that encloses the lower abdomen and lower extremities to maintain a controlled pressure differential below ambient (Figure 4-19). This device is used in conjunction with heart rate and blood pressure monitoring capabilities. It provides a continuous decompression and maintenance to -60 mmHg. Decompression from ambient pressure to -60 mm Hg can range from 10 seconds to 10 minutes (i.e., rapid to slow decompression). However, care is used in this approach since, if decompression is too fast, similar effects to postflight orthostatic intolerance can occur. An adjustable foot support, removable saddle, and knee fixation within the device provides skeletal "loaded" and "unloaded" LBNP. The decompression device is available not only for cardio-vascular research, but also for any other physiological research.

As in standing, the cardio-vascular system responds to LBNP by increasing blood pressure to maintain flow to the upper body and head. LBNP tests performed on-orbit provoke a larger increase in leg volumes, as fluid is shifted rapidly footward, than in control tests on Earth. There is also a much larger increase in heart rate (to maintain upper-body blood pressure) than on Earth. These results may indicate a loss of muscle tone in leg blood vessels and less resistance to expansion by fluids, as well as weakened ability to respond to short-term blood pressure changes in general (Charles et al. 1994). There is, however, significant inter- and intraindividual differences in the responses to in-flight LBNP tests, which make it difficult to use this test for predicting which astronauts will be more susceptible to orthostatic intolerance after landing.

Further in-flight orthostatic countermeasures and exercise equipment can also include whole-body elastic loading suits, such as the Russian "Penguin" suit (Figure 4-20), pharmacological preparations, and electromyostimulation.

Figure 4-20. The "Penguin" suit. The inside of the suit contains a system of elastic, straps, and buckles that can be used to adjust the fit and tension of the suit. This suit forces the subjects to use his extensor muscles in-flight to activate venous return.

5.1.3 Monitoring

The ISS Human Research Facility (HRF) and the Space Shuttle are outfitted with equipment necessary to make a variety of measurements of the cardio-vascular function. Most of these measurements may be made in conjunction with exercise equipment. The capabilities available for on-orbit research are summarized in Table 4-03.

Blood Pressure   *Capabilities include noninvasive monitoring and collection of blood pressure data, both extended duration and intermittent, on human subjects. The data can be collected by manual or automated methods during periods of rest or exercise*
Electrical Stimulation of Muscle   *Local noninvasive muscle stimulation on human subjects using a high current stimulator which provides trains of pulses up to 0.8 amps, according to pre-programmed protocols.*
ECG/EMG/EEG   *Acquisition of human physiological data such as ECG, EMG, EEG, temperature, and skin Galvanic responses. Multichannel data (16 differential channels) can be collected by means of portable, crew-worn devices over extended periods of time (24 hours), or via rack-mounted devices.*
Pulse/Blood Oxygen   *A pulse oximeter to monitor the percentage of hemoglobin oxygen saturation in the blood.*
Lung Volume   *Respiration of crewmembers can be studied by continuously monitoring lung volume using respiratory impedance plethysmography.*

Metabolic Activity/Pulmonary Physiology
 - Two gas analyzers will be available, one based on the use of mass spectrometry and the other on infrared gas analysis techniques.
 - Combined with ancillary equipment, including gas supplies for supplying special respiratory gas mixtures.
 - The following measurements will be possible:
    1. Breath-by-breath measurements of VO_2, VCO_2,
    2. Diffusing capacity of the lung for CO_2
    3. Expiratory reserve volume
    4. Forced expired spirometry
    5. Functional residual capacity
    6. Respiratory exchange ratio
    7. Residual volume
    8. Total lung capacity
    9. Tidal volume
    10. Alveolar ventilation
    11. Vital capacity
    12. Volume of pulmonary capillary blood
    13. Dead-space ventilation
    14. Cardiac output

Table 4-03. List of equipment available on the ISS for cardio-vascular research.

5.2   End of Mission

It is well known that drinking about one liter of a balanced salt solution leads to an increased blood plasma volume loads, by about 400 mL for at least 4 hours. Early Shuttle space missions verified that this technique of temporarily increasing plasma volume could be used by astronauts to ease the orthostatic intolerance on landing. The fluid loading protocol consists in ingesting about 1 liter of water or juice and 8 salt tablets about 1 hour before leaving orbit (Figure 1-20). This produces 1 liter of isotonic saline in the digestive track, which then leads to absorption and subsequent increase in plasma volume. This technique proved effective for short-duration mission. For example, for 26 astronauts, those who had practiced "fluid loading" had lower heart rates, maintained blood pressure better, and reported no faintness (compared to 33% astronauts having faintness in a control group). However, the effectiveness of fluid loading is reduced with longer time in orbit. May be other factors than cardio-vascular deconditioning become more important on longer flights with regard to causing orthostatic intolerance (Buckey et al. 1996a).

In the critical period of reentry and landing, Shuttle astronauts routinely may wear anti-gravity suits. These suits contain balloon-like pressure bladders in the pants, which can be inflated with air by the astronaut. When the astronaut inflates the bladders in his pants, the bladder presses against his legs, forcing body fluid into the upper body. This helps the heart pump the blood more efficiently by pushing the blood out of the lower extremities. The Russians wrap the lower body tightly with elastic strapping (*Karkas*) to achieve the same effect as the anti-gravity suit.

Anyone who has been on orbit for more than 30 days is required to be returned to Earth in the supine position (+Gx acceleration) to reduce the risk of orthostatic intolerance during reentry and landing. The Space Shuttle is equipped with recumbent seats for returning long-duration crewmembers from the ISS (Figure 4-21). There is, however, a concern that a long-duration flight crewmember could probably not egress from the recumbent seat system without any assistance.

In conclusion, despite the use of in-flight countermeasures, orthostatic intolerance remains a major, unresolved, clinical and operational problem (Churchill and Bungo 1997). The problem of long-term exposure to microgravity looms large; currently observed space effects may intensify or new ones may appear. At present, all cardio-vascular changes are entirely reversible upon return to normal gravity, and there appears to be no deleterious effect of spaceflight directly upon the heart. Might orthostatic intolerance become irreversible after long-term exposure? How will the time course of cardio-vascular re-adaptation to 1-g be affected by lengthier missions? Will long-term spaceflight bring irreversible myocardial degeneration or "hypotrophy"? These are just some of the questions that remain to be addressed.

Figure 4-21. This photograph shows three ISS crewmembers in their recumbent seats during return to Earth in the Space Shuttle (right), by comparison with the upright seat of the Shuttle crewmember (left). (Credit NASA)

6   REFERENCES

Buckey JC, Lane LD Jr, Levine BD, Watenpaugh DE, Wright SJ, Moore WE, Gaffney FA, Blomqvist CG (1996a) Orthostatic intolerance after spaceflight. *Journal of Applied Physiology* 81: 7-18

Buckey JC, Gaffney FA, Lane LD Jr, Levine BD, Watenpaugh DE, Wright SJ, Yancy CM Jr, Meyer DM, Blomqvist CG (1996b) Central venous pressure in space. *Journal of Applied Physiology* 81: 19-25

Busby DE (1968) *Cardiovascular Adaptations to Weightlessness. Space Clinical Medicine.* Dordrecht, Holland: Reidel Publishing Company

Charles JB, Bungo MW, Fortner GW (1994) Cardiopulmonary function. In: *Space Physiology and Medicine.* Nicogossian AE, Huntoon CL, Pool SL (eds) Philadelphia, PA: Lea & Febiger, Chapter 14

Churchill SE (1999) Response of cardiovascular system to spaceflight. In: *Keys to Space.* Houston A, Rycroft M (eds) Boston MA, McGraw Hill, Chapter 18.4, pp 1830-1834

Churchill SE, Bungo MW (1997) Response of the cardiovascular system to spaceflight. In: *Fundamentals of Space Life Sciences.* Churchill SE (ed) Malabar FL, Krieger Publishing Company, Volume I, Chapter 4

Eckberg DL, Fritsch JM (1992) Influence of 10-day head-down bedrest on human carotid baroreceptor-cardiac reflex function. *Acta Physioogica Scandinavia* 604 (Supplement) 69-76

Elert G (2002) Frames of Reference. *The Physics Hypertextbook*. Available at: http://hypertextbook.com/physics/mechanics/

Foldager N, Andersen TA, Jessen FB, Ellegaard P, Stadeager C, Videbaek R, Norsk P (1996) Central venous pressure in humans during microgravity. *Journal of Applied Physiology* 81: 408-412

Frey MA (ed) (1996) Proceedings of the International Workshop on Cardiovascular Research in Space. *Medicine and Science in Sports and Exercise* 28: S3-S8

Fritsch-Yelle JM, Charles JB, Jones MM, Beightol LA, Eckberg DL (1994) Spaceflight alters autonomic regulation of arterial pressure in humans. *Journal of Applied Physiology* 77: 1776-1783

Fritsch-Yelle JM, Leuenberger UA, D'Aunno DS, et al. (1998) An episode of ventricular tachycardia during long-duration spaceflight. *American Journal of Cardiology* 81: 1391-1392

Gharib C, Custaud MA (2002) Orthostatic tolerance after spaceflight or simulated weightlessness by head-down bed-rest. *Bulletin Academy National of Medicine* 186: 733-746

Hamilton D (2003) Cardiovascular disorders. In: *Principles of Clinical Medicine for Spaceflight*. Barratt M, Pool SL (eds) Chapter 18, in press

Herault S, Fomina G, Alferova I, Kotovskaya A, Poliakov V, Arbeille P (2000) Cardiac, arterial and venous adaptation to weightlessness during 6-month MIR spaceflights with and without thigh cuffs (bracelets). *European Journal of Applied Physiology* 81: 384-390

Kirsch KA, Rocker L, Gauer OH, Krause R, Leach C, Wicke HJ, Landry R (1984) Venous pressure in man during weightlessness. *Science* 225: 218-219

Levine B (1999) *Human Cardio-Vascular Adaptation to Altered Environments*. Transcript of a lecture given at the University of Texas Southwestern Medical Center

Lillywhite HB, Ballard RE, Hargens AR, Rosenberg HI (1997) Cardiovascular responses of snakes to hypergravity. *Gravitational Space Biology Bulletin* 10: 145-152

Linnarsson D (2001) Pulmonary function in space. In: *A World Without Gravity*. Seibert G (ed) Noordwijk: European Space Agency, ESA SP-1251, pp 48-57

Lujan BF, White RJ (1994) *Human Physiology in Space*. Teacher's Manual. A Curriculum Supplement for Secondary Schools. Houston, TX: Universities Space Research Association

Martin DS, South DA, Garcia KM, Arbeille P (2003) Ultrasound in space. *Ultrasound Medical Biology* 29: 1-12

Meck JV, Reyes CJ, Perez SA, Goldberger AL, Ziegler MG (2001) Marked exacerbation of orthostatic intolerance after long- vs. short-duration spaceflight in veteran astronauts. *Psychosomatic Med* 63: 865-873

Moore TP, Thornton WE (1987) Space Shuttle in-flight and postflight fluid shifts measured by leg volume changes. *Aviation, Space and Environmental Medicine* 58: A91-A96

Morey-Holton ER (1999) Gravity, a weighty-topic. In: Rothschild L and Lister A (eds) *Evolution on Planet Earth: The impact of the Physical Environment*, New York: Academic Press

Norsk P (2001) Fluid and electrolyte regulation and blood components. In: *A World Without Gravity*. Seibert G (ed) Noordwijk: European Space Agency, ESA SP-1251, pp 58-68

Reed R (1999) Cardiovascular Function and Fluids. Available at: http://www.spacebio.net/modules/

Rossum A, Ziegler M, Meck J (2001). Effect of spaceflight on cardio-vascular responses to upright posture in a 77-year-old astronaut. *American Journal of Cardiology* 88: 1335-1337

Sawin CF, Baker E, Black FO (1998) Medical investigations and resulting countermeasures in support of 16-day space shuttle missions. *Journal of Gravitational Physiology* 5: 1-12

Watenpaugh DE, Hargens AR (1995) The cardiovascular system in microgravity. In: *Handbook of Physiology*. Fregly MJ, Blatteis CM (eds) New York: Oxford University Press, Volume 1, pp 631-734

West J (1968) Regional differences in the lung. *Postgraduate Medicine Journal* 44:120-122

West JB, Elliott AR, Guy HJB, Prisk GK (1997) Pulmonary function in space. *Journal of the American Medical Association* 277: 1957-1961

Yates BJ (1996) Vestibular influences on cardiovascular control. In: *Vestibular-Autonomic Regulation*. Yates BJ, Miller AD (eds) Boca Raton FL, CRC Press, pp 97-111

Additional Documentation:

Integrative Physiology in Space (2000) *European Journal of Physiology*, Volume 441, No. 2-3

International Workshop on Cardiovascular Research in Space (1996) *Medicine and Science in Sports and Exercise*, Volume 28, Number 10 Supplement, S1-S112

Space Life Science and Space Sciences Flight Experiments Information Package (2001) A Companion Document to Agency Solicitations in Space Life Sciences and Space Sciences. Issued by the International Space Life Sciences Working Group

Chapter 5

THE MUSCULO-SKELETAL SYSTEM IN SPACE

Muscle and bone form as a result of mechanical forces exerted on the body. In microgravity, support muscles such as those in the calf and thigh decline in volume, strength, and mass. Similarly, bones lose calcium, the mineral from which they derive their structure and strength, through the process of demineralization. Is the reported loss of muscle and bone mass which occurs during spaceflight self-limiting or does it continue? Is it permanent or is it reversible? Could the parallel loss of muscular strength and coordination jeopardize the return of piloted spacecraft or limit work capability and performance for surface operations on Mars? This chapter examines the effects of spaceflight on structure and function of the musculo-skeletal system, what the implications of such changes might be for long-duration exploratory missions, and what countermeasures might be employed to prevent undesirable changes.

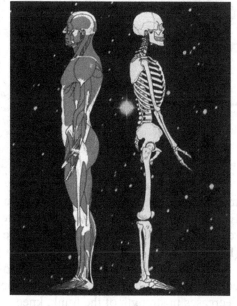

Figure 5-01. Muscle atrophy and bone demineralization are serious concerns for long-duration spaceflights. (Credit NASA)

1   THE PROBLEM: MUSCLE ATROPHY AND BONE LOSS

1.1   Muscle Atrophy

After a few days of exposure to microgravity, muscle atrophy begins and the urinary excretion of nitrogen compounds increases. This atrophy is characterized by structural and functional alterations. There is a decrease in muscle fiber size, with no apparent change in fiber number. Atrophy is considerably greater for postural muscles, i.e., those muscles that support activities such as walking, lifting objects, and standing on Earth, as compared to the non-postural muscles, which undergo only marginal changes. Astronauts lose 10 to 20% of their muscle mass on short missions. On long-term flights, the muscle loss might rise to 50% without using countermeasures.

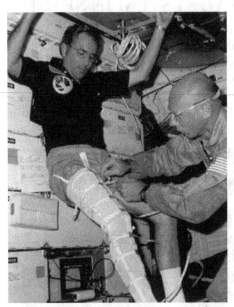

The visible reduction in the leg circumference has been used as an indicator of muscle atrophy (Figure 5-02). However, as seen in the preceding chapter (see Chapter 4, Section 3.2), this reduction is also influenced by the shift of fluids from the lower to the upper body in microgravity.

Figure 5-02. A "cast" placed on the leg of an astronaut is used to measure changes in leg circumference, as an indicator of changes in muscle mass during spaceflight. (Credit NASA)

The muscle loss is presumably caused by changes in the muscle metabolism, the process of building and breaking down muscle proteins. Experiments performed during long-term missions on board Mir have revealed a decrease of about 15% in the rate of protein synthesis in humans (Di Prampero et al. 2001).

In addition to pure muscle loss, the fibers involved in muscle contractions change their contractile properties and are weakened. Significant decreases in strength of the trunk, knee and shoulder muscles have been found within six days of a stay in microgravity. Extensor muscles are more affected than flexor muscles. Animal studies also revealed that muscle fiber regeneration is less successful in space. The associated continued excretion of nitrogen may also have deleterious hormonal and nutritional effects.

Spaceflight also results in increased susceptibility of skeletal muscle to contraction damage, which occurs in muscular atrophies on Earth-bound patients. These effects may compromise the ability of astronauts to do some of their activities in orbit. Likewise, they may not be able to withstand the stress of 1-g upon return to Earth. In fact, the muscle weakness, fatigue, faulty coordination and delayed-onset muscle soreness that the astronaut's experience after spaceflight mimics the changes seen in bed rest patients and the elderly. Finally, it is important to bear in mind that muscle atrophy caused by weightlessness also participates in the postural instability and locomotion difficulties seen after spaceflight (see Chapter 3, Section 3.3).

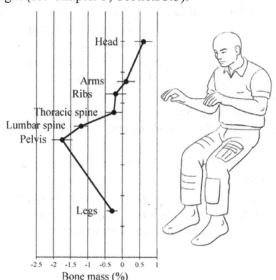

Figure 5-03. Changes in bone mass relative to preflight level in cosmonauts during long-term missions on board Mir. Bone loss seems to be a regional phenomenon in which the bone areas with the greatest decrease in load, i.e., the hip, lose the most bone. Interestingly enough, bone mass increases at the head level, presumably because of the increase in pressure due to the headward fluid shift in microgravity. (Credit MEDES)

1.2   Bone Loss

Bone loss during spaceflight is about 1-2% per month. The effect is especially marked in the weight-bearing bones of the legs and spine (Figure 5-03). Certain individuals on six-month flights have lost as much as 20% of bone mass throughout their lower extremities. There is no indication that this bone loss abates with longer flights. In addition, after return to Earth, bone loss continues for several months.

Bone loss of this magnitude leads to a significant increase in fracture risk, which may be as much as five-fold that expected with normal bone mass on Earth. Bones could fracture under the extreme stress of heavy work during extra-vehicular activity, for example, or upon return to 1 g.

Bones lose *calcium*, the mineral from which they derive their structure and strength, through the process of demineralization. This increased excretion of calcium may in turn affect various organs, especially the kidneys. For example, the risk of renal stone formation is increased and could have serious consequences during a mission. In addition to demineralization, changes in bone marrow (the site of blood-forming cells) have also been linked to bone loss.

Animal studies have indicated that the structure or "architecture" of the bone formed in space is different from that of animals left on Earth. Thus, for laboratory rats that have flown in space, strength does not increase proportionally to the increase in bone size as it does on Earth. If the same changes occur in humans, it is reasonable to ask what will be the new state for bone in microgravity after very long duration missions. The major health hazards associated with skeletal bone loss during these missions are accumulations of excess mineral in tissues such as the kidney, increased risk of fracture, and potentially irreversible damage to the skeleton.

Bone loss is a concern right here on our own planet as well. Millions worldwide suffer from bone loss, known as *osteoporosis*. Researchers hope that solving the issue of bone loss in space will reveal important clues about what causes osteoporosis on Earth.

Astronauts regularly perform weight-loading exercises that simulate the gravity of Earth. However, exercise alone has not prevented muscle and bone loss during spaceflight. Different types of exercise are required to build muscle strength and resistance to fatigue and injury, and maintain bone integrity. Studies are being conducted to address how muscles and bones should be loaded in microgravity in order to prevent these changes. A balance between healthy nutrition, therapeutic measures, drugs, and exercise is likely to be the most effective countermeasure.

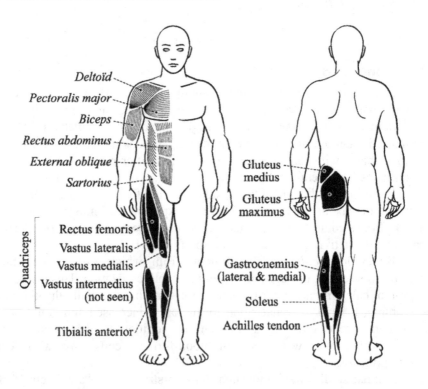

Figure 5-04. Major skeletal muscles in the body. The postural muscles (in black) are used to counteract the action of gravity during standing on Earth. (Adapted from Lujan and White 1994)

2 MUSCLE AND BONE PHYSIOLOGY

2.1 Muscle Physiology

The human body includes several types of muscle tissue. The muscles the most affected by spaceflight are those directly attached to the skeleton, i.e., the *skeletal muscles*. Skeletal muscles are the largest tissues in the body, accounting for 40-45% of the total body weight. These muscles are usually attached to the bones by tendons, and their contraction allows the movement of joints in everyday activities, like walking, lifting objects and standing. The anti-gravity muscles, also known as *postural muscles*, owe their importance and strength to the presence of gravity (Figure 5-04).

Skeletal muscle cells, called *fibers*, are cylindrical cells, about 50 microns in diameter. Each muscle fiber contains several hundred myofibrils, about 1 micron in diameter as well as many mitochondria (for adenosine triphosphate, or ATP, production), and a complex system of internal membranes called the *sarcoplasmic reticulum*, which regulates calcium ion levels in the fiber. A myofibril consists of many filaments of myosin and actin, the structural unit of contraction (Figure 5-05).

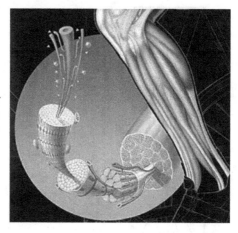

Figure 5-05. Muscles are composed of fibers, made up of smaller units, the myofibrils, which contain filaments that slide for contraction.

ATP is the basic source of chemical energy for muscle contraction. However, the amount of ATP present in the muscle cells is only sufficient to sustain maximal muscle power for 5-6 seconds. Consequently, new ATP must be formed continuously. Three processes can be used: a) the phosphagen system can sustain 10-15 more sec of muscle activity; b) the glycogen-lactic acid system (anaerobic step of glucose breakdown) allows another 30-40 sec "bursts" of energy; and c) the aerobic system provides muscle activity that is only limited by the oxygen and nutrients supplies.

Each muscle fiber is supplied by a motor nerve (axon), and contracts when that axon "fires" an action potential. Muscle action potentials are fast

(1-2 ms in duration) and are all-or-nothing, i.e., not graded. When a single stimulus is applied to the muscle fiber, it responds by a twitch. The twitch force is a weak force and is very slow compared to the duration of the action potential. There is a latent period between the start of the action potential and the time when the fiber begins to develop contractile force, during which the muscle fiber cannot be stimulated again. The duration of the twitch for any one muscle fiber is constant but it can be shorter (e.g., 10 ms in large, fast fibers) or longer (e.g., 50 ms in small, slow fibers). The latent period is about the same for both slow and fast types of muscle fibers.

Slow (oxidative) fibers, also called *Type I*, are characterized by a relatively slow development of force but are able to maintain this force relatively long. Marathon runners typically develop those in the Soleus muscle in the calf for prolonged lower leg muscle activity. Fast (glycolytic) fibers, also called *Type II*, are able to develop force faster. Sprinters and weight lifters typically develop those in the Gastocnemius muscle in the calf and in the biceps muscle for quick, powerful "bursts" of movement. The downside of fast fibers is that they fatigue rapidly.

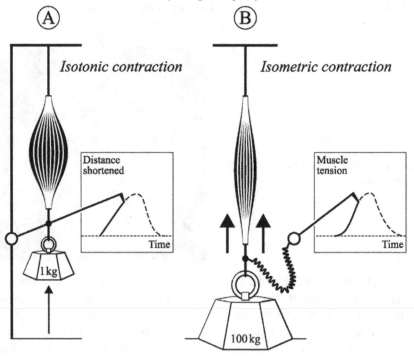

Figure 5-06. Isotonic contraction (A) is associated with constant tension, whereas isometric contraction (B) is associated with constant fiber length. (Adapted from Lujan and White 1994)

Contraction refers to the active process of generating a force in a muscle. The force exerted by a contracting muscle on an object is the muscle *tension* (Figure 5-06). The force exerted on a muscle by the weight of an object is the *load*. When a muscle shortens and lifts a load, the muscle contraction is *isotonic* (constant tension). When shortening is prevented by a load that is greater than muscle tension, the muscle contraction is *isometric* (constant length).

Another classification of muscle contraction is into *concentric* or *eccentric* contractions. Concentric contraction means that the muscle fibers decrease in length. Under the influence of external forces, muscle fibers can increase in length while contracting. This is called an eccentric contraction. An example of an eccentric contraction is walking downstairs, when the force of gravity causes the muscle to lengthen while contracting. During eccentric contraction, the force that is produced by the muscle is even greater than during isometric contraction. This greater production of force is still unexplained, but is surprisingly at the cost of hardly any ATP. When gravity is absent, eccentric contractions rarely occur, which has been suggested to be an important reason why muscles atrophy in microgravity (Convertino 1991).

During muscle contraction, there is a strict relationship between force and length. Because of this relationship it is important to standardize the angles of the relevant joints (i.e., standardization of the length of the muscle) when comparing muscle strength production before and after a certain period of time. In addition, the highest forces are developed at slower velocities of contraction. Consequently, it is also important to compare muscle strength production at identical angular velocities (i.e., standardization of the velocity of contraction).

The *power* a muscle can generate is largely dependent on the amount of actin-myosin filaments that can be used. More filaments mean more potential to generate muscular pull. The length and size of a muscle fiber can vary considerably between various muscles in the body and between individuals of different gender, fitness, and age. The length of a muscle fiber can vary between several millimeters and approximately fifteen centimeters, and is mainly responsible for the maximum velocity of contraction. The strength of a muscle is mainly determined by the size of myofilaments, which is often indicated by the surface area of a perpendicular slice of the muscle, the cross-sectional area. There is generally a high correlation between maximal strength and the cross-sectional area of a specific muscle.

"Eating alone will not keep a man well," said Hippocrates in 400 B.C. "He must also take exercise." Training increases the size of muscle fibers and even the number of muscle fibers, thereby increasing the maximal strength of a muscle. During exercise, the capacity of a muscle for activity can be altered: a) by transformation of one type of fiber to another: e.g., the muscles required to perform endurance-type activity will develop more Type I fibers and their number of blood capillaries will increase; b) or by the growth in size (hypertrophy) of the muscles fibers: e.g., weightlifting will induce hypertrophy in Type II fibers, with an increase in synthesis of actin and myosin filaments.

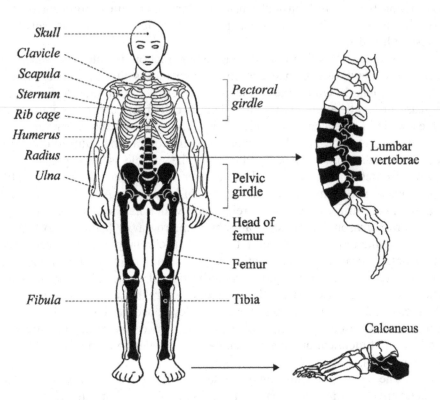

Figure 5-07. Major bones of the body with, in black, those primarily involved in counteracting gravity. (Adapted from Lujan and White 1994)

2.2   Bone Physiology

We tend to think of bones as something inert, but that's not the case. Bone is a living tissue. Bone tissue is constantly being broken down by certain cells, and built up by other cells to maintain its functional rigidity. Much of the activity from these specialized cells comes in response to the stress put on the bones, during walking or exercising. Even when in bed, there are still some muscular forces acting on the bone, providing the stimulus for the remodeling of the bone.

The *weight-bearing bones* provide a rigid support for the body in Earth's gravity (Figure 5-07). The porous structure of the bone is adapted to resist to mechanical constraints with a minimum mass. An estimated 80% of bone-strength is determined by pure bone mass. The remaining 20% are determined by the labyrinth-like structure of bone. In addition, the bones act as a mineral reservoir of calcium. An adult contains approximately 1000 grams of calcium, out of which 99% stays in the skeleton and only 1% in the extracellular space and soft tissues.

The major compartments of a long bone, such as the arm and leg bones, include: a) the periosteum or outer fibrous envelope of cortical bone, which contains the genes for locally acting growth factors; b) the compact bone, the outer bony layer, very strong and dense, which is the site of cortical bone remodeling; and c) the inner compartment or marrow space which contains both bone- and blood-forming cells (Figure 5-07). The cellular elements are contained within an interconnecting system of spongy bone also termed lamellar or trabecular bone.

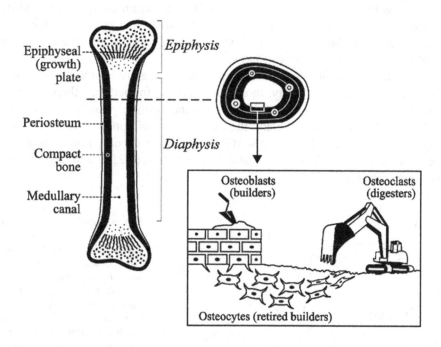

Figure 5-08. Structure of a long bone, showing the epiphysis, which forms a joint with another bone, the epiphyseal plate where elongation is achieved in the growth plate, and the three main compartments of cortical bone within the diaphysis. A cross-section of the compact bone shows the blood-forming cells on the outer surfaces, which continually build new bone to maintain its thickness and strength. (Adapted from Lujan and White 1994)

Bone contains a matrix of collagen fibers, which gives the bone a certain degree of elasticity. The matrix provides a milieu for the deposition of calcium crystals (hydroxyapatite). These crystals give its strength to the bone.

A layer of cartilage called the *growth plate* is where the bone grows longer by increasing its thickness. The cartilage is later calcified with hydroxyapatite. Bone is continually being remodeled under the influence of three types of highly specialized cells. Firstly, *osteoblasts*, or bone forming cells, synthesize the collagen matrix and control the mineralization of the bone. Secondly, *osteoclasts*, or bone resorption cells, secrete acids, which dissolve the minerals and act against the formation of bone-components (Figure 5-08).[*] Finally, *osteocytes* preserve the homeostasis of bone formation and resorption. Osteocytes are differentiated osteoblasts, which have become active to form bone, and are capable of both synthesis and resorption. Osteocytes are extremely sensitive to mechanical stress.

Bone remodeling is a continuous process throughout life: in adults, 20 to 30% of the bone is replaced each year. An, as yet, unknown trigger activates the osteoclasts to form holes and tunnels. These holes and tunnels are filled with a new matrix by the osteoblasts. Calcium compounds must be present for ossification to take place. Twelve to fifteen days after this, the mineralization of the newly formed bone starts. Complete mineralization has taken place after six to twelve months.

The balance of the osteoclast and osteoblast activity is not even. Until approximately the age of thirty, more bone is being formed than there is being dissolved, with an extra strong positive balance during puberty. Thereafter, the balance becomes negative and the total amount of bone decreases (Figure 5-09). The decrease is about 1-2% of bone mass per *decade*. For women this rate increases to 1-2% per *year* somewhere between three and eight years after menopause.

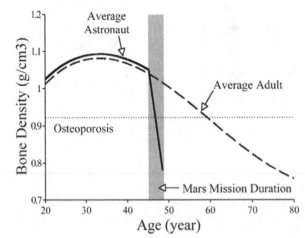

Figure 5-09. Under terrestrial conditions, a loss in bone mineral density of 0.5% per year would be considered normal past the age of 40, so bone loss during spaceflight cannot solely be attributed to that environment. During a mission to Mars, a 45-year-old astronaut could see bone deterioration reach the weakened state of severe osteoporosis (Adapted from National Geographic 2001).

---

[*] An easy way to memorize the function of osteoblasts and osteoclasts is the following: osteoblasts are bone-building cells, whereas osteoclasts are bone-crushing cells.

When humans lose bone density, some of this loss comes from cortical bone, but the main part comes from trabecular bone. Trabecular bone is mostly located next to joints at the ends of the long bones, such as the femur ball that fits into the hip socket, and in vertebral bones. Any loss of density at such locations, where the skeleton experiences the most stress, significantly increases the risk of fractures, hence the larger number of hip replacements among elderly people.

Osteoporosis is a bone disease in which the bone mass is reduced by 0.5-2.0% per year. It is a silent disease, which often leads to a fracture without any precursor symptoms (Figure 5-10). In Europe, this disease affects about 30% of the women and 6% of the men over 50 years of age, and costs approximately 9 billion Euros per year. With the overall aging of population, osteoporosis is an increasing public health issue. The main risk factors are genetic, hormonal and related to sedentary way of life with a lack of physical activity.

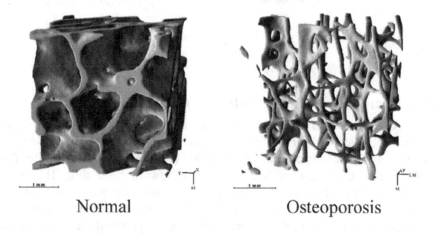

Normal   Osteoporosis

Figure 5-10. Scans of normal bone and osteoporosed bone. (Credit MEDES)

3   EFFECTS OF SPACEFLIGHT ON MUSCLE

3.1   Decrease in Body Mass

Since the muscles represent more than 30% of the body mass, changes in body weight during and after spaceflight are an indicator of muscle atrophy. In microgravity, body weight is measured using a special scale: the subject is seated in a device placed between two springs of known constant. When the seat is unlocked, the period of oscillation is proportional to the subject's mass. As a rule, body mass decreases by approximately 5% relative to preflight during the first two weeks of spaceflight (Thornton and Rummel

1977) (Figure 5-11). Part of this reduction is due to the fluid loss, as mentioned above (see Chapter 4, Section 3.3).

Interestingly enough, body weight tends to increase during bed rest or isolation studies. For example, when confined to a terrestrial Mir simulator for 135 days under conditions simulating a long-term spaceflight, three subjects gained between 5.1 and 9.3 kg. This increase in weight is thought to be due to an accumulation of sodium in extracellular space, leading to water retention and weight gain (Titze et al. 2002).

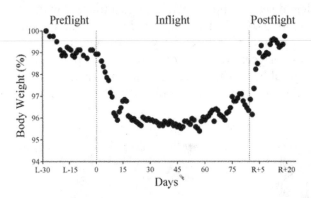

Figure 5-11. Changes in body weight in one astronaut during a three-week spaceflight. (Adapted from Thornton and Rummel 1977)

3.2   Decrease in Muscle Volume and Strength

A decrease in leg volume, also known as the "Chicken Legs Syndrome", is also observed in microgravity. By the end of a 3-month mission, leg circumference may decrease by 10-20%, mostly in the fleshier thighs. Part of this decrease is due to the headward fluid shift. However, although the fluid shift is virtually complete after one week, leg volume continues to decrease throughout the flight, suggesting that muscle loss significantly contributes to this decrease. During the postflight measurements, the leg volume does not immediately return to the preflight level, despite the quick re-hydration of the organism (Figure 5-12). This difference also indicates that leg volume is reduced in space partly because of muscle atrophy.

Another possible indicator of the reduction in muscle mass is the loss of nitrogen during spaceflight. Nitrogen is an essential element of every protein. Skeletal muscle is the largest active protein pool in the body, thereby being the major site of protein loss. Thus, the determination of nitrogen excretion in urine is an indicator of muscle tissue breakdown. The finding of increased excretion of the proteins 3-methylhistidine, creatinine, and sarcosine during spaceflight, which indicates muscle breakdown, confirmed this concept.

Significant atrophy was evident in human muscles after only 5 days in space (Edgerton et al. 1995). It has not been determined whether muscle deterioration reaches a plateau during long-duration spaceflight. The degree of atrophy is different for various muscles. The muscles of the arms and shoulders show smaller losses than the muscles of the lower back, abdomen, thighs and lower legs. These lower body muscles are critical to the maintenance of posture and balance on Earth, and suffer the most from the disappearance of gravity. The smaller losses in the upper limb may also be caused by an increased use of the arms during spaceflight. Indeed, under weightlessness, predominantly the arms are used to move within the spacecraft and during extra-vehicular activities.

The decrements in muscle strength resemble the decrements in muscle mass. Larger losses in the postural muscles and larger losses with increased flight duration are generally observed. Also, the decrease in muscle strength is commonly more profound in the extensor than in the flexor muscle groups in the legs. This may be due to the rest posture in microgravity (see Figure 3-12), which stretches the dorsal flexor muscles thereby maintaining size and strength of these muscles.

Besides losses in muscle mass and muscle strength, losses have also been found in muscular stamina and contractile endurance, both in humans and in rats (Baldwin et al. 1996). Again, these findings were made in the legs, but not in the arms. Although the underlying mechanism may be different, the reasons for these losses are thought to be identical to the reasons for maximal strength losses: disuse due to unloading and confinement in a small space.

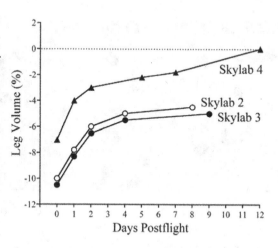

Figure 5-12. Changes in leg volume (both legs) relative to preflight in the Skylab astronauts. The variety of exercise devices available to the crew was increased for each mission. Skylab-2, astronauts used both a cycle ergometer and an isometric device for their 28 days in space. As a result of this flight, it was determined that additional exercise time and programs were required. During the 59-day Skylab-3 mission, ergometer time was increased, and a mini-gym provided additional exercises for the astronauts' trunks, arms, and legs. On the 84-day Skylab-4, ergometer time was further increased, and a treadmill was added. Typically, astronauts used the treadmill 10 min per day. The use of the treadmill considerably prevented the decrease in leg volume (Adapted from Thornton and Rummel 1977).

3.3   Changes in Muscle Structure

It is well known that, on Earth, if the nerve fibers to a muscle are severed or the motor neurons destroyed, the denervated muscle fibers become progressively smaller, their content of actin and myosin decreases, and connective tissue prolifers around the muscle fibers (*denervation atrophy*). A muscle can also atrophy with its nerve supply intact if it is not used for a long period of time. This phenomenon is known as *disuse atrophy*.

Muscle weakness following spaceflight is consistent with the reported 20-50% decrease in muscle fiber cross-sectional area and the loss of contractile proteins is space-flown rats (Riley et al. 1996). Biochemical and structural changes at the cellular and molecular levels have been seen in muscle biopsies collected on astronauts. However, these studies are very limited due to the painful character of such investigation.

Another, indirect method to evaluate structural changes consists in measuring oxygen consumption. Oxygen uptake and energy expenditure are closely related. When slow twitch muscles are exercised, they rely primarily on an aerobic process (one requiring oxygen) to extract the energy stored in carbohydrates, fats, and proteins. Fast-twitch fibers are more dependent on energy produced by the anaerobic breakdown of storages of glycogen. If a human's maximal oxygen capacity declines in space, the slow-twitch muscles may not be as efficient because of their increased dependence on anaerobic energy sources.

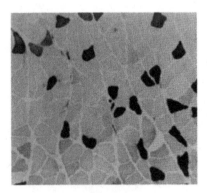 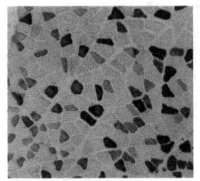

Preflight   Postflight

Figure 5-13. These light micrographs show the effect of microgravity on the size and type of muscle fibers in the leg muscles of rats. The larger cells (left) are from the muscle of a rat that remained on Earth and served as a control. The smaller cells (right) are from the identical muscle of a Spacelab-3 rodent that was in Earth orbit for 8 days. The dark-stained fast-twitch muscle fibers are more numerous in the muscle of the flight animal. (Credit NASA)

Another difficulty in interpreting data from human subjects comes from the fact that the test subjects participate in a wide range of in-flight payload activities (including EVA), which require variable and undocumented muscle use. Also, a further aspect that might confound the data is the unknown influence of the unreported nutritional status of the astronauts. It is likely that cosmonauts and astronauts are to some degree in an energy-deficit state during spaceflight (at least during the short-term missions with a heavy schedule), with the consequence that muscle protein will be lost. This problem becomes less important with increasing mission length, as in longer missions crewmembers have more time to prepare and consume food and consequently get closer to the recommended daily energy intake.

For these reasons, most spaceflight investigations on muscle have focused on animals, growing or mature. In-flight dissections of rodent skeletal muscle tissues have shown that antigravity slow-twitch fibers generally show the greatest deterioration following spaceflight (Figure 5-13). In fact slow muscle fibers seems to acquire fast fiber properties. This shift has the downside of rendering the muscle more fatigable (Figure 5-14). The greater reliance on anaerobic glycolysis contributes to the reduced endurance and increased fatigability.

Although reduced muscle's use, such as during spaceflight, decreases its size and strength, contractile proteins seem to adjust to maintain power output. Upon return to Earth, terrestrial motor strategies are rapidly restored and executed flawlessly. This occurs well before muscle fiber re-growth in cross-sectional areas and during the period of slow muscle fiber necrosis. It then appears that the central nervous system undergoes significant re-programming (plasticity) and performs compensatory activation of motor units that masks the deteriorated state of the muscular system (Riley et al. 1996).

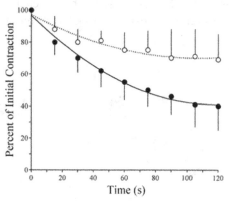

Figure 5-14. Results of a 2-minute isometric fatigue test of control and flight Soleus muscles of rat. The flight muscles were more fatigable than the control muscles, presumably because of their smaller size and their changes to fast-twitch properties. (Adapted from Baldwin 1996)

4   EFFECTS OF SPACEFLIGHT ON BONE

Due to the absence of gravity constraints on the body, astronauts can lose up to 2% of their bone mineral density each month. The bone loss observed after a spaceflight of a few months corresponds to that of several years on ground. By comparison with osteoporosis on Earth, astronauts could therefore be considered as "hyper-sedentary" persons. It has been recognized long ago that understanding the fundamental biochemistry and physics of bone mineralization on Earth is necessary to fully understand the potential effects in microgravity environments (Hattner and McMillan 1968). The opposite is also true.

4.1   Human Studies

The main mineral in bone is calcium, which makes calcium balance an important determinant of the status of bone mineral density (Figure 5-15). An increase in the fecal and urinary calcium excretion was first noticed after the Soviet Vostok missions. Calcium in urine and feces increased drastically in Skylab astronauts, parallel to a muscle and bone loss (Leach and Rambaud 1977). An aggressive exercise program was then implemented, with significant consequences for muscle volume (see Figure 5-12).

During the Gemini missions, bone mineral density was determined by X-ray densitometry, which measures the attenuation of two beams of X-rays by the calcium in the X-ray path. With this technique (which has a precision of 1-2%) a loss of approximately 2-4% of bone mass was detected in the heel bone after 4-11 days of spaceflight. The subsequent Apollo, Skylab, and Salyut data were obtained by single photon absorptiometry. After the Apollo missions lasting also 10 days, a 3-5% decrease in bone mass was observed. Therefore, 2-3 days spent on the Moon surface at 0.36 g did not prevent bone loss. After Soyuz missions, bone density had decreased by 8-10%, and Skylab measurements revealed a 1-3% per month loss in bone mineral.

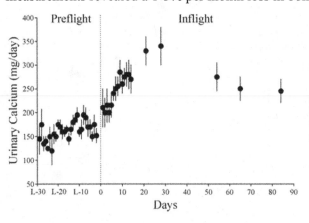

Figure 5-15. Increased urinary calcium excretion has been observed in astronauts in Skylab and other flights.

One cosmonaut of the third crew occupying Mir was examined with computerized tomography (CT), which gives a true (i.e., volumetric, in g/cm^3) density. Results showed a 10% loss of trabecular bone from lumbar spine after a 1-year mission. Other cosmonauts and astronauts flying on board Mir were examined both pre- and postflight using dual-energy X-ray absorptiometry (DEXA). DEXA is a specially calibrated X-ray device that provides a two-dimensional measurement of the mass of an entire bone, i.e., the trabecular bone and the cortical bone. These scans revealed that when using the onboard exercise countermeasures, there was a 5.4% decrease in bone density in tibia. Bone density did not return to preflight level in some individuals. Without countermeasures, there was approximately 1.3-1.5% per month decrease in bone density. In the worst case, a 15-22% decrease was measured in some bones after a 6-month mission (LeBlanc et al. 1996).

The most compelling data have recently been compiled from 15 cosmonauts who spent 1, 2, or 6 months on the Mir station (Vico et al. 2000). Bone mineral density was measured at the distal radius and tibia before, just after the spaceflight and up to 6 months after the mission. Neither trabecular nor cortical bone of the radius was significantly changed at any of the timepoints. On the contrary, in the weight-bearing tibial site, trabecular bone loss was noted after a 2-month flight, and was greater after a 6-month flight. Tibial bone loss persisted for at least 6 months after flight, suggesting that the time needed to recover is longer than the mission duration (Figure 5-16).

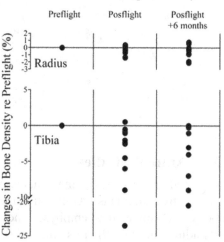

Figure 5-16. Mean loss of bone mineral density in the radius and the tibia relative to preflight values in cosmonauts following spaceflights of 6 months on board Mir. In some individuals the bone loss continues for about 6 months after flight. (Adapted from Vico et al. 2000)

No significant changes in bone regulating hormones (serum calcium, PTH, vitamin D, calcitonin, growth hormone) have been seen on astronauts after short-duration Shuttle flights (Stein et al. 1996). The duration of stay, calcium intake, and level of exercise performed in-flight all account for the wide range in average percentage losses of bone mineral density, as reported above. Individual values are even more variable, as changes in calcaneal bone density of one particular subject can range from 4 to 30% loss relative to preflight. A method allowing both identification of the recorded site and reproducible measurements are required for more accurate studies. Examinations using DEXA and CT techniques are being performed pre- and postflight with the ISS crewmembers. These studies will provide the first detailed information on the distribution of spaceflight-related bone loss between the trabecular and cortical compartments of the skeleton, as well as the extent to which lost bone is recovered in the year following return.

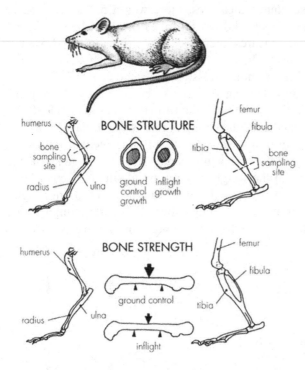

Figure 5-17. A change in bone strength, but not length, was observed in the limbs of growing rats in space. (Lujan and White 1994)

4.2   Animal Studies

One problem of using rodents as model is that their bone growth is different from humans. Although bone elongation ceases in humans after puberty, in mice the epiphyses never close and there is continuous longitudinal growth. In rats, the growth rate is significantly attenuated at about 12 to 14 months of age (they live to about 3 to 4 years old), but their cortical bone, which lacks vascular canals, is not similar to that of humans.

In early Soviet Cosmos flights, various types of muscle fibers were found in flown rats, together with histologic changes with random deletion of myofibrillar filaments (Ballard and Conolly 1990). There was a loss of muscle force and elasticity and some specific changes in enzyme activity. A reduction in the rate of bone formation was observed postflight, with a return to control levels in approximately 4 weeks. It is interesting to note, however, that these changes were largely prevented in the rats that were subjected to centrifugation in-flight (Nicogossian and Parker 1982).

Studies on young rats flown on board Spacelab missions revealed no changes in the length of antigravity bones, such as the tibia, the femur, and the humerus, compared to ground control animals. In other words, the rats grew at the same rate in microgravity as on Earth. However, the bone mass, hence its strength, was reduced (Figure 5-17).

An animal model used to study muscle atrophy and bone loss is the suspended rat (Figure 5-18). It incorporates the two features of spaceflight that might affect bone: the unloading of weight-bearing bones and the headward fluid shift. In this preparation, a harness raises the hind limbs off the cage floor and thus "removes" weight from the muscles of the hind limbs. The overhead pulley system has a swivel that allows the animal to move about the cage by only using its fore limbs. After a few days of adaptation, tail-suspended animals are active and eat and drink normally. This relatively benign technique is relatively rapid and represents an accurate simulation of the changes in muscle and bone occurring during spaceflight (Tischler et al. 1993, Piquet and Falempin 2003).

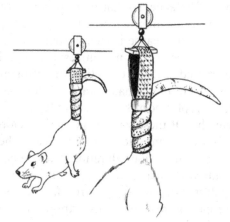

Figure 5-18. The rat hind limb suspension technique is widely used to simulate the unloading of muscle and bone as during spaceflight. (Lujan and White 1994)

Neither the communication system between muscle and bone nor the precise mechanism of bone loss is understood. Therefore, the characterization of the response to skeletal unloading at the tissue and cellular level is one of the major contributions of the use of the rat model (Morey-Holton and Globus 1998). For example, recording of Soleus muscle activity of suspended rats showed an immediate and persistent 75% reduction in contractile activity. After 7 days, the Soleus muscle showed a decrease in specific tension as slow fibers shifted toward fast fibers and thick, myosin-containing filaments were lost (Riley et al. 1990). These animals also lose about 25% of trabecular bone in the tibia, and showed a 30% reduction in the mechanical strength of the tibia shaft (Vico et al. 1991)

Unloading studies with immature rats indicate that gravity loading during the third and fourth weeks after birth is essential for normal development of locomotion. (Walton et al. 1998). Similarly, muscle development was disrupted when gravity-loading exercise was removed from immature rats flown on the Neurolab flight. In the absence of weight-bearing challenge, Soleus muscle fibers failed to grow in size and differentiate normally into slow fibers, and elaboration of the motor nerve terminals was retarded. Once mature, muscle tissue in unloaded animals is more prone to structural failure when reloaded because of fiber atrophy, and the ability to repair internal lesions is compromised (Arnaud et al. 1995). The role of fiber-type specific factors in regulating gene expression is being studied in transgenic animals (Capetanaki et al. 1997).

5   WHAT DO WE KNOW?

5.1   Muscle Atrophy

The human body constantly uses amino acids to build muscle protein, which then breaks down and must be replaced. When protein turnover gets out of balance, so that more protein breaks down than the body can replace, the result is muscle loss. But makes protein turnover to slow down?

One cause is lack of muscular activity. That is why bed rest is a good model because it minimizes activity. In fact, during bed rest there is an increased urinary excretion of nitrogen and muscle loss, as is observed in space, but these changes are variable and generally greater in degree. Most of the atrophy occurs in antigravity muscles, which are no longer bearing body weight. Of these various possible factors contributing to the excess excretion of nitrogen, muscle atrophy is clearly the main one.

In all long-duration astronauts, the high level of nitrogen excretion continued unabated for the duration of flight. This indicates a serious malfunction not likely to reach a new steady state until an extreme degree of atrophy is reached. This nitrogen loss was accompanied by losses of 15 to 30% of muscle mass and strength in the lower extremities. This poses a significant handicap to vigorous work in the gravity of Mars or on return to Earth.

Animal studies of muscle atrophy attempt to determine the physiological and biochemical mechanisms underlying muscle atrophy. Although the mechanism of the process of atrophy remains unknown, certain aspects have become evident. Muscle atrophy is accompanied by decreased synthesis of muscle protein and by some degree of increased degradation. As shown in rats that are suspended (hind limb unloaded), loading and stretching of otherwise inactive leg muscles prevented muscle atrophy and stimulated protein synthesis; the addition of electrical stimulation increased protein synthesis markedly. As shown in muscle cell cultures, stretching stimulates protein synthesis.

The uncertain value of physical exercise for suppressing muscle atrophy in human spaceflight has been noted previously. However, what is the signal and sequence of biochemical steps for initiating increased protein synthesis and deposition in muscle filaments, and what communicates a message to slow down protein synthesis? Answers to these questions would have an impact on muscle research far beyond spaceflight.

The effects of electrical stimulation of muscle have begun to be studied, but the possible combinations of frequency, voltage, and current are almost without limit. Stimulation of the sole cutaneous mechanoreceptors seems to reduce the muscular atrophy after hind limb unloading in rats (De Doncker et al. 2000).

A variety of techniques are available for muscle research: electron microscopy, electromyography, computerized tomography (CT) scanning, and stable isotope metabolic studies. In order to understand changes of muscle mass and strength, we must understand their underlying cellular and molecular mechanisms. Therefore, these existing technologies should be coupled with developing techniques in immunochemistry and in recombinant DNA and gene cloning. The genes encoding many major proteins of muscle, as well as their controlling elements, have been sequenced. The current goal is to relate mechanical stress, hormonal levels, and nutrition to the control of expression of these genes.

5.2 Bone Demineralization

Human bed rest studies correlating inactivity to factors such as diminished bone mass and increased urinary calcium have also proven to be useful models for potential changes during extended spaceflight. Studies of humans during long-term bed rest have shown that prolonged inactivity results in significant and continuing losses of calcium from the skeleton and nitrogen from muscle, and in considerable atrophy of both body systems. These changes were consistent, but quite different in degree from subject to subject. Genetic factors may account for these differences. However, as yet, no single gene has been convincingly proven to be a risk factor for osteoporosis. The genetic component has been attributed instead to the cumulative effects of a number of genes (including for example the vitamin D receptor gene) with small individual effects. Identification of such genes is currently under way (Tipton 1996).

In the severe paralysis of poliomyelitis, calcium losses led to X-ray visible osteoporosis in the bones of the lower extremities as early as 3 months after paralysis. While the overall rate of calcium loss in Skylab astronauts was 0.4% of total body calcium per month, the loss was estimated to be 10 times greater in the lower extremities than in the rest of the body (based on bed rest studies of calcium losses by metabolic balance compared with decrease in bone calcium density). This could lead in 8 months of flight to a decrease in bone density in the legs similar to that noted in paralytic poliomyelitis.

Studies of immobilized rabbits showed marked decrease in strength of tendons and ligaments after only one month. Thus, strains, sprains, and even ligament tears may be more likely to occur, and at an earlier time than bone fractures.

The cellular mechanisms of mineral loss are unknown. Excess excretion of calcium associated with increased hydroxyproline in the urine in humans is indicative of increased bone resorption. On the other hand, histologic examination of the bones of the rats on Cosmos showed suppressed bone formation. Many scientists believe that bone mass decreases in microgravity because the lack of stress on the bones slows the formation of osteoblast cells. Fewer bone-building cells, along with a constant level of bone-destroying activity, would translate into a net loss of bone mass. But why should microgravity inhibit the development of osteoblasts? A key chemical in the development of osteoblast cells from precursor cells is an enzyme called "creatine kinase-B". Investigators are trying to figure out which molecules in the body regulate the activity of this enzyme and how those chemicals are affected by reduced gravity, in the hope that this knowledge will point to a way to boost osteoblast formation in space.

In any case, the hypercalciuria associated with loss of mineral from bone in spaceflight might increase the potential for stone formation in the urinary tract. Although 75 to 80% of renal stones contain calcium, the likelihood of stone formation will depend not only on increased urinary concentration of calcium, but also on other factors such as urinary pH, concentration of inorganic elements (magnesium, potassium, and phosphorus), and concentrations of organic compounds (uric acid, citrate, and oxalate). Bed rest studies have shown a slight rise in urinary pH and a lack of change in urinary citrate, which in ambulatory states rises with increases in urinary calcium.[*] Both of these factors, if also noted in spaceflight, would favor decreased solubility of calcium salts. The likelihood of urinary tract stone formation during spaceflight may be small, especially if care is taken to maintain abundant urine volumes; nevertheless, such stone formation might be catastrophic to health and function for the astronaut involved, and thus to success of the particular flight.

Actually, NASA has developed a Critical Path Research Plan (http://criticalpath.jsc.nasa.gov/) to guide its bioastronautics research in systematically reducing or eliminating the risks to astronaut health, safety, and performance during and after spaceflight. Of the 55 risks identified in this

---

[*] Signs of renal stones were seen in several untreated subjects after a recent three-month bed rest (source MEDES).

Critical Path Roadmap, 11 are associated with altered musculo-skeletal function. Of particular concerns is the acceleration of age-related osteoporosis, the failure to recover bone lost after space missions, and the increased risk of fracture upon return to activity in 1 g. Critical questions to be addressed include:

- Will bone mass loss continue unabated for missions greater than six months in duration, or will it eventually plateau at some time consistent with absolute bone mineral density?
- What are the most important predictors for bone loss during prolonged exposure to hypogravity, especially with reference to ethnicity, gender, age, and bone morphometry?
- Is bone loss reversible and within what time frame?
- Does prolonged exposure to hypogravity lead to non-union of healing fractures? What evidence supports the alteration in vertebral morphometry during and after extended spaceflight?
- What practical diagnostic tools can be utilized during multi-year missions to monitor and quantify changes in bone mass and strength (e.g., biochemical markers, dual X-ray absorptiometry, ultrasound)?
- Are there important other mechanisms for bone loss with hypogravity that are critical to developing effective countermeasures (e.g., fluid shifts with altered hydrostatic pressure, changes in blood flow, immune system alterations)?
- Is there an optimal combination of exercise and a pharmacological countermeasure to minimize decrements in bone mass in hypogravity?

Figure 5-19. Interim Resistive Exercise Device (IRED) used on the International Space Station. This device provides a elastic based resistance for a variety of both upper and lower body exercises, such as squats, deadlifts, heel raises, bent over rows, upright rows, biceps curls, bench press, wrist curls, and shoulder press.(Credit NASA)

6   COUNTERMEASURES

6.1   Muscle

The considerable and time-consuming exercise activity of the astronauts on Skylab and Mir resulted in somewhat reduced loss of muscle mass and strength than on the earlier flights, but were obviously not adequate to be fully protective.

As mentioned earlier, flight surgeons recommend fifteen minutes of exercise daily on short-duration missions and two daily sessions of approximately one hour each on long-duration missions.

Several exercise devices will be available on ISS for research including a treadmill (to preserve an aerobic power), a cycle ergometer (to preserve aerobic capacity), a resistive exercise device (to preserve muscle strength and bone mineral), and hand grip equipment (to preserve hand strength for EVA). During exercise on the treadmill, the cycle ergometer, and the resistive exercise, status and data are controlled by the Human Research Facility rack.

The treadmill may be used for walking and running exercise. The device employs various strategies to simulate, as closely as possible, 1-g skeletal loading during exercise. Loads are exerted on the subject by restraint harnesses. The restraint system provides stabilization of the astronaut and load distribution on the body in a weightless environment. The treadmill can be motor-driven or passively operated. So, it is used as an ambulating trainer, endurance exercise of postural musculature, high impact skeletal loading (bone maintenance), and aerobic exercise. Moving air from a nearby duct is used to dry off the perspiration produced from exercising.

The cycle ergometer provides workload, driven by the hands or feet, which is controlled by manual or computer adjustment. It operates with the subject seated or supine, and provides time-synchronized data compatible with other complementary analyses. The data output consists of work rates (in watts) and pedal speed (in rpm) for use with a data acquisition system. The cycle ergometer is used as both aerobic and anaerobic exercise countermeasure, for the maintenance of lower body musculature endurance, for EVA arm exercise training, and as EVA 2-hour pre-breathe exercise countermeasure.

The interim resistive exercise device (iRED) includes a series of human-machine interface devices (e.g., handgrips, straps, curl bars, ankle cuffs, squat harness, etc.) that permit a variety of exercises to be performed by the astronauts. Cables on each side of shoulder straps are connected to two canisters, each containing a series of "flex packs" that can be dialed in sequentially to add greater resistance to the cables (Figure 5-19). The design of the hardware is such that the forces imposed upon a muscle group during an eccentric muscle action are less than the maximum concentric force that can be generated by the user. This device is used as training for muscle strength and endurance of all major muscle groups, to maintain skeletal muscle mass and volume, and to provide high-strain skeletal loading (bone maintenance). In addition, some core exercises are directed to emphasize the strength and endurance of postural muscles (Table 5-01)

Day 1	Day 2	Day 3
deadlift	shoulder press	squat
bent over rows	rear raises	heel raises
straight leg deadlift	front raises	straight leg deadlift
heel raises	hip abduction	bent over rows
	hip adduction	
Day 4	**Day 5**	**Day 6**
Biceps curls	deadlift	shoulder press
Triceps kickbacks	bent over rows	lateral raises
upright rows	straight leg deadlift	front raises
hip flexion	squat	hip abduction
hip extension	heel raises	hip adduction

Table 5-01. Resistive exercise workout recommended daily for the ISS astronauts. Note that lower body exercises are performed everyday.

As seen above, Russian cosmonauts wear the "Penguin" suit during long-duration missions (see Figure 4-20). Beside its effect on the cardio-vascular system, the elastic bands in the suit also simulate some of the gravitational effects on the musculo-skeletal system. Expanders (Figure 5-20) are also used occasionally. However, they do not provide sufficient force during axial loading for bone maintenance, and present a reduced range of motion against resistance compared to the interim resistance exercise device described above.

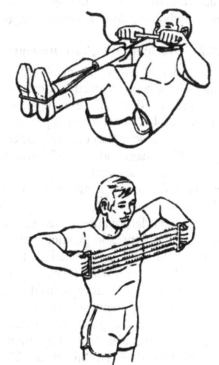

Figure 5-20. Examples of loading elastic devices for arm exercise in microgravity. (Nicogossian and Parker 1982)

Other countermeasure projects are attempting to increase protein synthesis rates with supplements of amino acids, which are the raw materials of protein. Indeed, early results during bed rest studies have suggested that the amino acid supplement was able to maintain synthesis rates and body mass.

6.2 Bone

A variety of studies are being conducted on the basic mechanisms of the effects of mechanical forces on bone dynamics and development. Such studies may give insight for countermeasures based on exercise, drugs, or diet.

6.2.1 Exercise

Evidence from bed rest studies and spaceflight suggests that bone loss is a regional phenomenon in which the bone areas with the greatest decrease in load lose the most bone (Oganov et al. 1992, LeBlanc et al. 1996). Skylab astronauts averaged 0.5% per month total body calcium loss despite exercising a number of hours a day through a series of exercises consisting of bungee cords for resistive exercises, bicycle ergometer exercise, and walking on a treadmill (Thornton and Rummel 1977). Exercise schedules typically required two hours of exercise daily. However, cosmonauts continue to lose bone selectively from the spine and lower extremities while maintaining upper body bone mineral density (Oganov et al. 1992).

6.2.2 Mechanical Countermeasures

All of the mechanical procedures tested thus far during bed rest studies have been ineffective. Correlative observations have indicated that the required procedure for use in-flight should provide the equivalent force on the skeleton of 4 hours of walking per day.

Some scientists currently believe that bone mass is not only controlled by the high-magnitude, low-frequency strain resulting from the mechanical loads on bones associated with vigorous exercise, but also by low-magnitude and high-frequency strain that musculature continuously places on bones while sitting or standing. It is well known that mechanical loads (stress) causes slight deformation called strain. The amount of strain is dependent on loading, elasticity, and geometry of the bone. An upper limit strain must be exceeded to provoke remodeling to increase bone mass. Mechanical strain below a lower limit will provoke adaptive remodeling to reduce mass.

Results of ground-based studies suggest that barely perceptible vibrations may generate enough strain to stimulate bone growth. For example, a group of sheep exposed to 20 minutes per day of vibrations experienced increased trabecular bone formation when compared to a control group without vibrations (Rubin et al. 2001) (Figure 5-21). In addition, when animals, prevented from regular, weight-bearing activity, were exposed to vibrations daily, bone formation remained at near-normal levels. However, animals not exposed to the treatment, but participating in weight-bearing activity each day, still exhibited signs of significant bone loss. If proven valuable for humans, low-level vibrations during spaceflight may offer an alternative for the current, time consuming astronaut exercise regimes for long-duration space missions.

Figure 5-21. After a year of daily 20-minute standing on a vibrating platform (0.3 g, 30 Hz), sheeps showed the robust striations of increased density (right). Control sheeps showed normal bone (left). (Rubin et al. 2002)

6.2.3 Nutritional Countermeasures

In addition to caloric intakes, protein and calcium, other nutrients that are associated with bone metabolism, phosphorus, sodium, potassium and magnesium have no limits or requirements specific for the microgravity environment. Nutritional recommendations for spaceflight are not different from the recommendations of the National Research Council for life on Earth (McCormick and Donald 2000).

1000 mg is a reasonable base figure for calcium in diet formulation for spaceflight to be taken daily with the principal purpose of "protecting" the skeleton. Among the countermeasures tested have been high calcium and high phosphorus intake in both bed rest subjects and Skylab and Mir astronauts. The study showed that this procedure maintained calcium intake and excretion level in balance for up to three months, following which the gradually rising fecal excretion of calcium caused a negative calcium balance. Hence, there is no basis at this time for recommending a higher intake level than 1000 mg/day.

Bed rest studies of the effects of high phosphorus intake showed some suppression of the tendency of urinary calcium to elevate, but the manipulation was ineffective because of gradually increasing fecal calcium excretion. Furthermore, the calcium to phosphorus ratio should not exceed a ratio of 1:1.8. Indeed, too high an intake of phosphorus will exert some binding effect on calcium in the intestine and tend to inhibit calcium absorption.

The current recommended level for magnesium is 350 mg/day for adult males. While studies of this element in relation to bone are far less numerous than studies of calcium, research to date indicates that deleterious effects apparently do not occur except possibly with low intake (as in an artificial diet) over a very long time. Some promise has also been noted in certain of the diphosphonates compounds that bind to bone crystal and tend to inhibit bone resorption. These countermeasure studies are being continued.

6.2.4   Pharmacological Countermeasures

Biochemical regimens have been studied during bed rest: a) synthetic salmon calcitonin, a hormone inhibiting bone resorption; b) phosphate supplements; c) oral calcium; and d) etidronate. All showed no beneficial effect (Fleisch et al. 1969, Hulley et al. 1971). However, during the last 3 weeks of bed rest, the usual progression of calcaneal mineral loss was no longer observed. Etidronate, however, has been associated with an accumulation of new bone tissue both in animals and man when given at high dose for extended periods of time (Meunier et al. 1987).

New bisphosphonates are being tested for treating global bone loss diseases such as post-menopausal osteoporosis. For example, alendronate (FOSAMAX, Merck, Inc.) was effective in preventing hypercalciuria and maintaining bone mineral density in the femoral neck, femoral trochanter, spine, and pelvis in humans during bed rest studies (Ruml et al. 1995). Moderate loss of bone mass density occurred in the calcaneus, but the loss was significantly less than in the control group. Other newer bisphophonates are likely to be even more effective.

Other studies are looking at the hormone glucose-dependent insulinotropic peptide, which is involved in insulin production that some bone cells have receptors for. They are trying to find out if the loss of bone in space can be prevented by modulating a person's own production of the hormone or by giving it by injection or tablet.

6.3   Aging and Space

There is a need for developing a practical, inexpensive, non-invasive way of making muscle and bone mass and strength measurements, a system sensitive enough to monitor and evaluate small changes. The need for such an instrument goes way beyond spaceflight. Since muscle and bone abnormalities affect a substantial portion of the population, such an instrument would offer broad utility as a tool for clinicians on Earth. For example, the information gained from this instrument may benefit the people here on Earth whose daily activities are affected by metabolic deficiencies, weakened muscles, or loss of bone mass. Some metabolic diseases, for example, result in debilitating muscular weakness, a condition that could be improved by advances in protein turnover research. Likewise, muscle wasting is problematic for senior citizens, patients confined to lengthy bed rest, patients with spinal nerve damage, and even burn victims recovering from traumatic accidents.

Older people also commonly experience a loss of bone mass, a condition often due to the age-related disease osteoporosis. Bone loss in space is not identical to osteoporosis on Earth, since there is a clear hormonal component in osteoporosis. As we age, we also lose muscle mass and strength, a phenomenon called *sarcopenia*. This continuous reduction of muscle strength is largest in the antigravity muscle. Aging effects differ from spaceflight in that: a) the entire body is concerned; b) muscle loss in the aging has no plateau; and c) is characterized by fast twitch (type II) to slow twitch (type I) fiber transformation (Rittweger et al. 1999). There is also a reduction of the number of muscle fibers and cross-sectional area. An imbalance in the natural cycle of protein turnover may be a contributing factor to decreased muscle mass.

Does spaceflight push the astronauts along the irreversible axis of aging? At the age of 77 when he flew on the Space Shuttle, John Glenn was the subject for a muscle loss experiment, whose aim was to investigate whether or how weightlessness can affect the elderly more than younger astronauts. Samples of blood and urine were collected during the flight after Glenn swallowed pills containing amino acid N-15 alamine. The study compared the amount of amino acids absorbed into the body to the amount passing out in urine, and calculated how quickly proteins are built up and broken down. The results of this experiment are currently being analyzed.

But the answer is not so simple, since aging is also associated with changes in hormones, activity levels, nutrition, and often, disease. Nevertheless, by exploring the interaction of aging and spaceflight, research will undoubtedly contribute to our knowledge of the aging process. A better understanding of bone and muscle changes in spaceflight will also lead to treatments for astronauts and Earth-bound patients alike.

202 *Fundamentals of Space Medicine*

7 REFERENCES

Arnaud SB, Harper JS, Navidi M (1995) Mineral distribution in rat skeletons after exposure to a microgravity model. *Journal of Gravitational Physiology* 2: 115-116

Baldwin KM et al. (1996) Musculoskeletal adaptations to weightlessness and development of effective countermeasures. *Medicine and Science in Sports and Exercise* 10: 1247-1253

Ballard RW, Connolly JP (1990) US/USSR joint research in space biology and medicine on Cosmos biosatellites. *FASEB Journal* 4: 5-9

Capetanaki Y, Milner DJ, Weitzer G (1997) Desmin in muscle formation and maintenance: Knockous and consequences. *Cell Structure and Function* 22: 103-116

Convertino VA (1991) Neuromuscular aspects in development of exercise countermeasures. *The Physiologist* 34: S125-S128

De-Doncker L, Picquet F, Falempin M (2000) Effects of cutaneous receptor stimulation on muscular atrophy developed in hindlimb unloading condition. *Journal of Applied Physiology* 89: 2344-2351

Di Prampero PE, Narici MV, Tesch PA (2001) Muscles in space. In: *A World Without Gravity*. Fitton B, Battrick B (eds) Noordwijk, NL: ESA Publications Division, SP-1251, pp 69-82

Edgerton VR et al. (1995) Human fiber size and enzymatic properties after 5 and 11 days of spaceflight. *Journal of Applied Physiology* 78: 1733-1739

Fleisch H, Russel RG, Simpson B, Muhlbauer RC (1969) Prevention of a diphosphonate of immobilization "osteoporosis" in rats. *Nature* 223: 211-212

Hattner RS, McMillan DE (1968) Influence of weightlessness upon the skeleton: A review. *Aerospace Medicine* 39: 849-855

Hulley SB et al. (1971) The effect of supplemental oral phosphate on the bone mineral changes during prolonged bed rest. *Journal of Clinical Investigation* 50: 2506-2518

Leach CS, Rambaut PC (1977) Biochemical responses of the Skylab crewmen: An overview. In: *Biomedical Results from Skylab*. Johnston RS, Dietlein LF (eds) Washington: DC. National Aeronautics and Space Administration, NASA SP-377, Chapter 23, pp 204-216

LeBlanc A et al. (1996) Bone mineral and lean tissue loss after long duration spaceflight. *Journal of Bone and Mineral Research* 11: 323-332

Lujan BF, White RJ (1994) *Human Physiology in Space*. Teacher's Manual. A Curriculum Supplement for Secondary Schools. Houston, TX: Universities Space Research Association

McCormick, Donald B (2000) Nutritional recommendations for Spaceflight. In *Nutrition in Spaceflight and Weightlessness Models*. Lane HW, Schoeller DA (eds) CRC Press, Boca Raton, Florida, pp 253-259

Meunier Y, Chapuy MC, Delmas P (1987) Intravenous disodium etidronate therapy in Paget's disease of bone and hypercalcemia of malignancy. *American Journal of Medicine* 82: S71-S78

Morey-Holton ER, Globus RK (1998) Hind limb-unloading of growing rats: a model for predicting skeletal changes during spaceflight. *Bone* 22: 835-885.

Nicogossian AE, Parker JF (1982) *Space Physiology and Medicine*. Washington, DC: US Government Printing Office, NASA SP-447

Oganov VS et al. (1992) Bone mineral density in cosmonauts after 4.5-6 month long flights aboard Orbital Station Mir. *Aerospace and Environmental Medicine* 5: 20-24

Picquet F, Falempin M (2003) Compared effects of hindlimb unloading versus terrestrial deafferentation on muscular properties of the rat soleus. *Experimental Neurology* 182: 186-194

Riley DA et al. (1990) Skeletal muscle fiber, nerve, and blood vessel breakdown in space-flown rats. *FASEB Journal* 4: 84-91

Riley DA et al. (1996) In-flight and postflight changes in skeletal muscles of SLS-1 and SLS-2 spaceflown rats. *Journal of Applied Physiology* 81: 133-144

Rittweger J, Gunga HC, Felsenberg D, Kirsch KA (1999) Muscle and bone— Aging and space. *Journal of Gravitational Physiology* 6: 133-135

Roer RD, Dillaman R.M (1990) Bone growth and calcium balance during simulated weightlessness in the rat. *Journal of Applied Physiology* 68: 13-20

Rubin C, Turner S, Bain S, Mallinckrodt C, McLeod K (2001) Extremely low level mechanical signals are anabolic to trabecular bone. *Nature* 412: 603-604

Rubin C, Turner AS, Mallinckrodt C, Jerome C, McLeod K, Bain S (2002) Mechanical strain, induced noninvasively in the high-frequency domain, is anabolic to cancellous bone, but not cortical bone. *Bone* 30: 445-52

Ruml LA, Dubois SK, Roberts ML, Pak CYC (1995) Prevention of hypercalciuria and stone-forming propensity during prolonged bedrest by alendronate. *Journal of Bone and Mineral Research* 10: 655-662

Schneider VS, LeBlanc A, Rambaut P (1989) Bone and mineral metabolism. In: *Space Physiology and Medicine*. Nicogossian A, Huntoon C, Pool S (eds) Philadelphia, PA: Lea & Febiger, pp. 214-221

Schneider V, LeBlanc A, Huntoon C (1993) Prevention of spaceflight induced soft tissue calcification and disuse osteoporosis. *Acta Astronautica* 29:139-140

Stein TP, Leskiw MJ, Schluter MD (1996) Diet and nitrogen metabolism during spaceflight on the Shuttle. *Journal of Applied Physiology* 81: 82-97

Thornton WE, Rummel JA (1977) Muscular deconditioning and its prevention in spaceflight. In: Johnston RF, Dietlein LF (eds) *Biomedical Results from Skylab*. Washington, DC: US Government Printing Office, NASA SP-377, Chapter 21, pp 191-197

Tipton CM (1996) Animal models and their importance to human physiological responses in microgravity. *Medicine and Science in Sports and Exercise* 28: S94-S100

Tischler ME et al. (1993) Spaceflight on STS-48 and Earth-based unweighting produce similar effects on skeletal muscle of young rats. *Journal of Applied Physiology* 74: 2161-2165

Titze J et al. (2002) Long-term sodium balance in humans in a terrestrial space station simulation study. *American Journal of Kidney Diseases* 40: 508-516

Vico L, Novikov VE, Very JM, Alexandre C (1991) Bone histomorphometric comparison of rat tibial metaphysis after 7-day hindlimb unloading vs. 7-day spaceflight. *Aviation, Space and Environmental Medicine* 62: 26-31

Vico L et al. (2000) Effects of long-term microgravity exposure on cancellous and cortical weight-bearing bones of cosmonauts. *Lancet* 355: 1607-1611

Walton K (1998) Postnatal development under conditions of simulated weightlessness and spaceflight. *Brain Research Reviews* 28: 25-34

Additional Documentation:

Integrative Physiology in Space (2000) *European Journal of Physiology* 441, Number 2-3, Supplement

International Workshop on Bone Research in Space (1999) *Bone*, Official Journal of the International Bone and Mineral Society, Volume 22, Number 5, Supplement

Muscle Research in Space: International Workshop (1997) *International Journal of Sports Medicine*, Volume 18, Supplement 4: S255-S334

Review of NASA's Biomedical Research Program (2000) Committee on Space Biology and Medicine, Space Studies Board, National Research Council. National Academy Press

Space Research. NASA Marshall Spaceflight Center (2002) Office of Biological and Physical Research. Volume 1, Number 4, September 2002

Chapter 6

PSYCHO-SOCIOLOGICAL ISSUES OF SPACE-FLIGHT

This chapter emphasizes the importance of mental and social well being in the success of both short and long space missions. What are the psychological and sociological issues, which must be addressed, especially for international missions? This section reviews the factors that may have a critical impact on the success or failure of a space mission, in terms of interactions of the crewmember with his habitat, with the space environment, and with the other crewmembers.

Figure 6-01. Fight scene of the movie "Lady Killer" (1933) by Director Roy del Ruth, featuring James Cagney, Douglas Dumbrille, Mae Clarke, Raymond Hatton, Russell Hopton. (Credit Warner Bros)

1   THE PROBLEM: REACTION TO STRESS

According to the media, the Russians once launched a rescue mission to the Mir space station. Among three men aboard the capsule, one replaced one cosmonaut in the two-person Mir space station, and the others escorted him back to Moscow. While cardio-vascular problems were the official reason for the Mir cosmonaut early homecoming, some U.S. space experts say there might have been additional difficulties from the stress of prolonged weightlessness, isolation, and confinement (Burrough 1998).

This example illustrates the problem with psychological issues during space missions: although they presumably exist, most of the reports are anecdotal. In fact, psychiatric problems during space missions, such as anxiety, depression, psychosis, psychosomatic symptoms, and postflight personality changes, have been rare or not methodically documented. Known negative psychological reactions to spaceflight have included sleep problems, reduced energy levels, mood and thought disorders, alteration in time sense, and poor interpersonal relations. Interpersonal issues include interpersonal tension, decreased cohesiveness over time, need for privacy, and task versus emotional leadership (Kanas and Manzey 2003). None of these problems, however, seems to have seriously affected a space mission yet, probably due to the extraordinary motivation and commitment of the astronauts. However, Dr. Patricia Santy, a psychiatrist and flight surgeon who worked at NASA, and author of the book "Choosing the Right Stuff" (Santy 1994), said that even highly motivated people have a limit. She sets that limit at three or four weeks. "After that, if you have interpersonal conflict in that confined microsociety, things can get out of hand".

Space travel requires establishing and maintaining effective, stable interactions between individuals in small groups that are under microgravity conditions and are isolated and confined for prolonged periods. Individual behavior adjustment, interpersonal conflict, and group performance effectiveness are typically exacerbated in isolated and confined groups. Such phenomena have been repeatedly documented in operational settings such as remote stations in the Antarctica, undersea habitats, and most pertinently, in spacecraft. However, the fact that observations are made and observed by people who actually share the experience limits the reliability of data.

Crewmen "wintering over" for eight months in Antarctic stations (corresponding to the summer season in the Northern hemisphere) have shown an increase of 40% in stress-related symptoms of anxiety, depression, insomnia and hostility (Sandal 1996). In space, where the stress of isolation is compounded by the stress of prolonged weightlessness, problems can go beyond hostility and anxiety. Astronauts usually complain of various psychosomatic symptoms, including sleep disturbance, time disorientation and headaches. Time compression and heavy work schedules have led to fights between astronauts and ground control crews, with the astronauts in space feeling rushed and the controllers on Earth growing impatient.

These problems are amplified by the difficulties in living in microgravity: Hygiene routines are time consuming and laborious; food does not taste the same and spices must be added for flavor; privacy is limited; the environment is noisy; the countermeasures may require effortful and time-consuming activity (Figure 6-02); and motivation to do the required countermeasures becomes increasingly hard. Related to the question of exercising in space is the problem of limited bathing facilities (the water dispersed from perspiration during exercise must be collected, and there is no shower, just sponge bath). In addition, the physiological adaptations to microgravity described in previous chapters challenge and stress individuals, and thus impact interpersonal interaction, concentration, and ability to perform group and individual work.

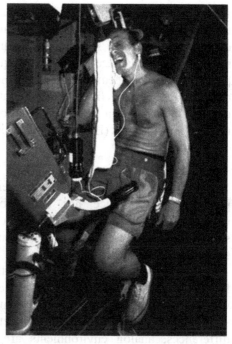

Figure 6-02. Astronaut Pete Conrad exercises in Skylab. During long-duration mission, daily exercise sessions require efforts and motivation (no shower!) and are very time-consuming. (Credit NASA)

As spaceflights become longer and more routine and the rewards less satisfying, the intense effects of being isolated take their toll and more problems in psychological adjustment are expected. Because of the technical and precarious nature of future, planetary, long-duration missions, the slightest upset in astronaut performance and behavior can have a deadly impact on the mission.

Typically, when an individual or small group of individuals is removed from a social environment and put into an isolated and confined environment, four groups of symptoms are to be expected (Harrison and Connors 1984). The first group includes mental deficiencies, decreased attention and concentration, learning problems, and hallucinations. These problems raise concerns when operating within dangerous environments such as space. In a worst case scenario, a warning light may be missed causing a catastrophe to occur. The second group is a decline in motivation when the individual's or the group's perceptions of the rewards inherent in the situation does not outweigh its costs. By comparison with the early space missions where astronauts were treated as heroes, spaceflight becomes more routine, thus creating a situation where reward is perceived as declining (Helmreich et. al. 1980). The third category of problems is somatic complaints, such as sleep disturbances, headaches, upset stomachs, and constipation. All of these complaints have some basis in the physical nature of the environment, but can also easily be exacerbated by or even caused by the unusual levels of stress found in that environment. Changes in mood and morale comprise the fourth category.

It is interesting to note that during the first several weeks of a mission in Antarctica, the interpersonal problems do not play a major part. When crewmembers first arrive, they don't have enough knowledge about their mission and surroundings to formulate their own ideas and so are willing to follow whatever the leader says. After the initial shock of the environment wears off, and crewmembers get to know their surroundings a little better, they begin to rebel against authority and each other. The increase in mood disturbance after the mid-point of winter isolation found in some studies suggests the existence of a "third quarter phenomenon" which is more psycho-social than environmental in nature and is independent of mission duration. It results from the realization that the mission is only half completed, and that a period of isolation and confinement equal in length to the first half remains (Bechtel and Berning 1991).

1.1   Analogs

In fact, much of our understanding of human behavior and performance in space has been obtained from the study of analog environments such as Antarctic research stations, polar expeditions, nuclear submarines, undersea habitats, oil drilling rigs, small rural communities, and space simulator experiments.

Analogs are not perfect simulations of the space environment. None, for instance, have the condition of microgravity. There are also differences with respect to characteristics of crewmembers, procedures for screening and selection, crew size, mission objectives and duration.[*] Despite these differences, analog environments are the only way to study behavioral impacts of isolation, confinement, and stress over long periods of time.

Nuclear submarines have long been the focus of analog studies for long duration spaceflight because of isolation and danger, as well as confined, close quarters. Whereas diesel submarines only stay below the surface a few days, nuclear submarines spend the better part of six months or more under water. The incidence rate of debilitative psychiatric illnesses among the crew is relatively low, with about 20-50 cases for 1000 men. Psychiatric symptoms generally observed include anxiety, interpersonal problems, sleep problems, performance decrement, and depression among a host of others (Weybrew and Noddin 1979). The low incidence may be due to the fact that submariners are some of the most thoroughly screened, tested, and trained individuals in the world. Also, the presence on board of a nuclear submarine medical officer who has specialized training in psychiatry may explain the low incidence rate for psychiatric symptoms.

Perhaps the closest *operational* analog of space occupancy is the undersea habitat, where aquanaut divers live and work on the ocean floor with a degree of isolation similar to that in space. Under these circumstances, and in Antarctic stations and submarine operations as well, observational measurements have focused upon critical individual and group factors that influence performance effectiveness and interpersonal relations.

---

[*] Space simulator experiments allow, however, to control some of these aspects.

Palinkas (1990) describes Antarctica as the best analog to the space environment. Because of the extreme nature of the environment, researchers must winter-over for 6-8 months out of the year. During this period, there is little contact with the outside world and groups are confined to their barracks because of the extreme temperatures. Several features of Antarctic research stations are particularly similar to outer space. Antarctic facilities and space facilities have similar scientific and political objectives. They are also similar in nature of work (science, exploration, and support), the heterogeneity of the crews (military and civilian men and women, Antarctic veterans and novices), the high level of skills, organizational similarities (division of labor, chain of command), and the rotational structure of tours of duty (Palinkas 1991). Outer space and Antarctica are also similar in that their environments are hazardous and stressful to work in and crews are heterogeneous, confined and isolated from the larger society. Because of these similarities, Antarctica has served as one of the primary means of gathering the psycho-sociological data for the ISS and interplanetary missions.

To some extent, analogs and space missions provides comparable report of psychological problems (Table 6-01). This is due mainly to the conditions of isolation and confinement. However, for space missions, most data comes from anecdotal reports (diaries, stories, personal books, and interviews) (Bondar 1994, Burrough 1998, Chaikin 1985, Collins 1990, Lebedev 1988, Linenger 2000), and methodically documented data is lacking.

Reported Problems	Mir	Shuttle	Submarines	Polar Expeditions
Interpersonal conflicts	Documented	Documented	Documented	Documented
Sleep disturbances	Anecdotal	Documented	Documented	Documented
Boredom, restless	Anecdotal		Documented	Anecdotal
Performance decrement	Anecdotal		Documented	Documented
Decline in group compatibility	Anecdotal	Anecdotal	Anecdotal	Documented
Substance abuse	Anecdotal			Documented

Table 6-01. Reported psychological problems during spaceflight and analogs.

One method used by psychologists to evaluate the effects of confinement and isolation during winter-over in Antarctica, for example, is to analyze the diaries of the crewmembers. The assumption is that the frequency that an issue is mentioned in a diary reflects the importance of that issue. Stuster and colleagues (2000) analyzed nine diaries from several expeditioners in French Antarctica (ranging from 69 to 363 days) written by the station leaders, the medical officers, and the technicians in charge of communications. These reports clearly indicate more negative experiences during the third quarter of isolation and confinement in an Antarctic station, regardless of duration of the expedition. It is also interesting to note that, when grouped into categories, the largest numbers of diary entries, and presumed importance of the issues, concern group interaction, communication, workload, recreation and leisure, and leadership (Figure 6-03).

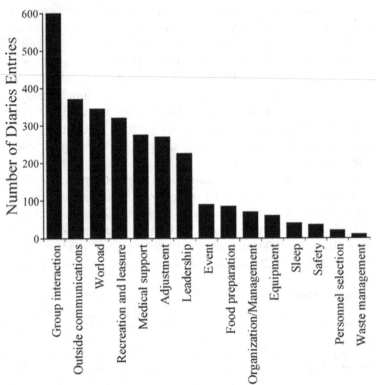

Figure 6-03. Number of diary entries classed by categories during expeditions in French Antarctica. (Adapted from Stuster et al. 2000)

1.2 Space Simulators

The Isolation Study for European Manned Space Infrastructure (ISEMSI) was a simulation experiment conducted by ESA in 1990 to provide psychological data on the day-to-day activities of astronauts during a long-term isolation and confinement inside a ground-based replica of a space station. The crew for ISEMSI consisted of six males, each from a different ESA member state. All subjects were civilians who had backgrounds in science and engineering. They were placed for a four-week period in hyperbaric chambers and monitored by a "ground" control team. Results from a battery of psychological tests performed showed no evidence of severe social or emotional conflicts during the experiment.

However, observations of social interaction and communication revealed considerable changes in the communication flow among the crewmembers (Figure 6-04). At the beginning of the confinement, all subjects participated in communication in a relatively balanced manner. At the end, subject D (the most dominant subject beside the commander) was totally isolated, and the communication of all other crewmembers remained limited to two-way communications with the commander. Despite these problems, the volunteers were seen to coalesce into a tightly knit group, even developing an aggressive attitude towards the "ground" control team (Sandal et al. 1995).

In a more recent experiment, social interactions of a mixed-gender crew from five countries (Russia, Canada, Japan, Austria, and France) were evaluated during a 240-day isolation study in a Mir simulator (SFINCSS-99). However, the crewmembers sometimes executed different flight programs and were housed in comparatively separate modules. As predicted, several incidents occurred that could be regarded as conflict situation between crewmembers. For example, a fistfight took place at the New Year's party, and a Canadian female crewmember accused a Russian crewmember of sexual harassment. Later, the commander informed the "ground" control team and insisted on withdrawal of two subjects from the study or proposed to close the hatch between the two modules. In accordance with his request, the hatch between the chambers was closed. From this point on, frustration grew up among crewmembers and toward the "passive support and insufficient management of crisis" by the ground control team. One volunteer expressed a wish to leave the chamber and did so (NASDA TM-02002).

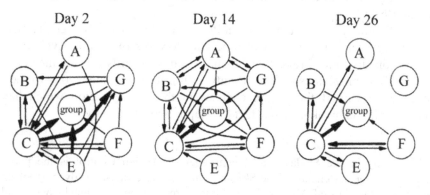

Figure 6-04. Pattern of communication at different days of the ISEMSI study. The thickness of the arrows indicates the frequency of occurrence. A, B, C, D, E, F represents the different crewmembers, with C being the commander. G represents the overall group, when, for example, one says, "let's go eat!". (Sandal et al. 1995)

The results of these experiments and others are somewhat consistent with what has been previously observed in actual spaceflights. The "Us vs. Them Syndrome" in particular has appeared at various times on spaceflights and in Antarctic studies.[*] Basically, the teams out in the field see those who stay at base as soft and weak. Their attitude seems to be "They're back there all warm and cozy! What do they know about what's going on out here?" The distance from a central authority seems to force the crew to assume a responsibility for themselves and their environment. As this happens, they begin to see the "outside authority" as unnecessary and conflict can occur if the central authority pushes itself on the crew (Nicholas and Foushee 1990).

1.3   Actual Space Missions

The former Soviet Union has had extensive experience in operating a crewed space station over long periods of time beginning in 1971 with the launch of the first space station, Salyut-1. Disagreements between the Russian crew and Mission Control over work overload, or regulations of crew activities imposed by Mission Control have been reported from several missions (Kelly and Kanas 1993).

Some of the problems that have been observed aboard Soviet stations include a general nervous or mental tiredness called "asthenia". Asthenia is comprised of hypersensitivty, irritability, and hypoactivity in its early stages. These symptoms are sometimes followed by psychosomatic illness and sleep disorders. Asthenia has been observed to occur primarily in the monotonous, later portions of the mission, and once recognized, it is treated through a manipulation of work schedules to provide more free time and stimulation, and through increased audiovisual communications with friends and family (Kanas 1991).

Another problem that has manifested itself on board Soviet stations is interpersonal tension. The Soviet proclivity towards the "collective" causes the cosmonauts to suppress their hostile feelings towards one another in the interest of the mission as a whole (Santy 1994). It may be interesting to see in the coming years if the abrupt changes in the Soviet Union which have occurred with its breakup will affect this suppressive tendency.

---

[*] The mutiny of the Skylab-4 astronauts against Mission Control is the perfect example. All the astronauts of the Skylab-4 mission were first-time flyers (rookies). Before they got adjusted, Mission Control transferred to them the same busy schedule as their predecessors in the space station. After their complaint about a heavy workload did not received enough attention by the ground controllers, the Skylab-4 astronauts declared an unscheduled day off to Mission Control and proceeded to turn off the radio while they got some rest. This mutiny led to much-needed workload adjustments. Perhaps as a result of this event, a rule states that at least one member of an Expedition crew on board the ISS should include a spaceflight veteran.

The conditions that will exist in long-duration space missions increase the potential for adverse effects already reported during relatively short-term missions (e.g., irritability, depression, sleep disturbances, and poor performance of both group and individual). During the joint NASA-Mir missions (Mars 1995-June 1998), several events led to tension among the crew and between the crew and personnel at Mission Control (Table 6-02).

- *Crew change at L-8 weeks*
- *Mission extended by 6 weeks*
- *Minimal control over in-flight work schedule*
- *Work overload / underload*
- *Social withdrawal*
- *Death of family member*
- *Dangerous atmosphere (ethylene glycol and contaminant leaks)*
- *Fire; decompression (loss of module); loss of power (free drift); communication system failures*
- *Anger with ground control / management ("us vs. them" syndrome)*
- *Crew friction*

Table 6-02. List of events leading to psychological disturbances during the NASA-Mir space missions. (Adapted from Ark and Curtis 1999)

1.4 Rules

Nearly all-essential human functions during long-term space missions will depend critically upon individual and group behavior. Selection, training, and organizational support are the focus of human behavioral initiatives (Table 6-03).

Selection
  Best psychological profile
  Appropriate skills
  Team player

Training
  Interpersonal communication
  Group dynamic and group problem-solving
  Multicultural sensitivity
  Performance feedback monitoring

Support
  In-flight counseling
  Communication with family/friends
  Scheduled breaks in routine

Table 6-03. The countermeasures for psycho-sociological issues of spaceflight are individual and crew selection, training, and in-flight support.

The first priority in considering the impact of individual factors upon personal adjustment to a space mission is the screening and selection of prospective participants. In Antarctica, a good "field person" is one who has a good knowledge of living and working in the new environment, and is committed to science and exploration. This individual is autonomous and can function without direct supervision. He also should have a certain amount of integrity and must accept one leader. Indeed, in a hostile environments, there is room for individuality, but not for "too many chiefs" (Stuster 1996).

In general these principles can be applied effectively to space missions. The astronaut should have an intimate knowledge of the environment he/she will be spending a good portion of time in and be committed to the mission he/she will be a part of. Palinkas (2000) in his review of the literature found that introverted individuals, those who are "more inner-directed, quiet, retiring types", tend to adapt and perform better in Antarctic situations than do extroverts.

Training should be both didactical and experiential. The goal is to sensitize both the crewmembers and monitoring ground personnel to the influences of socio-cultural factors, such as culture and language differences. Team building and conflict resolution exercises should also be included in preflight activities.

In-flight support is provided by the flight surgeons and psychologists on the ground, as well as family or friends, and includes surprise gifts and events to break the routine. Individuals are also encouraged to talk with one another to resolve interpersonal difficulties.

These countermeasures (selection, training, and support) are further detailed in the following sections.

2   INDIVIDUAL SELECTION

At the individual level, selection strategies have two-fold objectives: to eliminate unfit or potentially unfit applicants, and to select from otherwise qualified candidates those who will perform optimally. A distinction is therefore made between "select-out" and "select-in" criteria.

2.1   Select-Out Criteria

The first objective is to "select-out" or disqualify any candidate with a history of psychiatric disorder, current psychiatric symptoms, or other characteristics that place him or her at risk for a psychiatric disorder during a space mission. "Select-out" criteria are medical criteria specifying those psychiatric disorders, which would be disqualifying. These disorders include schizophrenia, major depression, and all the other psychiatric diseases listed in Table 6-04. To achieve this objective, selection procedures rely upon formal clinical evaluations and use of standardized psychometric tests.

L — *a validity scale; high values indicate evasiveness (e.g., different responses to about the same questions)*
F — *a validity scale; measuring the tendency to present one's self on an overly favorable light (low score = more favorable)*
K — *a validity scale; measures defensiveness (e.g., underreport, not completely honest in answering personal questions) (high score = more defensive)*
Hs — *Hysteria*
D — *Depression*
Hy — *Hypochondriasis*
Pd — *Psychopathic Deviation*
Ma — *Mania*
Mf — *Masculinity/Feminity*
Pa — *Paranoia*
Pt — *Psychasthenia*
Sc — *Schizophrenia*
Si — *Social Introversion*

Table 6-04. MMPI Scale. The first three measurements (validity scale) indicate how well the candidate responded to the test. The other measurements indicate the scores for each disqualifying psychiatric disorders. Note: The Mf score is not considered of any significance in defining sexual orientation: high scoring are described as sensitive, aesthetic, passive; low scoring are described as aggressive, rebellious, and unrealistic. (Adapted from Santy 1994)

Clinical evaluations generally are in the form of a structured psychiatric interview, with at least two independent psychiatrists. Each psychiatrist asks the same question in the same order and generally in the same manner, in order to avoid the problem known as interviewer bias. The interviews are conducted to counteract the tendency of applicants to minimize psychological symptoms ("staying clean"). Patricia Santy gives the following example in her book "Choosing the Right Stuff" (1994): instead of asking the question "Have you ever been depressed?" where most healthy subjects would realize it's not a good thing to be depressed and would probably answer "no", the question is formulated in the form of a request — "Tell me about the time where you have been most sad in your life"— which makes it hard for the subject to escape from giving some clinical information on the topic.

To assist the clinician in objectively determining whether or not an applicant is a "risk to flying safety", a series of psychometric tests is generally added to the psychiatric interview. These tests include self-reports questionnaires, such as the Minnesota Multiphasic Personality Inventory (MMPI), and the Million Clinical Multiaxial Inventory (MCMI). These tests have been standardized against a normal population and most of them have built-in scales that detect whether the applicants fake responses to test questions in trying to conceal pathology. Other tests try to create a situation in which the psychological issues of the applicant can be reflected. These projective personality tests include for example the Rorschach Ink Blot test, where the subject's association to ambiguous inkblots are observed and scored. In other projective tests the subject is asked to draw a person or complete a sentence.

Using these methods, Santy (1997) reports the incidence of psychiatric disorders in a study of 223 astronaut candidates to be 8-9%. The prevalence rates for these psychiatric disorders found in the applicant groups are very similar to the rates reported in the general population (0.4-8%) from a number of studies (Robins et al. 1984).

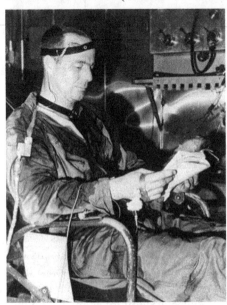

Figure 6-05. Astronaut Scott Carpenter during a stress test in a heat chamber prior to his flight on board a Mercury capsule. (Credit NASA)

2.2   Select-In Criteria

The second objective is to "select-in" or identify and select candidates with characteristics that predict for optimum performance in the isolated, confined, and hostile environment of space. This selection does not have specific medical or psychiatric implications. "Select-in" or psychological selection criteria identify those desirable personality traits or characteristics linked to a specific mission ("best person for the job"). Typically, these traits would be those required when applying for a qualified job, i.e., aptitude for the job, intelligence, "team player". In addition, given the stress and difficulty of the space environment, qualities such as the ability to tolerate stress, trainability and flexibility, and motivation are most important. Finally, sensitivity to self and others, emotional stability, maturity, ability to form stable quality interpersonal relationship, are prerequisites for dealing with sociological issues (Santy 1994).

"Select-in" criteria were easy to identify for the early space missions, where the astronaut requirements were limited to high piloting skills, good stress tolerance (high acceleration, reduced pressure, high temperature, and other stressors), an ability to make decisions, and a strong motivation for the success of the mission rather than personal objectives. However, when the space program later also required astronauts with engineering, scientific, or medical background, and not necessarily piloting skills, the definition of "select-in" criteria became more complex. In addition, there has been little evaluation of astronaut performance during space missions, which is a requisite for the validation of "select-in" criteria. Psychological tests used in the past have also failed to find significant personality predictors of performance. As a result, "select-in" criteria used by the psychiatrists were reduced from the Gemini to the Apollo program, and are basically not used for the selection of Shuttle astronauts (Table 6-05).

Procedure	Mercury	Gemini/Apollo	Shuttle
Number of hours for the psychiatric evaluation	30	10	3
Screening method	2 psych interviews 25 psych tests 5 stress tests	2 psych interviews 10 psych tests 1 stress test	2 psych interviews
"Select-in" criteria used by psychiatrists	1. Intelligence 2. Drive and creativity 3. Independence 4. Adaptive motivation 5. Flexible 6. Motivation 7. Lack of impulsivity	1. General emotional stability 2. High motivation 3. Adequate "self" concept 4. Quality of interpersonal relationships	None documented
Validation of criteria	Data not available	Not done	Not done

Table 6-05. Summaries of psychiatric and psychological selection procedures in the U.S. space program (1959-1985). (Adapted from Santy et al. 1991)

It is interesting to note that during the psychological evaluations of the candidates for the Mercury space program, two psychiatrists spent over 30 hours on each candidate. This included the time for taking the psychometric (for evaluation of motivation and personality) and performance (for evaluation of intellectual functions and special aptitudes) tests. Psychological reactions of the Mercury applicants were also monitored during stress experiments simulating some conditions of the mission, such as change in pressure, isolation, noise and vibration, and heat (Figure 6-05).

During the Gemini, Apollo and early Shuttle missions, the psychological evaluation of candidate was reduced to 10 hours, and the number of psychometric tests decreased from 25 to 10. For Shuttle candidates, only "select-out" criteria are used to eliminate possible disruptive behaviors, and the duration of the evaluation does not exceed 3 hours. Selection of those individuals evidencing the highest proficiency (select-in criteria) based on the results of psychological tests is absent after the success of the early space missions.

2.3   Psychological Profile of Astronauts and Cosmonauts

Using data collected over 30 years of candidate psychological evaluations at NASA, Santy (1994) has compared the results of a commonly used psychometric test, the Minnesota Multiphasic Personality Inventory (MMPI) test. This personality test consists of 566 questions for which a subject is asked to respond true or false. This test is used to identify psychiatric disorders, and it includes validity scales to detect if the applicant was honest in answering the questions (Table 6-04). What is remarkable is the similarity of all four groups of applicants (Mercury, Gemini, Apollo, and Shuttle) over the thirty years (Figure 6-06).

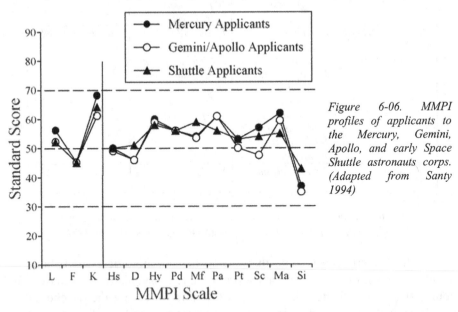

Figure 6-06. MMPI profiles of applicants to the Mercury, Gemini, Apollo, and early Space Shuttle astronauts corps. (Adapted from Santy 1994)

All groups are extremely defensive and present themselves in the best possible light (LFK scales). The low Si (social introversion) scales in all groups and the comparison with the normal population (Figure 6-07) suggest that applicants are much more socially extroverted than the normal population.

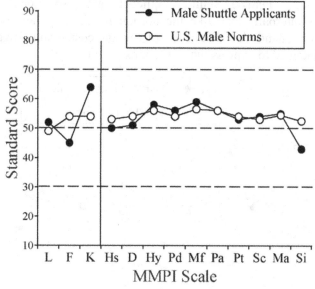

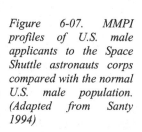

Figure 6-07. MMPI profiles of U.S. male applicants to the Space Shuttle astronauts corps compared with the normal U.S. male population. (Adapted from Santy 1994)

Interestingly enough, non-U.S. astronauts show a similar personality profile (Figure 6-08), even though the MMPI scores of the general population in various countries can be somewhat different due to cultural differences.

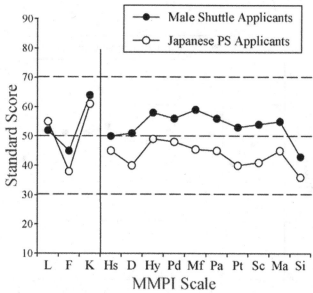

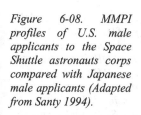

Figure 6-08. MMPI profiles of U.S. male applicants to the Space Shuttle astronauts corps compared with Japanese male applicants (Adapted from Santy 1994).

Female Shuttle applicants are also much more like their male counterparts than like the normal female population (Figure 6-09). On the other hand, the LFK scales of Russian cosmonauts suggest that they are more inclined to express their emotions (Figure 6-10).

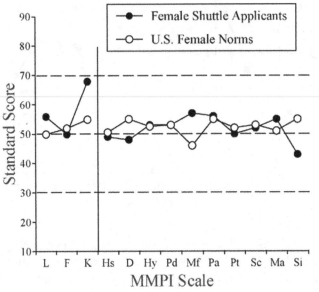

Figure 6-09. MMPI profiles of U.S. female applicants to the Space Shuttle astronauts corps compared with the normal U.S. female population. (Adapted from Santy 1994)

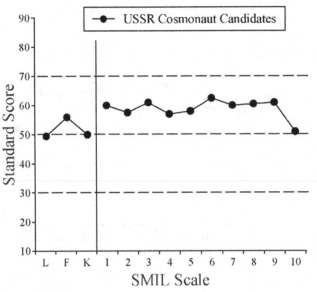

Figure 6-10. SMIL profiles of USSR cosmonaut candidates. The SMIL test is similar to the MMPI test, although the scales have been redefined. The numbers in the SMIL scales roughly correspond to the pathology described in Table 6-03 for the MMPI. (Adapted from Santy 1994)

Therefore, although psychometric tests are used to identify psychopathology and select out candidates, they also suggest both commonalties in personality traits and socio-cultural differences. These commonalties and differences will ultimately contribute to the psychological and sociological problems that will develop during a space mission among a group of highly selected individuals.

However, a psychiatric examination per se is not particularly helpful in determining which applicants to actually select. This is done though the characterization of those personality traits the most adequate for a mission, and through the validation of these criteria by subsequent behavior analysis and performance during training and space missions. This aspect will be developed in the Section 4. However, since today's space missions are composed of more than one single individual, let's review the social and cultural issues of group interaction and the process of crew selection.

3   CREW SELECTION

3.1   Sociological Issues

There are many variables that can affect the cohesiveness and performance of a group: culture, leadership, gender, age, personal attractiveness, emotional stability, competence, cooperativeness, and social versatility (Connors et. al. 1985). It is important to note here that these studies were conducted in analog environments with control groups, and that most of them include subjects from Western culture.

3.1.1   Confinement and Personal Space

Ground-based studies have determined that psychological impairment started to occur when the available volume was restricted to 1.42 m³ per person for 1 or 2 days on confinement, 7.36 m³ per person for 1 or 2 months, and 17.0 m³ per person for more than 2 months (Fraser 1968). Interestingly, except for space stations, the habitable pressurized volume in most spacecraft is less than these values (Table 6-06).

The number of individuals sharing confinement is believed to be another important variable affecting the amount of space needed per individual. More space per individual is needed as the number of individuals increases (Smith and Haythorn 1971). However, confined individuals tend to place heavy emphasis on assigned work and little emphasis on recreational opportunities. When recreation is sought, it tends to be passive in nature. This might account for the fact that astronauts and cosmonauts are not pursuing exercise programs enthusiastically (to say the least).

Spacecraft	Cabin volume	Crew	Volume per person
Mercury	1.53	1	1.53
Gemini	2.52	2	1.26
Apollo	9.1	3	3.0
Soyuz	10.2	2	5.1
Shuttle	74.3	7	10.6
Salyut	99	3	33.0
Skylab	361	3	120.3
Mir	378	3	126
ISS	1200	6	200

Table 6-06. Habitable pressurized volume (in m³) in past and present spacecrafts.

When two people are talking to each other, they tend to stand a specific distance apart. Each person has an invisible boundary around his/her body into which other people may not enter (Figure 6-11). If someone pierces this boundary, the "invaded" person will feel uncomfortable and move away to increase the distance with the "invader". (The major exception is family members and other loved ones.) This personal distance is not due to body odor or bad breath. Closeness lends a sense of intimacy, which degree varies with the distance between individuals.

Figure 6-11. Each person has an invisible boundary around his/her body into which other people may not come.

The average personal distance varies from culture to culture. Latin American, French and Arab interact at closer distances than U.S., English, Swedish or German individuals (Hall 1966). During two sessions of the International Space University (ISU), students from the Space Life Sciences Department performed a study trying to determine the personal space of students from various countries (Bui and Wong 2002). During a fake interview, the distance and the angle between the subjects and the interviewer were measured. Results indicated that this distance varied from 150 cm in Asian (e.g., Japanese) students, to 40 cm or less in Latin (e.g., Italian) students. There was also a strong tendency to not directly face the interviewer, but be at an angle. This angle seemed also strongly correlated with the subject's cultural origin (Figure 6-12).

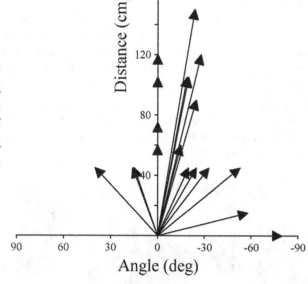

Figure 6-12. Mean distance and angle between two individuals during a sitted interview. Catherine Beaulieu, Julielynn Wong and Linh Bui compiled these results during the ISU summer session program, on a population of 23 students from 13 different countries.

The use of physical or eye contact also varies by culture. Although in some cultures, eye contact is a way to communicate, in other cultures physical or eye contact may lead to discomfort, and may indicate sexual overtones.[*]

When comparing the required personal distance in various environmental conditions, it was found that greater distances are required for personal space: a) in small rooms versus large rooms; b) inside locations versus outside locations; c) in high-anxiety settings versus in low-anxiety settings; d) with people who you expect to be interacting with over a long period of time versus a short interaction (Bishop 1997). Interestingly, all these factors (small rooms, inside, high anxiety setting, long interaction) are present during space missions. Consequently, the personal distance of the astronauts should increase in the conditions of a space mission. It would be interesting to actually measure the personal distance of crewmembers on board the ISS and compare for example that distance between astronauts and cosmonauts.

---

[*] Some astronauts have reported that the swollen face in weightlessness, due to the headward fluid shift, creates problem to communicate by eye contact.

On the other hand, the dimensions of a spacecraft are necessarily small, and there is little privacy (Figure 6-13). Ground-based studies indicate that privacy only seems to alleviate anxiety and stress in short-duration missions. In long-duration missions with privacy, it was noted that even with access (for conversation, social interaction) to another person in the group, stress levels are higher (Taylor et. al. 1968). This also seems to be true for space missions. Indeed, issues of personal hygiene and housekeeping alone account for about 40% of incidents during space missions with U.S. only crews (Santy 1994).

It is well known that under crowded conditions, men and women react differently. Women tend to perceive small, crowded places as friendly and sociable, while men tend to respond to such environments as irritating and uncomfortable. Men are more likely to feel their personal space violated in crowded places and perceive a continuing challenge to patterns of male dominance. Thus, men respond with greater irritation and hostility to crowded conditions than do women. The interesting note is that mixed-gender groups tend to respond nearly as well to crowded circumstances as do groups of women only (Bishop 1997).

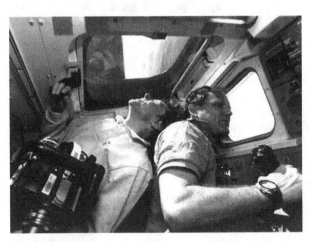

Figure 6-13. Two astronauts looking at the Earth from the observatory windows of the Space Shuttle. (Credit NASA)

3.1.2   Mixed Gender Issues

Women have performed equally or superior to male counterparts in Antarctic stations and underwater habitat studies (Connors et al. 1985). Though there is little question about the competence of women to handle space missions, the total number of female astronauts and cosmonauts represents less than 20% that of male astronauts. Hopefully, the increasing number of women astronauts will translate into a large participation in the ISS and interplanetary missions. However, factors associated with long-duration missions may be affected by the presence of women on board.

Antarctic expeditions started including women since 1979, but many stations still refuse to allow women to winter-over. Nevertheless, the results of studies in analogs suggest that the presence of women exerts a positive influence and discourages certain behaviors (e.g., drinking and fighting) that could lead to injury or group conflict (Palinkas 1991). However, other studies found that the introduction of a female into a male group caused destabilizing effects (Harrison and Connors 1984).

With the increasing length of future missions, sexual tensions and prejudices may be forced to the surface creating friction and this could impact both crew cohesiveness and performance. The concern about an affair occurring on board ISS is probably unwarranted, and if it does happen, the question will be what effect, if any, this new level of interaction will have on the performance of the crew. Helmreich et al. (1980) consider that more harm than good can come out of a policy which regulates the moral conduct of the space station crew. Such a policy will not make desire go away, only perhaps frustrate it into another, more dangerous form. He suggests that the space agencies should take a "hands off" approach (no pun intended!) to this issue and see what happens.

Gender stereotyping by members of the crew can also have a destabilizing effect. There is some evidence that male astronauts and cosmonauts (Oberg 1981) still hold on to outmoded views of women. These stereotypes should be avoided by having crewmembers work together on projects before a space mission in order to demonstrate each crewmember's, not just the women's, competence and technical proficiency.

3.1.3 Multi-Cultural Issues

Today the ISS, and tomorrow the exploratory missions, present the opportunity to have greater representation of different nationalities in space. This will bring problems that were not present in the more homogenous missions, which have occurred up to now. The Space Shuttle has had international crews and women crewmembers on a somewhat regular basis (Figure 6-14), but the short duration of these missions can not compare to the long-term missions we will be facing in the future.

Culture refers to widely shared beliefs, expectancies and behavior of members of a group on an organizational, professional or national level. Beside personal space described above, studies on pilot populations have shown significant national differences in attitudes, such as acceptance of hierarchical leadership and the necessity of adhering to rules and procedures. Another well-known cultural difference is time perception: for example, Anglo-Saxons typically emphasize schedules, appointments, segmentation, and promptness, whereas Middle East and Latin's cultures are more flexible and feel more at ease with several things going on at once. Such differences have obviously the potential to cause problems in safety, performance, and interaction between crewmembers.

The lessons learned from the NASA-Mir program with multi-cultural crews illustrate the difficulties that can be encountered. Several U.S. astronauts have commented that under conditions of high stress such as during long-duration missions, cultural differences disrupted the harmony among the crew. They suffered from the facts that: a) the language differences led to misunderstandings; b) they were the sole members of a cultural group; c) they had prolonged periods of no contact with English-speakers, even less with family; d) they had very restricted food selection; e) and they were not allowed to operate equipment. Some also complained they were treated as guest rather than working member (Burrough 1998).

Campbell (1985), in a review of cultural integration literature, identified a three-stage theory of adjustment to a new culture called the U-curve theory. The first stage of this theory is entry. In this stage the novelty of the environment precludes adjustment problems from manifesting themselves. The second stage is adjustment. In this stage the individual learns new ways of thinking and acting, and frustration with the environment is high. The third stage is adaptation. This stage sees the individual reconcile his/her expectations with the reality of the situation.

Figure 6-14. The crew of STS-51G (June 1985) included astronauts from the U.S., France, and Saudi Arabia, both male and female, with military or civilian background. (Credit NASA)

It can be expected that the adjustment of astronauts to the culture of a long-duration crew will be similar. The initial awe of being in space will give way to cultural reactions which may include difficulties in understanding non-verbal cues, difficulties in adjusting to new work regimes, and technical language difficulties. For this reason training in each other's culture and lifestyle is an essential part of any long-duration, international space effort.

3.2   Selection Issues

Crews are, in fact, small social systems shaped by multiple determinants, none of which, considered in isolation, can necessarily account for the variations in behavioral interactions or performance effectiveness.

Reviews of literature on Antarctica focusing on its relevance to long-duration spaceflight have identified the leader as the single most important role in the isolated group (Stuster 1996). Leaders organize, direct, and coordinate followers. They also exert their influence to: a) help the group maintain harmony and stability; b) interpret the conditions that confront it; c) set goals; and d) meet challenges posed from outside. The most effective leader let crews do work with minimal interference, but recognize when group activity is needed and arrange that activity. Good leaders are also able to swing between autocratic (i.e., make decisions without soliciting suborbinates's inputs) and participative (democratic) styles of leadership as needed. Prescriptions for good leadership often dwell upon the selection and training of leaders. However, such prescriptions could also involve the selection and training of followers, and the structuring of the social settings and the group's tasks.

One potential source of conflict in today's space missions, as for analogs, is that the leader's right to exert his/her influence is conferred through appointment by a higher authority, not by the group itself. To counter-balance this, another key factor of long-duration missions is to have open communication among the crew and allow feedback channels. Communication is essential for it provides updated knowledge of other people's attitudes and views, which is necessary for social comparison processes and for conflict management. Also, as seen above in the ISEMSI experiment, miscommunication (i.e., failure in the communication process) can contribute to interpersonal friction and conflict within the crew or between the crew and the ground personnel.

Another factor is to have clearly defined contingencies for achieving goals. Research findings are unambiguous in showing that a clear, engaging set of objectives is a powerful means for orienting members toward achieving overall organizational goals. Group goals encourage people to coordinate their activities for mutual gain, and hence are likely to affect the tone of interpersonal relations within the group. However, crewmembers must feel personally committed to these goals; it will not suffice to simply impose them from above. In addition, for long-duration missions, means must be found to maintain astronauts' interest in distant goals over prolonged periods of time. It may thus be desirable to establish a number of interim goals, which can be pursued and savored.

Finally, groups of more than two must have boundary role persons who act to interpret interests and concerns of all sides to allow activity to progress smoothly. Boundary role persons serve as agents for purposes of bargaining and negotiation. These boundary persons, with the help of the leader (see above) and the authority in Mission Control, should manage the inevitable problems and disputes that occur in real time and that threaten the overall integrity of the group, and arrange cooperative ventures with equitable outcomes for both sides (Connors et al. 1985).

3.2.1 Compatibility

In the context of multinational missions, one of the more important challenges is to ensure that the individuals are compatible and can work together effectively. Crewmembers may be considered compatible in that each member demonstrates qualities and behaviors that other crewmembers consider desirable and appropriate. This is challenging in the space program because individuals may have very different educational backgrounds. For example, scientists prefer their autonomy and tend to not interact and not to work well in a hierarchical command structure. On the other hand, pilots-astronauts often have military backgrounds, which lead them to prefer a more ordered command structure.

Evaluation of compatibility might be based on the results from psychological performance tests and personality questionnaires. Another more behavior-oriented approach, which has been developed in the context of industrial applications, includes the combination of a variety of behavioral exercises (e.g., role-plays, group discussions, group exercises). The objective is to select individuals who demonstrate capabilities for effective team functioning (team player) and problem solving.

The Soviet/Russian space program has spent considerable effort developing methods to assess interpersonal compatibility for long-duration missions. These methods include attitude assessments, psycho-physiological tests, and specific group exercises (Figure 6-15). The Russian psychologists believe that biorhythms are useful in selecting specific cosmonauts for space missions. For example, when the crew works together on a complex task during a training session, they monitor their pulse. As soon as the crewmembers start helping each other their pulses synchronize to some extent. It is believed that the higher the biorhythmicity, the greater the compatibility (Bluth 1987, cited by Santy 1997). However, these compatibility tests have not been validated.

Figure 6-15. The Mercury 7 astronauts during survival training in the desert. (Credit NASA)

3.2.2   Crew Composition

Obviously, the larger the group, the more chances of interpersonal conflict. In fact, it was found that larger groups react better to confinement situations. Irritations in these groups are not directed at other group members but at "things or non-personal aspects of the situation" (Smith and Haythorn 1972). Obviously, for space missions, especially interplanetary, the ultimate decision will be made upon propulsion considerations after calculating the total weight needed in terms of life support system mass per person. However, one rule of thumb is that crew size and heterogeneity must be as small as possible, since the complexity of interpersonal interaction increases with crew size and heterogeneity. (On the other hand, increasing crew size increases the number of possible social relationships and, among other things, options for social stimulation and developing friendships, which are favorable factors for group cohesiveness). Odd number of crewmembers is recommended over an even number to prevent the development of two equalized groups, which might hinder democratic problem solving.

A very important consideration is related to the occupational role of the crew. Traditionally all crews in human space exploration have included one or more astronaut pilots. Some of the designers of human missions (Zubrin 1999) state that given the demonstrated ability of guiding unmanned spacecraft safely to the surface of Mars taking a pilot would be an unnecessary waste of resources. They argue that a scientist trained minimally to override the automatic system in case of malfunction would be enough. All schools of thought consider essential to take on board a medical doctor to cope with medical problems during the trip. Should a mechanic be included in the mission to repair malfunctioning systems? How many scientists are necessary during a pioneer exploratory mission? Or should just extremely fit individuals integrate the crew on this occasion?

One sure thing is that the role of mission commander should belong to the most qualified individual in the crew, whether a pilot, engineer or scientist, that is one whose leadership style encourages group dynamics, group performance, and morale. Consequently, the role of mission commander will not automatically fall on the pilot-astronaut, as it has been the case for all space missions (with the exception of a couple of ISS increments) so far.

Of prime importance are the motivational issues raised by the prospect of long-term space occupancy. Motivation plays a critical role in maintaining individual performances and amicable social interactions over extended intervals of isolation and confinement. Duration and expected duration of missions also seem to be an important variable to consider. Taylor et al. (1968) found that when a group expects a long-duration mission there is more stress evident than a group expecting a short-duration mission.

For collective operations, the right people, well trained and properly configured (that is, with the right mix of skills, personal characteristics, task requirements, and work setting) are essential. However, this will ultimately bring differences in education, culture, and age, which may also contribute to issues in group cohesiveness and performance.

Even though there has been a significant advance of women astronauts lately (including several female Shuttle pilots), again in general all crews have been predominantly male. Long-duration planetary missions raise unique considerations. Having an all-male crew for such a long time is at least doubtful. Married couples are definitely a possibility, and mixed crews with non-married members are also being considered. But what if two crewmembers fall in love with each other or altercate due to their relationship? What about the other crewmembers that are "left out"? Men are believed to be physically stronger, but on the other hand women are more resilient psychologically. The fact that there is no quick way to get the crew back adds to the need to consider all possible alternatives. Experiments aimed at the understanding of the psychological and social consequences of sexuality and mixed-gender groups are also needed.

3.2.3   External Factors

It is interesting to note that there are many known factors that participate to group cohesion and performance or to group fission (Table 6-07). For example, during long-duration missions on board Mir, several crewmembers reported that the arrival of an international visiting crew, staying in the station for a period of one week, helped to neutralize tension between the crew. Cosmonauts on extended flights had letters and presents from home, along with special foods and fresh milk, delivered to them in space. They were also surprised by small, but apparently delighting toys, novelties, etc. The unexpected undoubtedly plays a very large role in what we see as the fullness of experience. Parties and group rituals (Figure 6-16) also help maintain moral and group interaction.

Factors that impact group fusion	Factors that impact group fission
• *Emergencies: when people are forced to work together for common survival*	• *Power and status (e.g., leader / followers)*
• *Arrival of outsiders (e.g., replacements, new personnel)*	• *Differences in work demands (e.g., shifts)*
• *Resentment towards outsiders (e.g., mission control, authorities)*	• *Differences in responsibility (e.g., pilots / non-pilots)*
• *Leadership: promotes performance and minimizes conflicts*	• *Differences in motivation*
• *Social events (e.g., surprises, parties, holidays)*	• *Differences in personal values*
• *Group rituals and habits*	• *Leadership (e.g., authoritarian / participative)*

Table 6-07. Factors that impact group fusion and fission.

It is interesting to note that *leadership* belongs to both categories, that is a good leadership helps group fusion, but bad leadership can be at the origin of group fission. Obviously, the ability to be a team-leader or a team-follower should be part of the "select-in" criteria. But how to define this criteria and others, and most importantly, how to validate them. We will review these issues in the following section.

Figure 6-16. As a group ritual, Shuttle crewmembers play poker on the launch day. They leave the crew quarters only after the commander wins. (Credit Douglas Hamilton)

4   ASSESSMENT OF CREW BEHAVIOR AND PERFORMANCE

So far, the psychological selection of astronauts has focused on selecting-out candidates with psychopathological disorders. In contrast to these well established "select-out" criteria, "select-in" criteria need to be developed in relation to specific aspects of the mission, such as mission objectives, mission duration and crew composition. Once determined, the "select-in" characteristics require validation against in-flight performance measures, and need to be explored in a mission-specific manner (Sandal et al. 1996).

To date, the absence of formal criteria for astronaut performance, and the limited research opportunities have made it very difficult to evaluate the efficiency of crew selection strategies. Such evaluation also requires that select-in criteria must not be used in the initial selection until they have been found to predict astronaut performance. One potential bias in validating selection criteria on astronauts and cosmonauts who have already gone through a formal selection process is related to the restriction of range in personality scores, as seen in the MMPI.

Another issue is a how to evaluate astronaut performance. Optimally, this evaluation should be performed during training and actual space missions. Performance would then result in a reevaluation of crew composition. This assessment is likely to increase the program of missions. For example, controllers in Mission Control in Russia were systematically keeping track of the number of errors performed by the crew on board Mir or by the ground personnel (Figure 6-17). Although useful for determining a change in performance throughout a mission, some astronauts might see such evaluation (and the psychologists who perform them) as a possible threat because they fear of being grounded or removed from a mission.

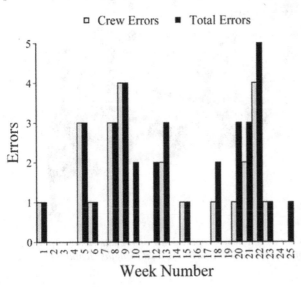

Figure 6-17. Number of crew and ground personnel errors during the course of a 28-day Mir mission. (Oleg Atkov, personal communication)

Over the recent years, however, various techniques of performance analysis are being developed. Test batteries consisting of one or a number of discrete individual tasks to measure such factors as vigilance, reaction time, tracking, limb steadiness, coordination, and perceptual speed have been used during space missions. Comparison between the time required to perform the same task in weightlessness compared to ground, for example, proved useful to determine "work-efficiency ratios", i.e., the total estimated time divided by the total number of hours available to work (Kubis et al. 1977).

The evaluation by peers proved to be an effective tool to assess both technical skills (job performance), as well as the ability of an individual to live and work with others (group living). In this type of evaluation, one crewmember is asked, for example, the following questions: "Name the 5 best crewmembers ", "Name the 5 worst crewmembers", and «If you can't go on a mission, who should replace you?" Using this technique in analogs, it was found that veterans of isolated and confined situations generally picked an effective crew if given a large applicant pool and time to conduct thorough interviews (Natani 1980). Preliminary results of peer evaluation performed by astronauts on each other during training on nine dimensions of performance (e.g., job competence, leadership, teamwork, group living, personality, communication skills) showed remarkable agreement with the ratings done by the astronauts supervisors (and decision-makers) (Rose et al. 1993). [*]

However, more objective and reliable methods for observing and recording the effect of space stresses upon complex performance processes must be developed. To date, no actual data have been published on astronaut performance. As Patricia Santy (1994, p. 152) wrote: "Such data are essential if behavioral scientists are to understand individual and group psychological factors as they relate to individual and crew performance in space. Without objective data to clarify these relationships, even the best guesses about what psychological criteria are critical in selecting astronauts remain only guesses."

5   PSYCHOLOGICAL TRAINING AND SUPPORT

Let's assume the selection process picked out the best individuals, and that the mission director, with the helps of the psychologists, then picked out the best crew for a given, long-duration mission. The hazards of such a mission, and the unknown implications of stress on human behavior under stress, will undoubtedly result in group interaction issues. Psychological training is required for preparing the crew to react to these situations. When conflicts arise, psychological support from the ground also enters into play.

---

[*] Mike Collins (1990) wrote: "As I used to tell John Young before Gemini-10, I was happy I was making my first spaceflight with him, but I wanted to fly so badly I would have gone up with a kangaroo!"

5.1   Training

One rule of thumb is that group dynamics and group problem-solving techniques should be dealt with prior to the mission. Astronauts not only have to be technically proficient in their area; they will also have to be aware of interpersonal dynamics and intercultural differences. Both the crew and their ground control personnel should be trained together preflight to use interactive techniques.

Optimally, once a crew has been selected for a certain mission, a training oriented for that specific crew should focus on the following issues: a) support of a team-building process within the crew (e.g., establishment of a stable crew structure, development of common behavioral norms, identification of common mission goals); and b) "anticipatory problem-solving" (i.e., making the crewmembers aware of how to deal with specific psychological problems most likely to arise in the course of a mission).

Three phases of training can then be identified: a) a phase of awareness where the crew learn the basics of group dynamics and interpersonal relations and their effects on performance; b) the group gets feedback on their newly-learned concepts by putting them into practice in role-playing and simulation exercises; c) these concepts are reinforced regularly to prevent backsliding (Nicholas and Foushee 1990).

Using this type of training, issues that caused tension between crewmembers and outside monitoring personnel have been ameliorated through "bull sessions" both in simulations and during actual space missions (Sandal 2001). Experts in group dynamics who work with, and are trusted by the crews, are available on the ground to assist in conducting such sessions during the mission if the need occurs (Palinkas et al. 2000).

Work tasks and schedules might be planned to minimize social and psychological issues, and to ensure and maximize individual and crew performance. Future crews should also be consulted on habitat function and design, including clothing, food, layout, decor, waste management, personal hygiene, privacy, tool and equipment design, and computer hardware and software.

Psychological training is of prime importance for Russian crews. Cosmonauts are involved from the moment of their selection in a series of psychological training designed to prevent the occurrence of severe adjustment problems for a space mission. Cosmonauts are tested in simulators, but also in real-world stress situations such as parachute jumping and remote survival missions (Figure 6-18). These missions are planned so that they are as real and dangerous as possible. The Russians say that this type of training develops self-confidence, discipline, and steadiness during an unexpected or emergency situation.* Training in stressful situation is also intended to make sure the crew works together as a harmonious, well-coordinated unit.

Figure 6-18. Training of ISS crewmembers during a survival mission after a Soyuz capsule landing in a remote area in Russia. As part of a generalized stress training, Russian crews are deposited in extremely hostile environments and survive only by their own wit and endurance. There are no rescue teams to help out if trainees go into trouble. (Credit NASA)

5.2   Support

5.2.1   Soviet/Russian Experience

The Soviet/Russian experience in space is extraordinary from a psychological point-of-view. The level of support which cosmonauts receive from start to post-finish is undoubtedly a factor in the success of their long-duration mission program. On the other hand, the U.S. psychological support has been minimal until the problems encountered by the U.S. astronauts during the NASA-Mir program (see Table 6-02). The psycho-social support program of the Russian may not be used exactly as is by the other partners of the ISS because of the culture gap, but it is definitely a starting point.

---

* Every cosmonaut makes day and night jumps from different altitudes, while performing tasks that become successively more difficult. For example, they may be required to carry on a radio conversation, identifying locations on the ground, *before* opening their parachute. This interest for parachuting can be traced back to Yuri Gagarin's day where the cosmonauts bailed out of the Vostok before landing (cited by Collins 1990).

From beginning of training to postflight, cosmonauts are constantly monitored for stress and psychological symptomology. They are given a battery of psychological tests, psychiatric interviews and are thoroughly tested for compatibility. The Psychological Support Group, a specialized cadre of psychologists, military, and civilian space personnel conduct the testing and monitoring. Drugs to regulate behavior, biofeedback, self-hypnosis, and relaxation strategies are all used in support of this effort (Kanas 1991).

During the flight, there is an ongoing monitoring of voice communications by psychologists in Mission Control in order to assess crew tension, cohesion and morale, and to look for potential interpersonal problems. An analysis of voice patterns of the cosmonauts is first performed on Earth during both stressful and non-stressful activities. These are compared to vocal patterns taken on the space station to check for stress. Two-way video observations also are used to interpret facial expressions and body language for signs of stress.

In addition, cosmonauts are routinely sent personal items and different recreational materials via Progress re-supply capsules. They also receive constant reminders of Earth via news, books, audio, and video material,[*] and have frequent contact with family and friends over private communication links (Figure 6-19).

The psychological support does not end with the flight, either. Returning cosmonauts are helped to adjust to new fame and to reintegrate with their families after such a long absence. This is achieved through counseling, drugs, and debriefings as required (Kanas 1991).

5.2.2   ISS Psychological Support

The NASA Psychological Services Group started in 1994 for supporting the stay of the U.S. astronauts on board Mir during the NASA-Mir program. Before this, NASA had focused far too little attention on psychological problems and their ramifications. This is partly due to the fact that there has been little need because of the short-duration of the space missions and partly because of a "technology and hardware first" attitude (Santy 1994). This has been changing recently though, and NASA is beginning to realize the implications of an astronaut succumbing to the

---

[*] After six months spent on board the ISS, the Expedition-6 crewmembers made an interesting description of their first sensations after their Soyuz landing. "When the hatch first cracked open, the smells of spring on the steppes and the sounds of birds overwhelmed us. Real earthy smells because we'd stirred up a fair amount of dirt when we landed and then we rolled and were dragged a bit. So you had this fresh dirt smell, which was just a beautiful smell... and it had a little bit of crushed grass in it, too. Then the next thing that hit us were all the birds chirping... It was just music to our ears."

stresses of her/his environment. The NASA Psychological Services Group is composed of behavioral scientists and psychologists, who learned significantly from the analog environments, from the many years of experience of the Russian psychologists, and from sharing lessons and experience with other International Partners (CSA, ESA, NASDA).

Figure 6-19. U.S. Astronaut Norm Thagard in his sleeping quarters on board Russian Space Station Mir. (Credit NASA)

Preflight psychological support begins when crewmembers deploy for training in Star City. Once there, tasks include advocating for improved conditions and resources, assisting with family contacts, and organizing off-hours recreation. This help is also directed toward all support personnel that follow the crewmembers with deployment.

The Group is also involved in "vehicle" issues that could later affect in-flight psychological support, such as habitability and stowage, acoustics and vibration, food variety and storage, and crew quarters. Similarly, it is involved in "human" issues, such as the work and rest schedules, language training, and culture training. Crewmembers are also familiarized with potential in-flight psycho-sociological issues. Immediate family is involved in these trainings, and informal meetings serve in the preparation of off-hours onboard activities by selecting movies, books, and hobbies of interest for each crewmember.

A Family Support Office is also created, which includes representatives from the Psychological Services Group, the Astronaut Office, and the Astronaut Spouses Group. This structure has a critical role for supporting families during all phases of a mission, and serves as a liaison role with the space agency. Through this organization, the space agency maintains regular contact with family (providing for example the assistance of a family member of another astronaut, or "escort"), and provides information about the mission (especially contingencies)

In-flight monitoring of onboard activities by the Psychological Service group ensures that there is a balanced regime of meaningful work and rest. Monitoring of the astronauts includes a questionnaire on mood, sleep and stress, and countermeasure usage and effectiveness. There are daily private communications between crewmembers and flight surgeons and, less frequently, with psychological support personnel. Some cognitive assessment methods, and well as behavioral and fatigue assessment tools are being developed to look for potential interpersonal problems.

In-flight support activities include surprise presents and favorite foods sent up via Progress and Shuttle, two-way communication with family and friends on the ground via private audio-video links, special family conferences on holidays (e.g., birthdays, Mother's Day), communications with friends, scientists, actors, and artists, audio and video news and sports news, e-mail, amateur radio communication, and onboard recreational software and audio-video material for leisure time use. A computer-based family picture album of spouses, friends, and co-workers is also proposed.

After the flight, debriefings are implemented with both the crewmembers and their family to help them readjust to their life on Earth.[*] These meetings are also used to assess the overall psychological health of the individuals and as a countermeasure for addressing residual intra-crew, crew-family, and ground-crew tensions that may develop during long-term missions. The other objective of these meetings is to assess the practical value of the current psychological preparation and support, and to obtain recommendations for improving the psychological support for following crews (Ark and Curtis 1999).

---

[*] Susan Helms, a crewmember of the ISS Expedition-2 said: "Before I went up on the ISS for five and half months, I moved out of my place, put all my possessions in storage, and moved into the astronauts crew quarters earlier than most people do. I didn't want telephone or credit-card bills, or anything except a bank account where my paycheck could go. I figured if I didn't have a home back here to worry about, the ISS could become my home. (…) I wanted it to be like a military deployment, like Navy guys who go out on a ship for 6 months and put all their stuff in storage."

5.2.3   Unsolved Issues

Despite their experience of long-duration missions, and the attention given to selection of cosmonauts, psychological training and support, numerous reviewers have pointed out that cosmonauts have faced periods of depression (Kanas 1991). For example, Cosmonaut Valentin Lebedev (1988), after 116 spent in orbit, wrote the following depressed thoughts in his diary: "Humming to myself, I float through the [Salyut] station. [...] Is it possible that some day I'll be back on Earth among my loved ones, and everything will be all right?" This example shows that there are serious psychological and social disturbances during long-duration spaceflight (Figure 6-20). No psychological selection strategy by itself will exclude that possibility.

Another source of concern is that we don't know what is going to happen when astronauts will embark on a 2- to 3-year trip to Mars or another planet. The isolation during such a mission is unique, since even the Earth will not be visible by the crew, and there will be long delays in communication. Crewmembers will have to deal with psychiatric problems themselves with no possibility of evacuating an affected individual. One thing that we know, or should know, is that the future of space exploration will require increased input from the psychological and social science community.

Careful attention to selection, training, and organizational functions should permit small groups of individuals to live and work effectively in space for continuous periods of several months or years. But there are enormous gaps in our understanding of how the multiple, complex behavioral factors operate independently to influence the behavior of individuals and groups.

It is therefore necessary to continue the research in this area, as recommended in a report by the Space Studies Board of the National Research Council entitled "A Strategy for Research in Space Biology and Medicine in the New Century" (1998).

It will be also necessary to develop more effective countermeasures to address the individual, group and cultural issues involved in these space missions. In particular, the following countermeasures need to be addressed further:

- Maintaining the presence of behavior and performance specialists through all phases of space mission design,
- Selecting full mission crews and critical ground personnel as a team,
- Training the astronauts in psychological method,
- Embedding tests of cognitive, emotional and behavioral performance in functioning mission hardware and experiments,
- Greater use of simulators for training on board the mission,
- Further development of self-report tools (diaries, personal logs, computer files),

- Developing virtual environments and telescience to address behavior and performance aspects of missions,
- Using ground-based analogs and simulators for selection and training,
- Further training of mission and ground crews together.

Behavioral and social problems are regarded to date as obstacles or "show stoppers" to long-term space missions. Adequate psychological selection, training and support can minimize these problems. Some research is needed, however, since many factors involved in personal and group dynamics in a hostile environment are still unknown. Even more important from a life science perspective, it seems likely that entirely new principles of human interaction and group dynamics will emerge as a result of such research to ensure effective human behavior in space environments and its analogs.

Figure 6-20. Cosmonauts prior to launch on a Soyuz vehicle. (Credit CNES)

6   REFERENCES

Ark SV, Curtis K (1999) *Spaceflight and psychology.* Psychological support for Space Station missions. Behavioral Health and Performance Group, NASA Johnson Space Center

Bechtel RB, Berning A (1991) A. The third-quarter phenomenon: Do people experience discomfort after stress has passed? In: *From Antarctica to Outer Space: Life in Isolation and Confinement* Harrison AA, Clearwater YA, McKay CP (eds) New York: Springer Verlag, pp 261-266

Bishop S (1997) *Psycho-Sociological Issues of Spaceflight.* Advanced Lecture at the International Space University, Summer Session in Houston, Texas

Bluth BJ, Helpie M (1987) *Soviet Space Station as Analogs* [with Mir update]. NASA Grant NAGW-659. Washington, DC: NASA Headquarters

Bondar R.L. (1994) Space qualified humans: The high five. *Aviation, Space and Environmental Medicine* 65: 161-169

Bui L, Wong J (2002) Intercultural study of personal space. Presentation at the International Space University Summer Session 2002, Pomona, CA

Burrough B (1998) *Dragonfly.* New York, NY: Harper Collins

Campbell AE (1985) *Multi-Cultural Dynamics in Space Stations.* Paper presented at the 36th International Astronautical Congress, Stockholm, Sweden

Chaikin A. (1985) *The loneliness of the long-distance astronaut.* Discover, February 1985 issue

Collins M (1990) *Mission to Mars.* New York, NY: Grove Weidenfeld

Connors MM, Harrison AA, Akins FR (1985) *Living Aloft: Human Spaceflight Requirements for Extended Spaceflight.* Washington, DC: NASA Scientific and Information Branch

Fraser TM (1968) Leisure and recreation in long duration space missions. *Human Factors* 10: 453-488

Hall ET (1966) *The Hidden Dimension: Man's Use of Space in Public and Private.* London: Bodley Head

Harrison AA, Connors MM (1984) Groups in exotic environments. In: *Advances in Experimental Social Psychology.* Berkowitz LZ (ed), New York, NY: Academic Press, Volume 189, pp. 49-87

Helmreich RL, Wilhelm JA, Runge TE (1980) Psychological considerations in future space missions. In: *Human Factors of Outer Space Production.* Cheston S, Winter D (eds) Boulder: Westview Press, pp. 1-23

Kanas N (1991) Psychosocial support for cosmonauts. *Aviation, Space and Environmental Medicine* 62: 353-355

Kanas N, Manzey D (2003) *Space Psychology and Psychiatry*. El Segundo, CA: Microcosm Press, and Dordrecht, The Netherlands: Kluwer Academic Publishers

Kelly AD, Kanas N (1993) Communication between space crews and ground personnel: A survey of astronauts and cosmonauts. *Aviation, Space and Environmental Medicine* 64: 795-800

Kubis JF, McLaughlin EJ, Jackson JM, Rusnak R, McBride GH, Saxon SV (1977) Task and motor performance on Skylab missions 2, 3, and 4: Time motion study, experiment M151. In: *Biomedical Results from Skylab*. Johnston RS, Dietlein LF (eds) Washington, DC: NASA SP-377, pp 136-154

Lebedev V (1988) *Diary of a Cosmonaut: 211 Days in Space*. College Station, TX: Phytoresource Research Inc, Information Service

Linenger JM (2000) *Off the Planet*. New York, NY: McGraw-Hill

Natani K (1980) Future directions for selecting personnel. In: *Human Factors of Outer Space production*. Cheston S, Winter D (eds) Boulder, CO: Westview Press, pp. 25-47

Nicholas JM, Foushee HC (1990) Organization, selection, and training of crews for extended spaceflight: Findings from analogs and implications. *Journal of Spacecraft and Rockets* 27: 451-456

Oberg JE (1981) *Red Star in Orbit*. New York, NY: Random House

Palinkas LA (1986) Health and performance of Antarctic winter-over personnel: A follow-up study. *Aviation, Space and Environmental Medicine* 57: 954-959

Palinkas LA (1991) Group adaptation and individual adjustment in Antarctica: a summary of recent research. In: *From Antarctica to Outer Space: Life in Isolation and Confinement*. Harrison AA, Clearwater YA, McKay CP (eds) New York, NY: Springer-Verlag, pp 239-251

Palinkas LA, Gunderson EKE, Johnson JC, Holland AW (2000) Behavior and performance on long-duration spaceflights: Evidence from analogue environments. *Aviation, Space and Environmental Medicine* 71, Suppl 1, A29-36.

Palinkas LA, Gunderson EKE, Holland AW, Miller C, Johnson JC (2000) Predictors of behavior and performance in extreme environments: The Antarctic Space Analogue Program. *Aviation, Space and Environmental Medicine* 71: 619-625

Robins LN, Helzer JE, Weissmann et al. (1984) Lifetime prevalence of specific psychiatric disorders in three sites. *Archives in General Psychiatry* 41: 949-958

Rose RM, Helmreich RL, Fogg L, McFadden T (1993) Assessments of astronaut effectiveness. *Aviation, Space and Environmental Medicine* 64: 789-794

Sandal GM, Vaernes R, Ursin H (1995) Interpersonal relations during simulated space missions. *Aviation, Space and Environmental Medicine* 66: 617-624

Sandal GM, Vaernes R, Bergan T, Warncke M, Ursin H (1996) Psychological reactions during polar expeditions and isolation in hyperbaric chambers. *Aviation, Space and Environmental Medicine* 67: 227-234

Sandal GM (2001) Psychosocial issues in space: Future challenges. *Gravitational and Space Biology Bulletin* 14: 47-54

Santy PA, Holland AW, Faulk DM (1991) Psychiatric diagnoses in a group of astronauts. *Aviation, Space and Environmental Medicine* 62: 969-971

Santy PA (1994) *Choosing the Right Stuff. The Psychological Selection of Astronauts and Cosmonauts*. Wesport, Connecticut: Praeger

Santy PA (1997) Behavior and performance in the space environment. In: *Fundamentals of Space Life Sciences*. S Churchill (ed) Malabar, FL: Krieger Publishing Company, Volume 2, Chapter 14, pp 187-201

Smith S, Haythorn WW (1972) Effects of compatibility, crowding, group size, and leadership seniority on stress, anxiety, hostility, and annoyance in isolated groups. *Journal of Personality and Social Psychology* 22: 67-79

Stuster JC (1996) *Bold Endeavors: Lessons from Polar and Space Exploration*. Annapolis, MD: Naval Institute Press.

Stuster JC, Bachelard C, Suedfeld, P (1999) *In the Wake of the Astrolabe: Review and Analysis of Diaries Maintained by the Leaders and Physicians at French Remote-Duty Stations*. Technical Report 1159 for the National Aeronautics and Space Administration. Santa Barbara, CA: Anacapa Sciences, Inc.

Stuster JC, Bachelard C, Suedfeld P (2000) The relative importance of behavioral issues during long-duration ICE missions. *Aviation, Space and Environmental Medicine* 71: A17-A25

Taylor DA, Wheeler L, Altman I (1968) Stress reactions in socially isolated groups. *Journal of Personality and Social Psychology* 9: 369-376

Weybrew BB, Noddin EM (1979) Psychiatric aspects of adaptation to long submarine missions. *Aviation, Space and Environmental Medicine* 50: 575-580

Additional Documentation:

A Strategy for Research in Space Biology and Medicine in the New Century (1998) Space Studies Board, National Research Council. Washington, DC: National Academy Press

From Antarctic to Outer Space: Life in Isolation and Confinement (1991) A Harrison, Y Clearwater, C McKay (eds.). New York, NY: Springer-Verlag

International Workshop on Human Factors in Space (2000) *Aviation, Space and Environmental Medicine* 71, Number. 9, Section II, Supplement

Lessons Learned from SFINCSS-99 and its Application to Behavioral Support Program (2002) NASDA TMR-02002

Review of NASA's Biomedical Research Program (2000) Committee on Space Biology and Medicine, Space Studies Board, National Research Council. National Academy Press

Chapter 7

OPERATIONAL SPACE MEDICINE

Despite careful screening, extensive training, and aggressive countermeasures against the physiological challenges encountered during spaceflight, incidents of ambulatory illness and medical emergencies are a certainty for both short and long duration space missions. In this chapter, strategies that evaluate the probability of medical events and the medical hardware, procedures, and physician and surgeon resources needed to mitigate them are explored. These risk mitigation strategies are used as the basis for design of health care procedures and facilities in space.

Figure 7-01. The roles of Nurse Christine Chapel (Majel Barrett) and Dr. Leonard 'Bones' McCoy (DeForest Kelley) were a vital part of the success of the original Star Trek series. (Credit Paramount Pictures)

1   WHAT IS IT?

As humans establish a permanent presence in space, whether it is on a space station or planetary bodies, it will be imperative that health care be provided to workers, scientists, and astronauts. The required medical facilities, procedures, and expertise needed to treat these crewmembers are unique to the constraints and unique stresses of space (Figure 7-02), and to the physiological changes described in the preceding chapters.

One essential aspect of a health maintenance facility is its interrelation to other life sciences activities. Experience over the past century in the development of modern medicine has shown a strong correlation between optimal medical care and scientific investigation. This concept also applies to space medicine. Doing so positively affects not only the quality of care but also the quality of research in space physiology.

Figure 7-02. One astronaut makes a dental examination on one of his crewmates during a Skylab mission. (Credit NASA)

As with any other medical field on Earth, space medicine involves proactive and reactive care of humans to optimize physical, physiological, and mental well being, although within the unique constraints of an extreme and unique environment. Unlike on the ground, the first priority is to support the mission, in this case the task of spaceflight. Ensuring the health and safety of crewmembers is in a way a secondary objective, but is necessary for fulfilling the first. To put the principle in context, it must be remembered that if the primary goal were to keep a select group of individuals as healthy and safe as possible, they would be kept safely on the ground. Spaceflight is inherently risky. The closest analog to space medicine would be aviation medicine as practiced by the military flight surgeon. Again, mission assurance is the first priority of all operational support (Barratt 1995).

Therefore, space *medicine* is different from space *physiology*: many of the physiological changes to weightlessness are just adaptive, not necessarily pathologic. Certainly there are some adaptive processes, such as bone demineralization, and environmental exposures, such as radiation, that may be considered pathologic, since without countermeasures they may eventually compromise health during a mission. Also, many of the adaptive responses to weightlessness, such as cardio-vascular, muscular, and neuro-sensory deconditioning, become maladaptive on return to normal gravity. It has often been said that if one did not have to return to Earth, low Earth orbit would be a great place! Finally, even a non-maladaptive and non-pathologic change may alter the way in which a given illness might present and be managed, causing space medicine practitioners to creatively reassess their diagnostic and therapeutic processes (Barratt 1995).

1.1 Objectives

"Il vaut mieux prévenir que guérir" ("an ounce of prevention is worth a pound of cure"). The wisdom of this adage is profound with regard to planning activities in the hostile, isolated environment of spaceflight (Holland and Marsh 1994).

Selection of the best-fit individuals is the first step of a health maintenance program. The medical requirements for the selection of astronauts will be detailed in subsequent sections of this chapter.

Once the personnel have been selected, the second component is *Prevention*, that is, the maintenance of physical and mental health. Considerations include physiological status monitoring, nutrition and stress management, safe waste management, hygiene, medical record keeping, environmental monitoring, exercise devices and medical research facilities, assurance of a suitable sleep environment, recreation and entertainment, social support aids, and communication with family and friends.

When Selection and Prevention are unable to mitigate the deleterious effects of spaceflight, *Countermeasures* are used. Countermeasures for motion sickness, postflight orthostatic intolerance, bone demineralization and muscle atrophy, and psychological issues have been described in the previous chapters.

When Selection, Prevention, and Countermeasures are unable to prevent or mitigate illness or injury, then *Treatment* is used.

In addition to prevention and treatment, *Rehabilitation* should be considered so as to enhance optimal crew productivity and return to operational capability.

Consequently, the types of care that a health maintenance facility must provide on a minimal basis fall into these five categories: selection, prevention, countermeasures, treatment of disease and injury, and rehabilitation.[*]

---

[*] Countermeasures and rehabilitation have already been covered in previous chapters, in particular regarding the bone and muscle systems (see Chapter 5, Section 6) and the psychological issues (see Chapter 6, Section 5).

1.2   Risk Assessment

Risks are the conditional probability of the occurrence of an adverse event from exposure to the space environment. Such exposure can result in dysfunctional physiological or behavioral adaptation that could lead to increased injuries, illness, loss of life, or loss of mission objectives.

Injury is the most likely debilitating or potentially life-threatening process, if personnel are young, healthy individuals. There are, however, certain medical and surgical emergencies that affect even young people, such as appendicitis, perforated ulcer, renal stones, and subarachnoid hemorrhage. The Polaris submarine experience from 1963 to 1973 revealed 269 surgical cases in 7,650,000 man-days. There have been approximately 21 cases of appendicitis, 17 of which were successfully treated with antibiotics, and 4 of which resulted in death (Hamilton 2002).

When reviewing the medical experience of the U.S. Navy, the Antarctic winter-over statistics, and the Soviet/Russian space experience on long-duration missions, it can be reasonably expected that a critical medical event requiring surgery (e.g., appendectomy) in a permanent space station of seven astronauts would occur every 14 years (Hamilton 2002). Given the fact that there has been a quasi-permanent human presence in space since 1986 (first Mir and then ISS), surgery in space could soon be a necessity. Surgery is the only alternative when antibiotics fail and is the primary treatment on Earth in a non-remote setting. If the health maintenance facility is incapable of providing surgical care, the workers and scientists, as well as the public must be aware of such a possible decision.

The key issue is to prevent surgeries by detecting the presence of potential diseases. However, most today clinical tests are used to screen pathology in patients. These tests have a poor ability for detecting disease in very healthy individuals. In addition, there are no tests that are designed to select-out the occurrence of pathology over the next three years.[*] This is a serious problem for selecting a crew for a Mars Mission.

The Institute of Medicine of the National Academy of Sciences (2001) has recently published a report on the medical risks of long missions. The top, number one conclusion by the institute is sobering: "Space travel is inherently hazardous. The risks to human health of long duration missions beyond Earth orbit, if not solved, represent the greatest challenge to human exploration of deep space", the report says. Furthermore, the development of solutions "is complicated by lack of a full understanding of the nature of the risks and their fundamental causes."

---

[*] As Dr. Arnauld Nicogossian, NASA's director of life sciences, has pointed out: "Never before has medicine been called upon to certify that an individual will be healthy enough to perform for three years following the examination" (cited by Collins 1990).

The authors of this report, a panel of 14 medical doctors, clinical psychologists, and health care specialists, are well aware that during such missions, all risks cannot be predicted. They point out: "The successes of short-duration space missions may have led to misunderstanding of the true risks of space travel by the public. Public understanding is necessary both for support of long-duration missions and in the event of a catastrophe". The public must be prepared for the possibility that "all countermeasures may tragically fail, that a crew may not return from a prolonged mission, or that individuals may not be able to function physically or mentally upon their return", the study group warned.[*]

2   ASTRONAUT SELECTION AND TRAINING

2.1   Crew Position

The first astronauts and cosmonauts were jet aircraft pilots. Physical and psychological screenings were intense, due to the hazardous nature of pioneering missions. Then the prime emphasis shifted away from flight experience toward superior academic qualifications. Some applications were invited on the basis of educational background alone. These were the scientist astronauts, so called because the applicants who met minimum requirements had a doctorate or equivalent experience in the natural sciences, medicine, or engineering.

The first group of Astronaut Candidates for the Space Shuttle program was selected in January 1978 and included 20 Mission Scientist astronauts and 15 pilots. Six of the 35 were women. Since then, candidates are selected as needed, normally every two years. Both civilian and military personnel are considered for the program. Pilot astronauts serve as both Space Shuttle commanders and pilots. During flight, the commander has onboard responsibility for the vehicle, crew, mission success, and safety of flight. The pilot assists the commander in controlling and operating the vehicle and may assist in the deployment and retrieval of satellites using the remote manipulator system (RMS), referred to as the *robotic arm*.

Mission Specialist astronauts work with the commander and the pilot and have overall responsibility for coordinating Shuttle operations in the following areas: Shuttle systems, crew activity planning, consumables usage, and experiment operations. Mission Specialists are trained in the details of the Shuttle onboard systems, as well as the operational characteristics, mission

---

[*] The recent *Columbia* disaster seems to reveal a trend in the public acceptation of space failures. The grief and sadness were immense at the news of the tragedy. However, compared with the *Challenger* tragedy that occurred 17 years earlier, and which also took the lives of seven astronauts, much less interest was given to the investigation by the media and the public during the weeks that followed.

requirements and objectives, and supporting equipment for each of the experiments conducted on their assigned missions. Mission Specialists perform extra-vehicular activities (EVAs), operate the remote manipulator system, and are responsible for payloads and specific experiment operations.

Payload Specialists are persons other than NASA astronauts (including foreign nationals) who have specialized onboard duties; they may be added to Shuttle crews if activities that have unique requirements are involved and more than the minimum crew size of five is needed (Figure 7-03).

Participants are invited or paying guests for a given mission. They have included politicians, journalists, scientists, teachers, and tourists.

First consideration for additional crewmembers is given to qualified NASA Mission Specialists. When Payload Specialists are required, NASA, the foreign sponsor, the scientific investigators, or the designated payload sponsor nominate them. All applicants must have the appropriate education and training related to the activities that they would perform while in space.

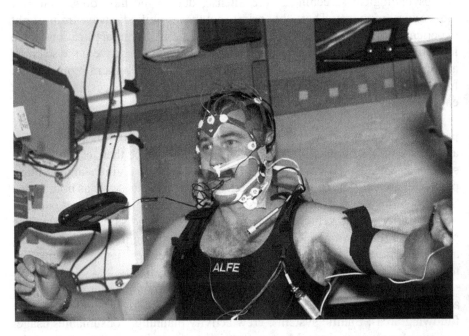

Figure 7-03. Mission and Payload Specialists are performing specific intra- or extra-vehicular activities. This photograph shows Mission Specialist Dave Williams being equipped for a recording session of brain activity during the Neurolab STS-90 mission. (Credit NASA)

2.2 Physical Requirements for Astronaut Selection

All applicants must meet certain physical requirements and must pass NASA space physical examinations with varying standards depending on classification (Table 7-01).

Pilot astronaut applicants must meet the following requirements prior to submitting an application:

- At least 1,000 hours pilot-in-command time in jet aircraft; flight test experience is highly desirable,
- Ability to pass a NASA Class I space physical, which is similar to a military or civilian Class I flight physical,
- Height limitation due to the size of the Space Shuttle and Soyuz flight decks.

Item	Pilots (Class I)	Mission Specialist (Class II)	Payload Specialist (Class III)	Participants to Spaceflight (Class IV)
Distant vision	20/50 or better uncorrected; correctable to 20/20 each eye	20/150 uncorrected; correctable to 20/20 each eye	Correctable to best eye	Same as Class III
Near vision	Uncorrected <20/20 each eye	Uncorrected <20/20 each eye	Not specified	Not specified
Hearing	Each ear: 30 dBA @ 500 Hz 25 dBA @ 1,000 Hz 25 dBA @ 2,000 Hz 50 dBA @ 4,000 Hz	Same as Class I	Better ear: 35 dBA @ 500 Hz 30 dBA @ 1,000 Hz 30 dBA @ 2,000 Hz	Must hear whispered voice at 1 m (hearing aid allowed)
Height	162—191 cm	152-191 cm	Not specified	150-190 cm (Soyuz)
Refraction/ astigmatism	Specified	Specified	Not specified	Not specified
Contraction visual field	15 deg	15 deg	30 deg	Not specified
Phorias	eso>15; exo>8 hyper>2	eso>15; exo>8	Not specified	Not specified hyper>2
Depth perception	No errors in 16 presentations of the Verhoeff stereopter Test	Same as Class I	Not specified	Not specified
Color vision	Pass Farnsworth lantern Test	Pass Farnsworth lantern Test	Not specified	Not specified
Blood pressure	140/90	140/90	150/90 allowed	150/90 allowed
Radiation exposure	<0.05Sv/year	<0.05 Sv/year	Not specified	Not specified

Table 7-01. Medical requirements for NASA Class I, II, III and IV astronaut applicants.

Mission Specialists have similar requirements to Pilot astronauts (Table 7-01), except that the qualifying physical is a NASA Class II space physical, which is similar to a military of civilian Class II flight physical. Height requirements for Mission Specialists correspond to the limits that the space suits (for extra-vehicular activities) can accommodate. Medical requirements for Payload Specialists are even less stringent. Medical guidelines for the selection of "participants to spaceflight" or "space tourists" are currently being developed by space agencies or expert panels (see *Additional Documentation* in the Reference list, Section 6).

Discipline panels evaluate applicants who meet the basic qualifications. Those selected as finalists are screened during a weeklong process of personal interviews, thorough medical evaluations (Table 7-02) and orientation. The recommendations of the Aerospace Medicine Board (which consists of up to 15 physicians qualified in aerospace medicine) and of the Astronaut Selection Board (mostly composed of astronauts), which includes a list of the final candidates, are sent to the NASA Administrator. The Administrator then makes the decision of who will actually join the program.

Selected applicants are designated *Astronaut Candidates* ("Ascans" in short) and assigned to the Astronaut Office at the NASA Johnson Space Center for a 2-year training and evaluation program. Civilian Astronaut Candidates who successfully complete the training and evaluation are expected to remain with NASA for at least 5 years. Successful military Astronaut Candidates are detailed to NASA for a specified tour of duty.

The medical evaluation process for Russian cosmonauts is carried out in three phases: a) an evaluation by various specialists of detailed medical history from a questionnaire decides whether a candidate is fit/unfit for hospital examination; b) a hospital evaluation with the aim to detect any latent pathology, early disease and functional endurance capabilities of various systems, through clinical and laboratory investigations including psychological evaluation; c) a final selection carried out at the Yuri Gagarin Cosmonaut Training Center, in Star City near Moscow. Basically all test results conducted so far are reviewed and involves one-week in-patient evaluation in Moscow at the Central Military Aviation Hospital for military Cosmonaut Candidates and at the Institute of Bio-Medical Problems for civilian Cosmonaut Candidates.

Similar medical evaluations apply to the astronauts (Table 7-02). In addition, after selection, both astronauts and cosmonauts undergo annual certification examination.

Examination	Astronaut Candidates	Cosmonaut Candidates
Medical history	NASA medical survey Questionnaire	Includes surgical history
Physical examination	General physical Anthropometry Muscle mass Pelvic exam and pap smear Procto sigmoidoscopy	General physical Anthropometry Rectal exam Pelvic and uterine exam
Cardio-pulmonary evaluation	History and examination Pulmonary function test Exercise stress test Blood pressure Resting and 24 hr ECG Echocardiogram	History and examination Pulmonary function test Exercise stress test Blood pressure Resting and 24 hr ECG Echocardiogram Phono- and mechano-cardiogram Cardiac cycle analysis
Ear-Nose-Throat Evaluation	History and examination Audiometry Typmanometry	History and examination Audiometry Typmanometry Exo- and endoscopy Vestibular function Optokinetic stimulation
Ophtalmological evaluation	Visual acuity, refraction and accommodation Color and depth perception Phorias Tonometry Perimetry and retinal photography Endoscopy	Visual acuity, refraction and accommodation Color and depth perception Night vision Tonometry Extra ocular muscles Slip lamp exam and endoscopy
Dental examination	Panorex and full dental X-rays within last 2 years	Orthopantomography Electro-odontodiagnosis Vacuum test
Neurological examination	History and examination EEG at rest EEG with photic stimulation EEG with hyperventilation, valsalva and sleep	History and examination Doppler study of cranial vessels
Psychiatric and Psychological evaluation	Psychiatric interviews Psychological tests	Psychiatric interviews Psychometry Personality inventory Sleep monitoring
Radiographic evaluation	X-ray DNS Mammography Medical radiation exposure history and interview Abdominal and urogenital USG	X-ray abdomen, cranium Spine IV punction Genital system Liver scan and biliary scan Excretory urogram Abdominal and urogenital USG

Examination	Astronaut Candidates	Cosmonaut Candidates
Laboratory investigation	Complete heamogram Blood biochemistry Immunology Serology Endocrinology Urine analysis 24 hour chemistry Renal stone profile Urine endocrinology Stool RE Occult blood Ova and parasites	Complete heamogram Blood biochemistry Immunology Serology Urine analysis 24 hour chemistry Stool RE, ova, parasites Analysis of duodenal and intestinal secretions
Other tests	Drug screen Montoux Test Microbiological, fungal and viral tests Pregnancy test Screening for sexually- transmitted diseases	Decompression and hypoxia Centrifuge for +Gz and +Gx resistance Tilt table studies LBNP Heat stress Parabolic flight

Table 7-02. Medical evaluation for astronaut and cosmonaut applicants.

Figure 7-04. Water survival training. (Credit NASA)

2.3 Astronaut Training

Astronaut Candidates receive training at the Johnson Space Center near Houston, Texas. They attend classes on Shuttle systems, in basic science and technology. Mathematics, geology meteorology, guidance and navigation, oceanography, orbital dynamics, astronomy, physics, and physiology are among the subjects.

Candidates also receive land and sea survival training (Figure 7-04), training in scuba diving, and using space suits. All Astronaut Candidates are required to pass a swimming test during their first month of training. They must swim 3 lengths of a 25-m long pool in a flight suit and tennis shoes. The strokes allowed are freestyle, breast, and sidestroke. There is no time limit. They must also tread water continuously for 10 minutes.

Candidates are also exposed to the problems associated with high (hyperbaric) and low (hypobaric) atmospheric pressures in the altitude chambers and learn to deal with emergencies associated with these conditions. In addition, Astronaut Candidates are given exposure to the microgravity of spaceflight. Modified KC-135 or Airbus A-300 jet aircrafts produce periods of weightlessness for 20 seconds (Figure 7-05). During this brief period, astronauts experience the feeling of microgravity. The aircraft then returns to the original altitude and the sequence is repeated up to 40 times in a day.

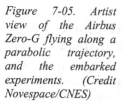

Figure 7-05. Artist view of the Airbus Zero-G flying along a parabolic trajectory, and the embarked experiments. (Credit Novespace/CNES)

Pilot astronauts maintain flying proficiency by flying 15 hours per month in NASA's fleet of 2-seat T-38 jets (Figure 7-06). Pilots build up jet aircraft hours and also practice Shuttle landings in the Shuttle Training Aircraft, a modified corporate jet aircraft. Mission Specialist astronauts fly a minimum of 4 hours per month.

The astronauts begin their formal Shuttle training program during their year of candidacy by reading manuals and by taking computer-based training lessons on the various spacecraft systems ranging from propulsion to environmental control. Then, they begin training in simulators (both fixed-base and motion-base) using generic training software until they are assigned to a particular mission, approximately 10 months before flight. Once they are assigned to a flight, the astronauts train on flight simulators with actual flight-specific training software.

During the last weeks prior to a mission, the astronauts also train with the flight controllers in the Mission Control Center (MCC). Computer links the simulators and MCC in the same way the spacecrafts and MCC are linked during an actual mission. The astronauts and flight controllers learn to work as a team, solving problems and working nominal and contingency mission timelines. Total hours in the simulators for the astronauts, after flight assignment, is about 300 hours.

Figure 7-06. Flight training on board the T-38 jet aircraft. (Credit NASA)

In parallel with the simulator training, several other part-task trainers are used to prepare astronauts for Shuttle and ISS missions. These trainers have varying degrees of fidelity and each serve a particular purpose. For example, the Neutral Buoyancy Laboratory is a large water tank that helps the astronauts in becoming familiar with planned activities and with the dynamics of body motion during extra-vehicular activities.

In addition, all Mission Specialist EVA crewmembers are trained to performed the following Shuttle contingency tasks (if necessary) for each flight: a) failure to close payload doors; b) failure of airlock hatch latches; and c) failures of the robotic arm. There are, however, limitations of the Neutral Buoyancy Training. For example, the viscosity of water produces a different EVA environment than actual space: in water it is hard to initiate motion, but easy to stop; in space it is easy to initiate motion, but hard to stop. Also, gravity is still present in water; it is uncomfortable for crewmembers to work upside-down, and some tools and other items that cannot be made neutrally buoyant are "heavy". Finally, it is not possible to effectively simulate the transitions between day and night. Training is done in a fully-lighted pool, whereas there are 90-min light-dark cycles on orbit. In addition, the thermal environment is constant in a pool, by contrast with the wide temperature extremes in space. Because of these differences, predicting the amount of time to perform an EVA in space is difficult.

Full-sized Shuttle and ISS mockups are also used to train astronauts for onboard systems orientation and habitability training. In these trainers, astronauts practice meal preparation, equipment stowage, trash management, use of cameras, and experiment familiarization. Virtual reality training is also used. Stereo video goggles and headphones allow the astronauts to "see" inside the modules of the ISS and hear in a computer-generated world, and the gloves allow them to move around and grasp objects. This technology seems particularly useful for EVA training to assist in proper positioning for operation tasks while on the end of the robot arm. Crewmembers are also trained to move the robot arm to desired locations using virtual reality.

Another mockup of the forward section of the Shuttle can be tilted vertically, to train for on-orbit procedures, emergency launch pad egress, and bailout operations. When in the horizontal position, this mockup is also used for practicing emergency egress after Shuttle landing (Figure 7-07).

Figure 7-07. Training for emergency egress after landing of the Space Shuttle. (Credit NASA)

Pilots training for a specific Shuttle mission receive more intensive instruction on board a jet airplane modified to perform like the Shuttle during landing. Because the Shuttle approaches landings at such a steep angle (17-20 deg) and high speed (over 500 km/hr), this plane approaches with its engines in reverse thrust and main landing gear down to increase drag and duplicate the unique glide characteristics of the Shuttle. Assigned pilots receive about 100 hours of landing training prior to a flight, which is equivalent to 600 Space Shuttle approaches.

Astronauts who participate in the ISS program receive Russian language training before transferring to the Yuri Gagarin Cosmonaut Training Center for approximately 13 months. Four weeks prior to the Shuttle launch that will deliver them to the ISS, the astronaut returns to NASA Johnson Space Center to train and integrate as part of the Shuttle crew during the final phase. Russian language courses continue at the Gagarin Training Center until the astronaut reaches the level required to begin technical training. The Russian technical training for U.S. astronauts includes theoretical training on the Russian vehicle design and systems, EVA training, scientific investigations and experiments, and biomedical training (Figure 7-08).

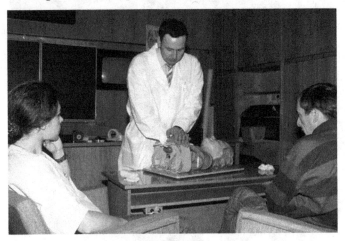

Figure 7-08. During their training, all cosmonauts and astronauts learn the basics of physiology and medicine. (Credit CNES).

Cosmonauts assigned for flight to the ISS using the Soyuz are generally trained over an 18-month period. This training first includes theoretical and active hands-on sessions on Soyuz and ISS simulators, and on the onboard life support systems. It is interesting to note that this phase of training also includes 300 hours of lectures and practical sessions on space physiology and medicine.

When the crew is selected for a specific mission, then the assigned cosmonauts learn how to work as a team. Joint training sessions include space navigation, flight theory, onboard systems of Soyuz and ISS, onboard computers and audio-visual equipment, as well as survival training. All cosmonauts undergo an annual medical evaluation at the Gagarin Cosmonaut Training Center in Moscow and a final pre-launch medical evaluation at the launch site in Baikonur.

3 PREVENTION: HEALTH HAZARDS IN SPACE

In efforts to predict which medical problems might occur in orbit along with their frequency, one can look at the data from populations in analog environments which have the same age range, remoteness, and limitations of available medical resources as astronauts. Investigations of data from surface ship crew, submarine crews and Antarctica studies, however, cannot account for the specific environmental risks unique to spaceflight. The most useful data is prior experience during human space missions. Over the years, in both the U.S. and Russian space programs, several medical problems have arisen, usually with minimal impacts on the mission objectives and timeline.[*] Many of these were successfully treated using the onboard medical facility. Consequently; all of these events have influenced the selection of medical hardware for the current space program, that we will review in the subsequent section.

1. Anorexia (loss of appetite)
2. Space motion sickness
3. Fatigue
4. Insomnia
5. Dehydration
6. Dermatitis (skin inflammation)
7. Back pain
8. Upper respiratory infection
9. Conjunctival irritation (eye irritation)
10. Subungual hemorrhage (bruises under fingernails suit gloves)
11. Urinary tract infection
12. Cardiac arrhythmia (abnormal heart beat)
13. Headache
14. Muscle strain
15. Diarrhea
16. Constipation
17. Barotitis (ear problems from atmospheric pressure difference)
18. Bends (decompression-caused limb pains)
19. Chemicals pneumonitis (lung inflammation from EVA)

Table 7-03. Medical problems most encountered in-flight (from the most frequent to the least frequent).

---

[*] Conversely, more serious manifestations of illness have prompted the early return in at least two cosmonauts, one for high fevers (later diagnosed as chronic prostatitis) and another for cardiac dysrhythmias.

Condition	Frequency	Percent
Facial fullness	226	81.0%
Headache	212	76.0%
Sinus congestion	173	62.0%
Dry skin, irritation, rash	110	39.4%
Eye irritation, dryness, redness	64	22.9%
Foreign body in eye	56	20.1%
Sneezing/coughing	31	11.1%
Sensory changes (e.g., tingly, numbness)	26	9.3%
URI (common cold, sore throat, hay fever)	24	8.6%
Back muscle pain	21	7.5%
Leg/foot muscle pain	21	7.5%
Cuts	19	6.8%
Shoulder/trunk muscle pain	18	6.5%
Hand/arm muscle pain	15	5.4%
Anxiety/annoyance	10	3.6%
Contusions	10	3.6%
Ear problems (predominantly earaches)	8	2.9%
Neck muscle pain	8	2.9%
Stress/tension	8	2.9%
Muscle cramp	7	2.5%
Abrasions	6	2.2%
Fever, chills	6	2.2%
Nosebleed	6	2.2%
Psoriasis, folliculitis, seborrhea	6	2.2%
Low heart rate	5	1.8%
Myoclonic jerks (associated with sleep)	5	1.8%
General muscle pain, fatigue	4	1.4%
Subconjunctival hemorrhage	4	1.4%
Allergic reaction	3	1.1%
Fungal infection	3	1.1%
Hoarseness	3	1.1%
Concentrated or "dark" urine	2	0.7%
Decreased concentration	2	0.7%
Dehydration	2	0.7%
Inhalation of foreign body	2	0.7%
Subcutaneous skin infection	2	0.7%
Chemical in eye (buffer solution)	1	0.4%
Mood elevation	1	0.4%
Phlebitis	1	0.4%
Viral gastrointestinal disease	1	0.4%

Table 7.04. Medical events in the Space Shuttle program reported by frequency from postflight medical debriefings with the crewmembers. This Table include the data compiled for all Shuttle missions from 1988 to 1995 (STS-26 to STS-74). (Source NASA)

3.1 Medical Events during Spaceflight

The most common medical problems encountered during space missions (both U.S. and Russian, short- and long-term) are listed in Table 7-03. More detailed lists of those specific medical events which occurred during Shuttle missions between 1998 and 1995, and during the Mir missions between 1987 and 1996, as well as their frequency, are presented in tables 7-04 and 7-05, respectively. It is important to note that the events in these last two lists were collected from postflight medical debriefings and log files.

Medical Event	Initial Events (n=169)	Recurrences (n=135)
Superficial injury	34	2
Arrhythmia/conduction disorder	30	98
Musculoskeletal	29	NR
Headache	16	8
Sleeplessness	10	9
Tiredness	10	4
Conjunctivitis	4	2
Contact dermatitis	4	3
Erythema of face, hands	4	NR
Stool contents (preflight)	4	NR
Acute respiratory infection	3	NR
Asthenia	3	2
Surface burn, hands	3	NR
Dry nasal mucous	2	NR
Glossitis	2	1
Heartburn/gas	2	NR
Foreign body in eye	2	NR
Constipation	1	NR
Contusion of eyeball	1	NR
Dental caries	1	NR
Dry skin	1	1
Hematoma	1	NR
Laryngitis	1	5
Wax in ear	1	NR

Table 7.05. In-flight medical events for cosmonauts in the Mir program from February 7, 1987 to February 29, 1996. NR = None Reported. (Source NASA)

These data show that the experience so far is that of routine disorders, such as minor respiratory infections (toxic inhalation from chemicals or products involved in investigations, pyrolysis products from fire, propellants), skin disorders such as contact dermatitis, and minor trauma. All these events are extremely common in the industrial setting, and are observed in about the same proportion during Antarctic studies (Table 7-06). Spaceflight, however, is also characterized by microgravity-specific disorders, such as space motion sickness, foreign bodies in the eye (particles do not "settle out"), musculo-skeletal problems and cardio-vascular events.

Group	Number of Cases	Percent
Injury and Poisoning	3910	42.0%
Respiratory system	910	9.7%
Skin and subcutaneous tissue	899	9.6%
Nervous system and sense organs	702	7.5%
Digestive system	691	7.4%
Infections and parasitic disease	682	7.3%
Muscle, bone, and connective tissue	667	7.1%
Other illness	335	3.6%
Mental disorders	217	2.3%

Table 7-06. Illness and injury in Antarctica. Data compiled from stays in Antarctica between 1988 and 1997. Group categories are based on the International Classification of Diseases, 9th Revision. (Adapted from Lugg 2000)

It is interesting to note that sleep disorders, fatigue and insomnia rank respectively 3rd and 4th in the most commonly encountered disorders during spaceflight. Many astronauts experience sleep difficulties, averaging only about six hours of sleep a day in contrast to the seven or eight hours they get on the ground. These disorders could be related to the psychological stress related to isolation and confinement (see Chapter 6, Section 1.1) or to their heavy work schedules.

In space, sleep is fragmented, or otherwise disturbed. Interestingly enough, the change in sleep pattern that typically comes with aging is early waking and fragmented sleep. Optimal alertness during the day and sound sleep at night, valuable qualities on Earth and in space, require proper synchronizing of the human circadian pacemaker (the "body clock"). Thus, researchers seek to better understand how aging and spaceflight affect the mechanisms governing circadian rhythms. The examination of astronauts' circadian alertness using sleep diaries, brain activity (Figure 7-03), and oral temperature rhythms suggests that the endogenous circadian pacemaker seemed to function quite well up to 90 days in space (Monk et al. 2001). However, after about 3 months in space, the influence of the endogenous circadian pacemaker on oral temperature and subjective alertness circadian rhythms is considerably weakened, with consequent disruptions in sleep. While researchers think that aging changes the properties of the body clock, they are not precisely sure how these changes occur.

Shuttle missions typically operate on 23.5-hour days, and astronauts exploring Mars would experience a natural 24.65-hour day. Research has shown that bright light can reset the body clock. In a normal day/night sleep cycle, the level of melatonin, a hormone that regulates the body's sleep activities, will rise about two hours before an expected sleep period to help the body prepare for rest. The levels are even higher during sleep and of course, low during the day. When subjects are in a dim light, such as in spacecraft, and on a different day schedule (such as working night shift or during jet lag) the melatonin cycle loses its normal rhythm. Levels are often high when the person is awake and low when she is trying to rest. This factor makes it difficult to sleep at the scheduled time. A treatment to adjust the internal clock, originally developed for aging people, more recently has proven useful to astronauts preparing for spaceflight (Wright and Czeisler 2002).

By contrast with the Antarctic studies data (Table 7-07), the occurrence of serious infections in space has been very uncommon. It is possible that after a certain "incubation period" no new pathogens are introduced and the astronauts leave in a disease-free environment. In normal conditions, when challenged by pathogens, the immune cells in bone marrow and lymphoid organs initiate and regulate lymphocyte and antibody responses as well as control the production and function of cells in the blood and connective tissues. During spaceflight, the immune system is not challenged any more. Most studies on the immune system in space have tried to determine if cells of the immune system can proliferate in space and maintain a normal immune function. Spaceflight is known to result in significant reductions of both plasma volume and red blood cell mass within days (see Chapter 2, Section 2.2) as well as in abnormalities in human and animal lymphocyte numbers and morphology. Recent studies have shown that lymphocytes do not respond to stimuli that normally cause division, suggesting an impaired ability to proliferate in space. This could have profound implications for the immune and red blood cell formation systems.

Cardio-vascular function
- Increase in weight
- Increase in lipids
- Increase in blood pressure
Immune function
- Delayed reactivity to bacteria
- Increase in virus shedding
- Decrease in T-cell function
Thyroid function
- Increase in TSH
- Increase in T3 production/clearance
Other
- Decrease in hydroxylation of vitamin D (decrease in UV-B radiation)
- Increase in PTH
- Decrease in testosterone

Table 7-07. Physiological response to isolation and confinement in Antarctica.

By analogy, aging also depresses the human immune response (though the change in space is temporary while the change due to aging is not). It is not clear, however, whether aging or other factors that typically accompany aging, such as declining activity, cause this immune system depression. Models of age-related changes in immune function are difficult to find, so microgravity may be a very useful model system to use to increase our understanding of changes due to aging.

3.2   Medical Aspects of Extra-Vehicular Activity

Due to the complexity of the ISS and the need for repair or maintaining satellites in orbit, the need for extravehicular activity (EVA) is evident. The EVA astronaut becomes analogous to the commercial deep-sea diver or the Antarctic field researcher, each facing their respective environmental hazards protected by technology in what are now routine excursions. One difference, though, is that the organism of the astronauts at the time of the EVA is underlying some adaptive changes to weightlessness, as seen in the previous chapters, which may alter their physical condition. Also, the design of the space suit influences many secondary decisions, such as spacecraft or station cabin pressure, medical hardware inventory, and power and consumables requirements (Hills 1985).

The requirements, design and operations of the EVA suits are described at length in Peter Eckart's book published in this series (Eckart 1996). In the present section, we will speculate on the physiological and medical effects of partial or complete failure of this life support system.

To detect and prevent any adverse trends as early as possible during an EVA, the space suit and physiologic parameters are monitored by crewmembers within the spacecraft and controllers on the ground. These parameters include suit pressure, temperature, O_2 consumption, CO_2 partial pressure, electrocardiogram, heart rate, and radiation exposure. However, in the event of a medical emergency, the patient is not immediately accessible to medical treatment. He/she must be moved to the airlock and reenter in the spacecraft, possibly requiring the aid of a fellow crewmember, undergo the re-pressurization cycle, and finally have the bulky space suit removed to whatever degree is necessary to accommodate emergency treatment (Barratt 1992).

Barratt (1992) explains that "a simple but vital concept when discussing closed gas system is that the biological responses of most gases are dependent on their partial pressures, not their overall concentrations. At sea level (where total pressure equals 760 mmHg or 101.3 kPa), with an O_2 concentration of 21% and a partial pressure of O_2 (ppO_2) of 21 kPa (158 mmHg), the respirable atmosphere is said to be normoxic. The same 21% are hypoxic at altitude, where ppO_2 diminishes in step with total pressure, and hyperoxic in hyperbaric atmospheres. Either of these conditions may be detrimental. Similarly, the toxic effects of CO_2 are partial pressure dependent; thus, what may be an acceptable concentration at sea level, e.g., 3%, may be toxic at hyperbaric pressures of a few atmospheres".

In general, the operating pressure inside a space suit is 30-40 kPa, whereas the cabin pressure is 101.3 kPa. The O_2 concentration is 100% in the suit, whereas it is only about 30% in the cabin. The use of lower pressure in the space suit has the advantage that the joints are more flexible. However, this system requires extensive pre-breathing with higher cabin pressures (for more details on this procedure, see Eckart 1996).

A malfunction of the space suit or a failure in the pre-breathing procedure could have severe consequences on the crewmember. For example, a slow leak and partial depressurization of the space suit could result in hypoxia, with such symptoms as loss of color vision, followed by confusion and eventual loss of consciousness.[*] Telemetry will of course alert the controllers before serious symptoms occur, but the onsite medical facility must be prepared to deal with the consequences.

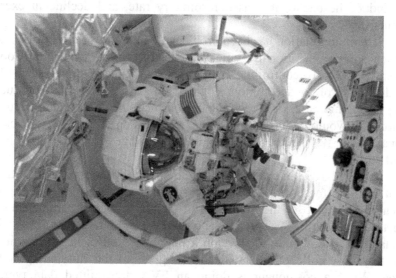

Figure 7-09. Astronaut in the airlock of the ISS preparing for an EVA. (Credit NASA)

The transition from the Shuttle or ISS to the suit pressure is the equivalent of ascending from sea level to approximately 9144-m altitude in an unpressurized aircraft (Heimbach and Sheffield 1985). There is therefore the potential for nitrogen bubbles to move in the tissue and generate localized limb and join pain, a symptom known as *decompression sickness* or "the

---

[*] During the STS-37 mission in April 1991, an EVA crewmember who had returned inside the Shuttle after an EVA with no noticeable events noticed a blood spot on the inside of a glove where a pinhole leak had developed and induced a small skin injury.

bends". Pre-breathing with 100% O_2 prior to initial depressurization during exercise on the cycle-ergometer, along with a final in-suit pre-breathe just prior final depressurization is currently used to limit this problem (Figure 7-09). The increased cardiovascular circulation while breathing 100% O_2 rapidly purges the blood stream of excess nitrogen. A rather short period of this exercise replaces many hours of the standard oxygen pre-breath. However, should the bends occur in orbit, the cabin atmosphere would need to be re-pressurized to maximum cabin atmosphere (110 kPa) immediately and continuing on 100% O_2 (Newman and Barratt 1997). The suit pressure would be increased to 160 kPa to provide some hyperbaric oxygen. This would prevent the suit from being used again. Return to Earth would be performed as soon as practical if symptoms did not resolve (Hamilton 2002).

At sea level, prolonged exposure to 100% O_2 eventually leads to pulmonary O_2 toxicity, manifested by chest discomfort, cough, decrease in tidal volume, and eventually serious pulmonary and respiratory problems. Also, following loss of suit ventilation, high levels of CO_2 (greater than 2 kPa) could induce headache, increased respiratory rate, and decline in exercise performance.

Thermoregulation has been problematic in the early days of EVA, but has been solved operationally with the introduction of the Liquid-Cooling Garment (LCG). By controlling water inlet temperature, this system offers individual control to accommodate the wide variation in heat production during changing workload requirements. Since the water temperature is monitored, it would be possible for the controllers on the ground to decide to terminate an EVA before detrimental heat storage occurs (Newman and Barratt 1997).

In 1992-1993, I participated with my colleagues at MEDES (the French Institute for Space Medicine, based in Toulouse) in the design of a mathematical model for the prediction of the physiological parameters of an astronaut during an EVA. This model was of special interest for ESA, which was planning at that time to develop a new, more efficient space suit. We worked in close collaboration with the designers of the Russian space suit, Zvezda. When a cosmonaut is doing an EVA, transmitted data typically include heart rate, respiratory rate, the rate of O_2 absorption, the CO_2 level, the water temperature in the LCG, and body temperature measured from a sensor located behind the ear. Based on this information, our model was able to predict additional parameters, such as total metabolic heat production (i.e., the efficiency of the LCG), CO_2 expiration rate, mean body temperature and the temperature at various skin location, e.g., head, trunk, arm, hand, leg and feet. The outputs revealed that the average metabolic expenditures of an astronaut during an EVA range from 175 to 250 kcal/h. However, during extensive effort for a limited duration the metabolic expenditures rate can go up to 400 kcal/h. Our model was divided in four submodels: the space suit, the thermal exchanges, and the cardio-vascular and pulmonary systems (these last two submodels used the well-known Stolwijk's model). It was validated using the data collected during 71 EVA simulations in vacuum chambers by 30 cosmonauts and during 8 actual EVAs. The outputs of our model proved quite useful in determining the actual efficiency of the space suit and the comfort of the crewmember (Bagiana et al. 1993).

As discussed in a previous chapter, cardiac dysrhythmias have been occasionally observed on crewmembers during EVA (see Chapter 4, Section 3.3). Although, none of these dysrhythmias have led to the interruption of an EVA, they signify alterations in cardiac function that were not detected prior to the spaceflight. The psychological stress associated with the EVA, the heavy workload, and the dehydration that follows could be responsible for these symptoms.[*]

The most common medical events associated with EVA, though, are mostly localized aches and pains, such as finger bruises, resulting from the rigidity of the suit and the physical work (see Tables 7-04 and 7-05). Perhaps some of these symptoms are related to decompression sickness. Fortunately, none serious injury occurred, despite the hazards of having people locomoting, or being moved on robotic arms between massive objects, some with sharp angles. Perhaps the forces that lead to such events terrestrially, such as vehicle accidents and falls, are not present up there. Crush injuries and ankle or knee ligament injuries are, nevertheless, a possibility.

3.3 Medical Problems of Radiation in Space

Our life on Earth has a most bittersweet relationship with radiation. Our human existence has been vitally shaped by solar radiation by providing us with food and energy. Visible and non-visible radiation from the depths of the universe is responsible for illuminating us with a glimpse of our origins. Radiation also has a deadly face as seen by human use of atomic bombs in past history, by nuclear reactor disasters, and by the fear of skin cancer caused by ultraviolet radiation. Perhaps for these reasons it is often claimed that radiation provides the greatest obstacle ("show stopper") to planetary missions.

In low Earth orbit, crewmembers are protected from ionizing radiation by the Earth's magnetic field. However, they will be exposed to significant heavy ion radiation during interplanetary missions or while inhabiting a Moon or Mars base. This exposure could have disastrous effects on the central nervous system, because heavy ion radiation has been shown to inflict "single hit" damage, even death, on non-dividing cells. These aspects are reviewed in the following section.

---

[*] During a space walk, astronauts can drink from a drink bag located inside the space suit by means of a straw.

3.3.1 Space Radiation Environment

Space radiations, including their physics and the concepts of radiation dose and protection, have been reviewed in details in Eckart's (1996) book. We will only summarize them here, to introduce the Section on the medical issues of space radiation (3.3.3).

There are basically three sources of naturally occurring space radiation that can be hazardous to human spaceflight: the geomagnetically trapped proton and electron environment in the Van Allen belts, galactic cosmic radiation (GCR), and solar particulate radiation (see Figure 2-22).

The Van Allen belts consist of high-energy protons (approximately 1 keV to several hundred MeV) and electrons (approximately 1 keV to several MeV) trapped in the geomagnetic field. The proton belt extends to an altitude of approximately 20,000 km, with peak intensities occurring at approximately 5,000 km. The electron belts extend to an altitude of 30,000 km, with peaks at about 3,000 and 15,000 km. Models of the trapped proton and electron environments have been developed from satellite measurements.

Galactic cosmic radiation (GCR) consists of extremely energetic (up to 1013 MeV) ionized nuclei (or HZE particles, for "high charge and energy ions") ranging from hydrogen to uranium and originating outside the solar system. Models of the GCR environment have been generated from geostationary satellite and high-altitude balloon measurements.

Solar radiation (or solar particle events, SPE) consists of high-energy particles, predominately protons, ejected from the Sun, usually during solar flares. Solar activity has an 11-year cycle, during which a tenfold variation in the frequency of SPE has been observed. No reliable physical model can predict the timing or magnitude of solar flares occurrence with acceptable accuracy. This feature makes SPE a significant hazard in long-duration space travel. Additionally, solar flare activity can substantially increase the fluence of HZE particles, at least up to energies of a few hundred MeV per nucleon.

Radiation exposure in low Earth orbit (LEO), where the Shuttle and ISS orbits lie, comes primarily from the proton and electron belts and GCR. Trapped-radiation exposure increases with altitude and varies with orbital inclination. GCR exposure also varies with orbital inclination. The geomagnetic field provides some degree of protection from SPE, depending on the orbital inclination; flux is almost totally eliminated for a 28 deg orbit and reduced to about 30% of the free space flux for polar orbit.

Exposure at geosynchronous (GEO) altitude will be primarily from bremsstrahlung (X-rays) created by the trapped electrons as they interact with spacecraft shielding (see Figure 2-23). The electron environment at GEO has a diurnal fluctuation, and intensities can increase by several orders of magnitude with magnetic storm activity. GEO is also susceptible to the full exposures from GCR and SPE, as are lunar and interplanetary missions.

3.3.2 Spacecraft Radiation Environment

Incoming radiation from space is modified as it passes through the body of a spacecraft and any additional shielding that may be present (see Figure 2-23). The biological effects of radiation must be determined, therefore, by starting with this modified spectrum. The physical principles by which radiation interacts with matter are well known, but the way to combine these principles to form a good model of the resulting secondary spectrum is not. HZE and SPE can also cause problems in electronic components (the so called "single event upset").

A substantial amount of data obtained from various forms of dosimetry on board Apollo, Skylab, Shuttle, Mir, and now ISS missions (such as the Phantom Torso described in Chapter 2, Section 5.2) has provided measurements of radiation exposures, but it is difficult to extrapolate these data to free space. Nevertheless, with available models and limited spacecraft data, the daily exposure for various mission configurations has been estimated (Table 7-08).

Mission	Absorbed Dose (mGy)	Dose Equivalent (Sv)
Space Shuttle (7-day mission orbiting Earth at <450 km)	2-4	0.005
Space Shuttle (8-day mission orbiting Earth at >450 km)	5.2	0.05
Apollo-14 (9-day mission to the Moon)	11.4	0.03
Skylab-4 (84-day mission orbiting Earth at 430 km, 28.5 deg inclination)	77.4	0.178
Mir (1-year mission orbiting Earth at 400 km, 51.5 deg inclination)	146	0.584

Table 7-08. Radiation dose exposure during spaceflight. The absorbed dose is a measure of the energy deposited in tissue by radiation. The standard unit (SI) of absorbed dose is the Gray (Gy) with 1 Gy = 100 rad. One Gray is the amount of ionizing radiation corresponding to 1 Joule absorbed by 1 kg of material. Note that 1 Gy from high-energy protons is the same as 1 Gy from x-rays (one chest x-ray = 1 mGy). Since the biological effects are different for the various types of radiation, the concept of dose equivalent unit has been introduced, which takes into account a quality factor (QF) depending on tissue interactions with various radiations. The Sievert (Sv) is the dose-equivalent SI unit for humans (1 Sv =100 rem).QF = 1 for most beta, gamma, and x-rays; QF= 2-5 for neutrons; QF = 10-20 for alpha particles. (Sources: Comet 2001, Durante 2002).

For the ISS, the dose-equivalent to the blood-forming organs (BFO) has been estimated to be approximately 0.5-0.6 Sv per year, of which approximately 90% will be from trapped protons. This dose represents about 10 times the maximum allowable dose to a terrestrial worker in a nuclear plant. During SPE, the absorbed dose would be mainly from high-energy protons. HZE from the GCR contribute about 0.2 Sv per year in free space, independent of the amount of shielding. Indeed, the inability to shield effectively against the GCR in free space will be a persistent problem for long-duration missions and for Moon or Mars bases.

As stated previously, solar flare activity can result in heavy ion fluences at a few hundred MeV per nucleon that are up to 10 times higher than the background GCR flux. This fluence may result in a 24-hr exposure in a significant dose (Table 7-09).

Time	Event
06:21	Solar flare observed
13:00	30-day limit exceeded for skin and optical lens
14:00	30-day limit exceeded for BFO; annual limit exceeded for lens
15:00	Annual limit exceeded for skin
16:00	Annual limit exceeded for BFO; career limit exceeded for lens
17:00	Career limit exceeded for skin
18:00	Career limit exceeded for BFO

Table 7-09. Effects of the August 1972 solar flare on a hypothetical crew in orbit at that time with 2 g/cm^2 shielding. BFO: blood-forming organs.

3.3.3   Medical Effects of Radiation Exposure

The next two sections are largely inspired from an essay written by a student from the Space Life Science Department of the International Space University (Bhardwaj 1997, with permission).

When highly charged particles, for example electrons or protons, contact living tissue, they have the ability to ionize molecules like water or oxygen. This reaction produces highly reactive products called free radicals, which can inflict much damage to cellular components. The most significant effects stem from interaction with DNA and other "controlling" macromolecules. There are primarily two effects of radiation, a short-term and a long-term effect. The short-term, also known as the "Acute Radiation Syndrome", may cause nausea, a decrease in blood counts, or even death if the dose is high enough. The long-term effect is known as a stochastic risk and predominantly involves cataract or cancer formation. Tissues vary in response to immediate radiation injury. The tissues with higher cell turnover are the most vulnerable: in descending order of vulnerability: lymphoid, gonadal system, bone marrow, epithelial cells of the gastro-intestinal system, epiderm, hepatic tissue, pulmonary alveloli and biliary epithelium. The DNA effects include decreased mitotic (division) rate, impaired synthesis with abnormal progeny cells and cell lines (cancer). However, a high enough dose will induce necrosis in any tissue.

Acute Radiation Syndrome

Although most space radiation doses will be low, a very large solar particle event can expose astronauts to high-dose radiation, which can produce clinically significant effects. These effects are non-stochastic: the severity of the effect increases with dose above some effective threshold. The Acute Radiation Syndrome (Table 7-10) at sub-lethal doses is characterized early on by transient anorexia, nausea, vomiting, and diarrhea. Later, the survivors may suffer temporary or permanent sterility and cataracts, as well as cancer. Lethal doses lead to bone marrow suppression and immune system malfunction, which leads to death in 30 to 60 days. These high doses lead to severe gastrointestinal disturbances in 1 day to 1 week. Extreme doses can produce central nervous system derangement in a matter of hours. The Acute Radiation Syndrome has been studied extensively in animal models, but the human clinical experience is extremely limited.

Dose (Sv)	Probable Medical Effects
0.1-0.5	No effects except minor blood changes
0.1-1	5-10% subjects experience nausea or vomiting; fatigue for 1-2 days; slight reduction in white blood cells
1-2	25-50% nausea and vomiting, with some other symptoms; 50% reduction in white blood cells
2-3.5	75-100% nausea, vomiting, fever, with anorexia, diarrhea, and minor bleeding; 75% reduction in all blood elements. 5-50% subjects will die
3.5-5.5	100% nausea, vomiting, fever, bleeding diarrhea, and emaciation. Death of 50-90% in 6 weeks. Survivors require 6 months convalescence
5.5-7.5	100% nausea and vomiting in 4 hours. 80-100% die
7.5-10	Severe nausea and vomiting for 3 days. Death within 2.5 weeks
10-20	Nausea and vomiting within 1 hour. 100% subjects will die within less than 2 weeks
45	Incapacitation within hours. 100% subjects will die within 1 week

Table 7-10. Symptoms and time course of Acute Radiation Syndrome.

Long-Term Effects

There are very severe implications when cellular DNA (the "blue prints" of the organism) is affected by either free radicals or by the radiation particle itself. If certain regions of DNA are damaged, then that particular cell may undergo uncontrolled cell division, which later manifests itself in the gross scale as cancer. There are two major compensatory mechanisms that attempt to avoid this outcome. The first method is that if the cell is damaged enough, then it undergoes a morphological set of events from nucleus shrinkage, condensation, and ultimately DNA fragmentation. This sequence, called apoptosis, is basically a mode of carefully orchestrated cell death.

The other mechanisms involve natural molecular level processes, such as the p53 gene. This gene is known as the "guardian of the genome" because it codes for a protein making the cell to pause before it undergoes mitotic division. This pause allows for the DNA repair mechanisms to function so that any damage can be fixed up before the ensuing DNA replication and following cell division. But if this gene is damaged, then cancer may result.

3.3.4   Exposure Limits

The biological effects of ionizing radiation have been extensively studied for almost a century. The data come from studies of controlled irradiation of cell cultures, small and large animals, and non-human primates, as well as from retrospective studies of humans exposed to nuclear weapons blasts, radiation used for medical treatment, and nuclear occupational hazards. However, most of the information has been obtained with low linear energy transfer (LET) radiation such as X-ray, gamma, and electron radiation. Separated clusters of ionization along the path of the primary photon or electron characterize the low-LET radiation. In contrast, high-LET radiation, such as stopping protons, secondary-stopping protons from neutrons, alpha particles, and energetic heavy multicharged particles, is densely ionizing.

It has been known for decades that a given amount of energy deposited by high-LET radiation could be several times more damaging than the same amount of energy deposited by low-LET radiation. Because of the higher relative biological effectiveness of high-LET radiation, a *quality factor* (QF) is applied to occupational doses (in physical units) to obtain a weighted unit for assessment of radiological health risk (or *dose equivalent*). For example, the QF for neutrons from a nuclear reactor would be about 10.

More generally, the assumed linear relationship between absorbed dose and observed biological effect has come into question for HZE particles or high-LET particles in general. Since the manner in which energy is deposited in tissue by HZE particles is so different from that of low-LET particles, this linearity may not apply to HZE particles. Of current interest has been the "microlesion" concept. This theoretical model of the interaction of heavy particles with biological tissue has raised the question of a whole new spectrum of biological damage, including damage to non-dividing cells, particularly the central nervous systems. It appears that the microlesion concept is worthy of further investigation, as there may be significant consequences in long-duration spaceflight (>3 years) if an accidental underestimation of the effect of HZE particles is made.

The assessment of the health risks for various missions and thus the operational limits for such missions are dependent on QF, which in turn will be greatly dependent on the evaluation of biological damages (life shortening, tumor induction, chromosome abnormalities, mutation, and so on). The database using space-type radiation for such assessments is very small. Also, the current knowledge of the GCR hazard is inadequate because of the poor understanding of the effects of HZE particles on biological tissue.

Gender	Age	A	B	C	D
Male	25	34.9	3.10	18.5	1.99
	35	35.2	1.84	18.7	1.20
	45	35.5	1.38	18.9	0.92
	55	35.4	1.12	18.7	0.75
Female	25	35.6	6.24	15.7	2.93
	35	35.2	3.50	15.5	1.70
	45	33.9	2.22	15.1	1.19
	55	30.8	1.73	13.9	0.99

Table 7-11. Cancer morbidity and mortality by age group and gender, with and without radiation. A: lifetime incidence (%) unirradiated; B: additional incidence (%) from 1 Sv; C: lifetime mortality (%) unirradiated; D: additional mortality (%) from 1 Sv. The risk of developing a fatal cancer increases by 1-3% with a dose of 1 Sievert (Sv) of radiation spread over 10 years. 1 Sv at 0.1 Sv/year for 10 years starting at indicated age. (Source: U.S. National Council on Radiation Protection, NCRP)

Table 7-11 shows the best currently available estimates carcinogenic risk for the effects of 1 Sievert (Sv) of radiation spread over 10 years. These radiogenic cancer risk estimates have served in part as the basis for the set of astronaut radiation exposure limits being recommended to NASA. These limits are shown in Table 7-12.

Depth	BFO	Eye	Skin-
30 Days	0.25	1	1.5
Annual	0.5	2	3
Career	1-4*	4	6

*Table 7-12. Astronaut ionizing radiation exposure limits (in Sv) (BFO: blood-forming organs). *The career dose-equivalent is based upon a maximum 3% lifetime risk of cancer mortality. The Total dose equivalent yielding this risk depends on gender and on age at the start of exposure. The career dose equivalent limit is approximately equal to 2 + 0.075 (age 30) Sv, for males, up to 4 Sv maximum; and 2 – 0.075 (age 38) Sv, for females, up to 4 Sv maximum.*

The female astronaut brings special concerns for several reasons. In general, her overall body size and organ sizes are smaller than those of her male counterpart (thus her radiation doses will be higher, given the same amounts of administered activity and similar biokinetics); her gonads are inside of her body instead of outside, and are located nearer to several organs often important as source organs in internal dosimetry (urinary bladder, liver, kidneys, intestines); her risk of breast cancer is significantly higher than that of her male counterpart; and in the case of pregnancy, very little is known about how much activity may cross the placenta and expose the embryo/fetus and the nursing infant.

Within these dose limits, the risk for crewmembers to develop a cancer from space radiation during their lifetime is 3%. However, the genetic effects for crewmembers of childbearing age (especially women) become increasingly possible.

In low Earth orbit, it is unlikely that any astronaut will receive 1 Sv over a career. However, in high orbit or during interplanetary travel, where a rapid evacuation is not possible, an Acute Radiation Syndrome could result from a solar particle event without adequate shielding. It is estimated that the 1 Sv value will be approached during a 2- to 3-year Mars mission, given currently used quality factors and average shielding of 10 g/cm^2. During a Mars mission the total dose could reach 0.8-2 Sv, including 0.2-0.8 Sv per year due to GCR and 0.3 Sv per year due to one large SPE.

It is obvious that the threat of cancer to astronauts after a prolonged mission is a serious question. In Zubrin's opinion, every 0.6 or 0.8 Sv (female and male values respectively) of radiation absorbed over extended periods of time only adds a 1% increased chance of fatal cancer later in life to a 35-year old adult (Zubrin 1996). Thus based on these results, a spacecraft with today's technology in terms of shielding and having a safe-haven interior shelter for the crew in times of solar flares, would be able to house a crew to Mars within acceptable radiation limits. On the other hand, experiments have shown that DNA damage repair occurs at a reduced rate in yeast in microgravity (Pross et al. 1994). These results magnify the radiation risk.

Other stochastic effects involve cataract formation in the eye lenses of the astronauts and central nervous system effects. To date no Russian cosmonauts who have undergone long-duration missions have had cataracts, but the problem is still a potential risk. As a countermeasure, antioxidants in the diet of the astronauts could be extremely helpful of warding off the ill effects of radiation. Vitamin E, vitamin C, and beta-carotene are well-known and effective antioxidants. Novel and new flavanoids, such as venoruton, have also been shown to decrease cataract formation in rat models (Kilic et al. 1996).

Calculations have been made which indicate that the cell nucleus in every single cell in the body would be hit by a proton once every three days, given a nuclear area of 100 mm^2, as a result of the background cosmic radiation (National Research Council 1996). The effects of this are unknown, but there is concern that there may be substantial effects on the nervous system due to changes in transcription rate and function of proteins.

Not much research can be done safely on Earth to investigate these radiation effects, since cosmic rays are difficult to generate, and no one would consent to being exposed to a theoretically fatal dosage. The ISS could provide a good testing ground, since large numbers of astronauts will be exposed to modest amounts of radiation in their 3-month tours of duty, but a full investigation might require waiting decades until these astronauts retire and die either of natural causes or of cancer. Obviously Mars mission advocates have no intention of waiting that long. It actually makes the most sense to accept the radiation risk on the Mars mission, since after all this is a journey into the unknown, and the risk of radiation is mild compared to the dangers that explorers on Earth have faced in the past, and overcome (Reifsnyder 2001).

3.3.5 Radiation Countermeasures

Prevention

Various solution strategies exist which attempt to counteract the deleterious effects of radiation. First of all, as a preflight measure, a bone marrow sample could be obtained from the astronaut so that it could be used to regenerate the bone marrow, should the astronaut be inflicted with cancer at a later time. Another possible problem is that children conceived postflight might have a larger risk of birth defects due to their parent's higher radiation dosage. Thus, a proposed solution could be to cryogenically preserve the ova or sperm, from the female and male astronauts, so that they could be used in the future if and when desired.

Another idea as a measure to decrease an astronaut's chance of getting cancer may be through astronaut selection. Given that certain oncogenes have hot spots that are especially vulnerable to damage and that cancer is not a one-mutation process, but is instead a multiple hit process, then the following potential solution could be considered. As molecular biology techniques are quickly advancing, it is not hard to imagine that certain loci on genes, which are especially vulnerable to becoming oncogenic, may be isolated. It then follows that astronaut candidates who have already had a mutagenic hit on this gene, may have a greater likelihood of getting cancer, and thus may be jeopardizing their future by embarking on an interplanetary mission.

Other solutions have included the use of chemoprophylactic drugs such as WR-2721. This was found to be a radioprotective drug, but the side effects caused vomiting and vasodilatation, which resulted in hypotension; thus being a problem for functional control in spaceflight. However, new drugs such as WR-33278 that binds DNA, and Lazaroid that protects neural tissue, may provide hope in terms of pharmacological solutions.

Monitoring

As already mentioned in Chapter 2 (Section 5.2), passive dosimeters are available on ISS to determine the space radiation dose at specific locations within ISS (Table 7-13). At present, the information from these readers is returned to the ground and analyzed in a laboratory to obtain the Linear Energy Transfer (LET) spectrum. The LET spectrum is then combined with the dose information to determine a corrected total dose.

Active dosimeter systems are also available on ISS. Incidences of charged particles detected by the Real-Time Radiation Monitoring Device are monitored on the ground in real time. Small chambers for biological specimens may be attached to the sensor unit. A tissue equivalent proportional counter and a charged particle directional spectrometer also have the capability for real-time data collection and viewing. They are mounted both inside the habitation module and outside.

	Shuttle	*ISS*	*Agency*
Tissue Equivalent Proportional Counter	X	X	NASA
Charged Particle Directional Spectrometer	X	X	NASA
Passive Dosimeters	X	X	NASA
Small-Size Passive Dosimeter Package	X	X	NASDA
Real-Time Radiation Monitoring Device	X	X	NASDA

Table 7-13. Radiation monitoring tools on the Space Shuttle and ISS. (Source NASA)

Satellite Solar Flare Advanced Warning

In the event of a solar flare, the protection of the crew is vital, and early detection is of extreme importance. However, the electromagnetic radiation (X and gamma rays) would travel at the speed of light from the Sun out towards the planets. If this made up the entire solar particle event, then satellites would be useless because by the time the satellite sensed the event, and sent the signal to the spacecraft at the speed of light, the radiation would have reached the spacecraft before the crew received the message. This is not the case, because the high-energy charged particles that inflict the bulk of the damage do not travel at the speed of light, so a satellite-warning signal would be most useful.

Solar observatory satellites (such as SOHO) observe particles of solar, interplanetary, interstellar, and galactic origins; thus solar winds and flares, as well as cosmic radiation. Since these satellites are closer to the Sun than the spacecrafts in Earth orbit or en route to Mars, a solar event would be first sensed by the satellite and then the warning could be relayed directly to the crew onboard. This would provide life preserving valuable time to the crew, which would then undertake the necessary precautions, as will be discussed in the next Section.

Radiation Damage Repair

The possibilities of modifying the biological damage by radiation deserve further attention. Recent evidence obtained by the cancer research community indicates that the multiphase process of carcinogenesis can be interrupted at various stages. For example, at the DNA damage or initiation phase, free radical scavengers, such as vitamin E and possibly vitamins A and beta-carotene, and vitamin C (ascorbic acid), can protect. Several of the trace elements also have an antioxidant effect; these include copper, iron, manganese, selenium and zinc. Some data indicate that the promotion phase, in which a radiation-damaged cell changes to a potentially cancerous cell cluster and then goes on to the progression phase yielding a tumor, can be interrupted by agents such as dimethylsulfoxide (DMSO) or protease inhibitors. Implementation of the results of studies directed toward early detection of cancer could help improve the prognosis for crewmembers unfortunate enough to contract cancer.

Before a commitment to a Lunar or Martian colony is made, mutagenesis and teratogenesis by high-LET radiation must be extensively studied. Mutation and developmental abnormalities are, like cancer induction, stochastic effects: the severity of the effect is independent of dose, but the probability of occurrence increases with dose. The mutation risk to future generations from expected space radiation doses is apparently fairly low, but the available information is largely inadequate for assessing the teratogenic risks to fetuses (Bhardwaj 1997).

3.3.6 Strategies in Radiation Shielding

In the prevention of high-dose acute radiation exposure, special shielding is the most commonly considered modality. Mass shielding is just the passive ability of bulk mass to inherently shield radiation (Table 7-14). A fundamental property of mass shielding is that the thickness must increase enormously as the energy of the radiation particle increases.

Shield Thickness	*BFO*	*Skin*	*Lens*
0.2 (g/cm^2 Al)	6.0	3.0	1.9
1.0 (g/cm^2 Al)	6.3	3.5	2.4
5.0 (g/cm^2 Al)	8.9	8.0	6.5

Table 7-14. Time (in hour) for radiation exposure to reach the 30-day limit in the blood-forming organs (BFO) and at the skin and eye lens levels, using various thickness of aluminum shielding. (Source NASA)

In situations where it is not possible to de-orbit or lower the altitude of the spacecraft to a protected region of space, such as during the mission to Mars, the vehicle will most likely include a "storm shelter" or "safe haven", where the crew will stay until the radiation has subsided to acceptable level.

For such shelter, the use of food racks and water tanks packed around the walls to absorb the radiation, or a water-filled collapsible cocoon has been proposed. Fortunately most of a solar flare's energy is in alpha- and beta-particles that can be stopped with a few centimeters of shielding.

Cosmic rays are a different story. They are constantly present, coming from all directions. The radiation consists of heavy, slow moving atomic nuclei that can do far more damage to more cells than alpha and beta particles. This radiation requires several meters of shielding for complete blockage, and since the nuclei come from all directions at all times, unlike the brief solar flares that last only a few hours or days, a storm shelter would be insufficient to protect the crew.

Another solution for shielding is to create a magnetic field around the spacecraft being able to deflect solar radiation. Technology using low temperature superconductor coils seems inadequate because too costly in energy. However, new high temperature superconducting coils are promising since they can produce a high energy, low intensity magnetic field (Goldman 1996). One possible concern with this mode is that there are still lingering concerns about the effects of magnetic field exposure to human tissue, especially neural cells.

For added protection in case a very large solar event occurs, partial body shielding of a small amount of bone marrow stem cells can be very effective in raising the lethal threshold. For example, in one study, monkeys that had 1% of their stem cells protected survived a dose that killed all unshielded animals. In the future, *ex vivo* cell storage techniques may allow a bank of shielded bone marrow to accompany astronauts on a long-duration mission.

It should be noted, however, that a 100% efficient radiation shielding system might not be desirable. It is possible that a minimum level of ionizing radiation is necessary to keep the biochemical repair cellular mechanisms in functioning order. This beneficial effect of a low-level exposure to an agent that is harmful at higher levels is called hormesis. Obviously, the shielding technology needs improvement for interplanetary travel, but there may be a non-zero optimum value (Bhardwaj 1997).

3.3.7   Conclusion on Radiation Issues

In space medicine, radiation is often seen as a "show stopper" for a mission to Mars. However, when evaluating all the risks involved in such a mission, it might not be the worst. In addition, technological leaps are being made in the fields of molecular biology and supraconductors, which could provide valuable countermeasure solutions in the near future. Bhardwaj (1997) concluded his essay by the following interesting thought: "the essence of our physical life form originated from matter ejected from supernovae, which is *radiation*. The engine of evolution, which transformed a unicellular organism into a human being, was fuelled by nothing but galactic cosmic rays, which also is *radiation*. The day might come where life will not be possible on planet Earth, because of the dreadful *radiation* of a nuclear bomb. When exploring other planets as a possible refuge, [humankind will] again be confronted with *radiation*. Ironically, the spacecraft carrying the human crew to Mars will also probably use some sort of nuclear propulsion!"

3.4 Conclusion on Space Health Hazards

At this time it is not possible to certify any physiological system to be unaffected by several years at microgravity or to preclude any as a fruitful area of research. At present, we cannot assume that as spaceflight increases from months to years, unanticipated malfunctions will not appear. For this reason, scientists recommend that a reliable database must continue to be established so that new phenomena can be recognized and addressed by research before proceeding to longer flights. To accomplish this, the approach of incremental exposure of humans to microgravity should be continued with careful surveillance during and after exposure.

So few data on space medicine are currently available that any projection for human space missions of several years is only tentative. The physiological effects of short-duration spaceflight are tolerated, or compensated for, by the state of current countermeasures. However, the long-term effects of microgravity or the reduced gravity of Mars on bone and muscle metabolism and on cardio-vascular function remain poorly understood.

The more general problem of the ability of human beings to interact and perform well in a closed, stressful environment assumes novel importance and exigency with extended spaceflights. In addition to the problems of weightlessness and heavy ion radiation, the crew may have to deal with increased microbial density in the cabin air, organic and inorganic toxins (outgassing products), nutritional limitations, and the problems of health care delivery in space. These physical stresses will exacerbate the severe emotional stresses associated with working and living in confined quarters. Many of these problems cannot be studied in terrestrial analogs, and many scientists think that they must be understood in much greater depth during low Earth orbit space missions before a human mission to Mars.

4   TREATMENT: SPACE MEDICAL FACILITIES

Since the onboard medical facility cannot be equipped for all possible eventualities, the supplies and equipment included in its design must be carefully selected for maximum utility. A major task involves ranking candidate diseases and injuries according to their potential impact during a space mission. Typically, the risk associated with an event is defined as the product of its frequency and its consequences. Where an event may have a variety of consequences, an aggregate risk is defined as the sum of the risks of each of the consequence types in common units (McCormick 1981).

A universal maxim in medicine is that "common things occur commonly". This maxim definitely applies to space missions, as indicated by the frequency of minor medical events reported during space missions (see Tables 7-04 and 7-05). This supports the premise that the bulk of the on-orbit medical care will be directed toward more "general symptoms" or routine disorders (Nelson et al. 1990). Space medicine here is analog to the environmental medicine for employees in industrial settings, where consideration is given to the environmental hazards (chemical, physical, biologic, and ergonomic) that may cause sickness, impaired health, or significant discomfort. It is obvious, however, that even minor medical problems may have a great impact considering the cost and risk of maintaining an orbital work force.

In examining the less common but more severe medical problems that might occur, more mission-specific parameters are evaluated. For instance, any toxic substance required operationally must be accompanied by the means to treat inadvertent exposure to it. This includes the toxicity of chemical substances (e.g., hydrazine) or their mixture, and the likely failures of environmental control life support systems and their medical implications. The onboard medical doctor or an individual trained for medical procedures (so called Crew Medical Officer, CMO) must be able to recognize, diagnose, and treat these disorders quickly. It must also be born in mind that each illness or injury is occurring on top of the physiological adaptive changes affecting the multiple body systems aforementioned. Consequently, signs, symptoms, or presentation of various diseases and their treatment may be altered by fluid shifts, electrolyte changes, hemodynamic changes, and so on. And finally, it is for fact that manpower is in short supply aboard spacecrafts and space station. The crew is on its own with the available equipment and a limited support from Earth (Barratt 1992).

There are basically two systems of health care during space missions. The first system is a facility that provides simple first aid, with one or all members of the crew trained in basic care, and minimal equipment. Medical kits used on board the Space Shuttle include an emergency medical kit and a medications-and-bandages kit. Each kit is contained in three fabric packages, and weighs less than 3 kg. Such items as a stethoscope, blood-pressure cuff, sutures, disposable thermometers, and injectible medications are in the emergency pack. Band-aids, adhesive tape, gauze bandages, and oral medicine are in the medications-and-bandages kit. The third kit is an instrumentation pack, which includes a respirator, an intravenous fluid system, and a defibrillator (Figure 7-10).

The second level of health maintenance facility is a dedicated area for first aid and exercise, which includes eventually equipment for treatment of hypobarism. The objectives is to stabilize the injured patient until rescue could occur, treat minor injury, and even carry out some minimal invasive diagnostic studies and simple diagnostic testing. Such a facility requires extended training of a crewmember. Symptoms and clinical signs can be described to physicians on the Earth, who direct treatment giving instruction to the medical doctor or the crew medical officer on board ISS.

Figure 7-10. Emergency medical instrument kit used on board Mir. (Credit CNES).

4.1   Crew Health Care System (CHeCS)

To support the medical needs of crewmembers during ISS assembly and operations, NASA has developed the Crew Health Care System (CHeCS). The CHeCS consists of three primary elements: the Health Maintenance System, the Environmental Health System, and the Countermeasures System. The latter includes the treadmill, the cycle ergometer, and the resistive exercise device described above (see Figure 5-20) for crewmember to exercise and minimize the effects of spaceflight on the body.

The primary purpose of CHeCS is to provide for and monitor the well being of the astronauts in orbit. However, components of CHeCS occasionally may be used to support life sciences research. Similarly, CHeCS may require occasional use of research equipment for periodic assessment of crew health. The CHeCS is thus complemented by the Human Research Facility (HRF), which houses equipment to investigate the effects of microgravity on human physiology. The HRF is composed of two racks, which provides services and utilities to experiments and instruments installed within it. These include electrical power, command and data handling, cooling air and water, pressurized gases and vacuum. For example, computers are used to transmit data from environmental experiments which measure radiation, such as the

Phantom Torso, Bonner Ball Neutron Detector and Dosimetric Mapping experiments. They also transmit data from life sciences experiments and crew psychological surveys.

The ultrasound imaging system (ATL 5000) located in the HRF provides image enlargement of the heart and other organs, muscles and blood vessels (see Figure 4-15). This generic diagnostic research tool is capable of high-resolution imaging in a wide range of applications, both research and diagnostic, such as:

- Echocardiography, or ultrasound of the heart,
- Thoracic ultrasound,
- Abdominal ultrasound, deep organ,
- Vascular ultrasound,
- Gynecological ultrasound,
- Muscle and tendon ultrasound,
- Ultrasound contrast studies,
- Small parts ultrasound.

Also housed in the rack is the Gas Analyzer System for Metabolic Analysis Physiology, or GASMAP. GASMAP is used for periodic assessment of crew aerobic capacity. It analyzes human metabolics, cardiac output, lung diffusing capacity, lung volume, pulmonary function, and nitrogen washout in subjects at rest and during exercise. Other equipments stowed in the HRF include sample collection kits, a continuous blood pressure device, a foot ground interface, and a lower body negative pressure device.

4.1.1   The Health Maintenance System

The above examination of the factors for health hazards during spaceflight has led to the development of specific facilities for the Health Maintenance System. The Health Maintenance System includes the following: a) a defibrillator; b) an ambulatory medical pack; c) a respiratory support pack; d) an advanced life support pack; e) a crew medical restraint system; and f) a crew containment protection kit.

During spaceflight, the accessibility and use strongly influence the success of a mission (or the survival of an individual). Some issues are analog to clinical medicine of Earth, For example, medical waste (e.g., sharp needles) must be carefully disposed, and medications must be tracked and discarded when shelf life is exceeded. In addition, the absorption of oral medications may be sensitive to the adaptive changes of body function, such as the fluid shift or the digestive function, to microgravity. Also, alternate routes of administration (intramuscular, intravenous, nasal spray) may be better than on Earth for some drugs.

In addition, due to microgravity, some of the equipment or procedures must be adapted to the environment of space. Let's now review some of the unique elements of the ISS Health Maintenance System.

Body Restraint System. If the medical hardware is not at proximity, then either it is transported to the patients, or the patient is transported. For transport and medical care in microgravity, proper restraint is required. For maximal efficiency, the restraint function is integrated with diagnostic and therapeutic systems. The schematic in Figure 7-11 illustrates the procedure for restraining Shuttle crewmembers in the event of spinal injury or illness requiring immobilization. A Crew Medical Restraint System can be secured to the ISS structure within two minutes, in order to provide a patient restraint surface for performing emergency medical procedures, such as during advanced life support (see below). It can also restrain two crewmembers during their delivery of medical care.

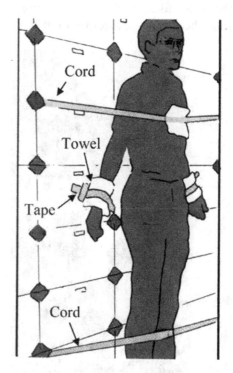

Figure 7-11. Procedure for immobilizing the body of Shuttle crewmembers in case of spinal injury. (Nicogossian and Parker 1982)

Cardiac Defibrillator. Early electrical defibrillation correlates best with survival in event of cardiac arrest. On Earth, applying the charged paddles to the patient's chest normally requires a weight of 11 kg, which is provided by the weight of the rescuer lining over the patient's chest. In microgravity, the rescuer has no weight, and self-adhesive defibrillator pads are used. In addition, insulation from delivered voltage and electro-magnetic interference (EMI) shielding must be considered to protect other crewmembers and sensitive avionics.

Advanced Life Support Pack. This large soft pack stores emergency life saving medications and medical equipment in easily accessible form. Solving the problems of providing emergency cardiac care during spaceflight sometimes requires new life-saving techniques. For example, under normal conditions, cardio-pulmonary resuscitation (CPR) involves repeated applications of force to the patient's chest while ventilating the lungs. In space, however, both patient and rescuer are free-floating. Unable to stabilize the patient on a surface and with no force of gravity to provide weight, the rescuer could not easily perform Earth-bound CPR (Figure 7-12). The problem has been bypassed by modification of the conventional technique. The patient is secured on a restraint and the rescuer is held in place over the patient with a simple harness attached to the restraint. The harness prevents the recoil force of the chest compressions from propelling the rescuer away from the patient, and conventional CPR can be performed.

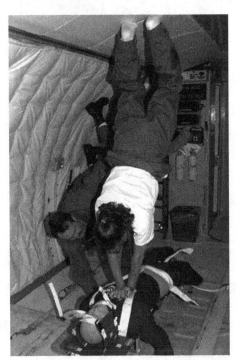

Figure 7-12. Two NASA flight surgeons are trying to perform cardio-pulmonary resuscitation (CPR) on a restrained dummy during the microgravity phase of parabolic flight. (Credit Douglas Hamilton)

4.1.2   The Environmental Health System

This system is used for monitoring the internal environment of the ISS. The Environmental Health System assesses toxicology, water quality, microbiology, and the radiation environments. As seen earlier, the radiation environment is monitored with a variety of dosimeters located inside and outside the ISS (see Table 7-11). The toxicology system includes a volatile organic analyzer, a compound-specific analyzer for combustion products, and a compound-specific analyzer for hydrazine. A water sampler and total organic carbon analyzer enables crewmembers to assess water quality. A surface sampler kit, a water microbiology kit, and a microbial air sampler enable microbiology assessments.

It is important to keep in mind that a substance might be physiologically ineffective at a very low dose, therapeutic at an intermediate dose, and toxic at a high dose. In addition, below a given threshold, a high dose of a substance usually has a greater effect than a low dose; this is known as the "dose-response curve". For space missions, a Threshold Limit Value (TLV) and Permissible Exposure Level (PEL) have been determined for each substance. Furthermore, there is a maximum allowable concentration authorized for each substance in human spacecrafts (Substance Maximum Allowable Concentration, SMAC).

Toxicological issues for human missions include the accidental contacts with the chemicals used in the life support system (leaks or spills) or the fuels for the orientation control systems (hydrazine and nitrogen tetroxide). These fuels are toxic when absorbed through skin or by inhalation, and provoke immediate and violent irritation of nose, throat, eyes, and the respiratory tract. Longer exposure can result in respiratory arrest and lung, kidney, and liver damage, and possible death. For example, 1 mL of hydrazine vaporized in the Shuttle cabin will produce 11.8 ppm (part per million), whereas the SMAC for a 7-day space mission is only 0.04 ppm. Decontamination procedures are especially difficult in microgravity, where the contamination has to be contained. Skin surfaces and eye must be washed thoroughly and the wash water must be contained as well (Figure 7-13).

Figure 7-13. ESA Astronaut Jean-François Clervoy washing his eyes with water in microgravity. (Credit NASA)

4.1.3 Typical ISS On-Orbit Medical Assessment

On 5 December 2002, NASA released the following ISS on-orbit status, which describes the medical operations performed by the ISS crewmembers (Source: http://www.spaceref.com/news):

"All ISS systems continue to function nominally [...]. After wake-up at 4:00am EST, the crew made preparations for more medical assessments, starting with a test of the crew medical restraint system by Commander Ken Bowersox and Flight Engineer Don Pettit. Throughout the day, all crewmembers completed their first general physical fitness evaluation, which checkups on blood pressure and the electrocardiogram during programmed exercise in the Lab. Commander Bowersox, a qualified crew medical officer, assisted with the examination. Readings were taken with the blood pressure/electrocardiograph (BP/ECG provides automated noninvasive systolic and diastolic blood pressure measurements while also monitoring and displaying accurate heart rates on a continual basis at rest and during exercise). Sox plugged it in the defibrillator outlet in the Lab, since it requires special voltage for operation, whereas the CHeCS heart rate monitor, which measures and records heart rate only, can operate on 24 Volts for use on the exercise equipment in the Russian segment, such as the treadmill."

"The first round of the renal (kidney) stone experiment program continued. Don Pettit, who started his turn yesterday, today began his urine sample collections, several times during the day. Flight Engineer Nikolai Budarin had his first day reviewing the computer-based training disk and starting his diet log. Bowersox will join in tomorrow."

"Budarin completed the periodic routine inspection of the condensed water separator in the Service Module. Later, he also performed the routine tasks of life support systems maintenance and in-fight medical system inventory update file preparation."

"Ken Bowersox reconfigured the Lab common cabin assembly air conditioner for the reduced crew size [after the departure of the Shuttle STS-111 crew]."

"Later, Bowersox and Budarin installed the remaining metal oxide canisters for regeneration and conducted a dry-run of the space suit systems, in preparation for next week's spacewalk (12/12/02) [...]."

4.2 Telemedicine

The on-orbit facility is part of a larger integrated health system, consisting of the medical facility itself, the onboard doctor or Crew Medical Officer, and ground medical personnel (flight surgeons, engineering support, and consultants), and a telemedicine link. All must function in a coordinated fashion. A general philosophy is that the onboard Crew Medical Officer serves more of a technician and procedural role, with ground personnel providing decision-making support; essentially a "brains on the ground, hands in orbit" approach. The telemedicine link not only connects the crewmember with the most qualified consultants; it also provides real-time downlink of video and diagnostic procedural images such as ultrasound images. Using the onboard system, video imaging of the eye, ear, nose, throat, and skin is possible. In case of illness or disease, continuous biomedical monitoring (ECG, ppO_2, blood pressure, heart rate) is performed and data are downlinked in real time to ease the requirements on the crew medical officer.

Such telemedicine systems have been and are being utilized in remote terrestrial settings, such as Antarctica stations or ships, during which medical specialists on the ground were able to review cases and provide therapeutic and procedure advice.

Computerized medical databases and medical diagnostic programs are being evaluated for incorporation into the onboard medical facility. Large toxicology databases are already utilized during space missions, accommodating rapid search and information retrieval to offer immediate advice in treating exposure to toxic substances. Future customized medical data management system could include medical databases, diagnostic software, and crew medical records with medication allergies, ill-effects history, and environmental exposures into an integrated health care tracking network (Barratt 1992).

However, the technology and sophistication mentioned above must be balanced against the capabilities and proficiency of the onboard Crew Medical Officer. New technologies may be very demanding in terms of time required, power needed, and complexity required to use them. A well-trained clinician may immediately dismiss an entire pattern of diagnostic considerations which the automated system feels it must process, and even the best video images may not transmit what the crew medical officer can detect at a glance. Technology can never completely substitute for skills in physical diagnosis. For this reason, the level of training of the Crew Medical Officer is a vital consideration in the integrated health system (Barratt 1992).

4.3   Emergency and Rescue

As in terrestrial medical care, the response to an orbital medical event depends on 5 factors: a) the severity of illness or injury; b) the capability of the onboard medical system; c) the ability of surgeon to assist during medical event; d) the level of skill and training of Crew Medical Officer; e) the ease and feasibility of medical evacuation to Earth.[*]

---

[*] An easy way to remember these 5 factors is "SCALE", with letter S for severity, C for capability, A for ability, L for level of skills, and E for ease (Hamilton 2002).

4.3.1   Crew Medical Officer (CMO)

For missions of a few months' duration, the trade-off between the capability for emergency transportation back to Earth and the capability for emergency treatment in space must be studied. If emergency rescue is found to be impossible or impractical, then emergency care capability must be improved. It is then necessary to provide a physician for in-flight care. Such a person could also be a trained astronaut capable of performing other duties including life sciences research. On long-duration missions, e.g., at a lunar base or on a Mars mission, the need for such personnel will increase greatly.[*]

The number of crewmembers aboard the ISS will range from three to seven (when a new habitat module will be delivered), for stays that can last between 90 and 180 days. One mission safety rule states that the crew shall be able to return to Earth in less than 24 hours. This delay is too short for launching a Space Shuttle in a rescue mission. For this reason a Soyuz vehicle is permanently docked to the ISS, and ready to leave in a few minutes (Figure 7-14).[†]

Figure 7-14. Landing of the X-38 capsule at the Kennedy Space Center in Florida. The X-38 was supposed to be the successor of the Soyuz vehicle as an emergency return vehicle from the ISS. (Credit NASA)

---

[*] When the Soviet Union decided to embark the first crew for a 8-month mission, Dr. Oleg Atkov, a cardiologist working at the Moscow Cardiology Institute, was "asked" by the authorities to participate in that mission as a crew medical officer. This was the first flight of a qualified medical doctor in the Soviet space program, presumably in anticipation for the potential medical problems during long flights.

[†] The Soyuz docked to the ISS needs to be replaced by a new one every 6 months, because of the lifetime of its propellants and other life support systems components. This rotation of Soyuz is called a "Taxi flight".

Two astronauts of an ISS Expedition crew are trained as Crew Medical Officers (CMOs). The CMOs on ISS are usually not physicians. They receive 34 hours of medical training completed at least 6 months prior to flight. The other crewmembers (non-CMO) only receive 17 hours of preflight medical training. When on board, the CMO have one hour per month of in-flight computer-based refresher on their medical training, and must complete one emergency simulation per mission. In addition they receive an optional "Space Emergency Medical Training" course at a local community college (20 hours in classroom, 50 hours in clinical setting).

As stated earlier, the CMO must recognize, treat, and stabilized acute injury, as well as prepare the patient for transport in case of an emergency return. CMOs receive cross training in different disciplines so that they could provide surgical assistance, anesthesia support, and diagnostic capability, such as in the laboratory or imaging areas. However, due to their limited training, it is imperative that these physicians have access to consultation with other medical specialists on Earth. One current problem is that communications from the ISS in its current state are limited to about 50% of coverage (Hamilton 2002).

The medical events during planetary missions obviously require a more substantial capability than low Earth orbit, where rapid return to Earth is a viable option. By any means, the concept of a safe haven is also imperative in planning any health maintenance facility. Such safe havens could provide temporary protection against radiation, but also fire, environmental toxins, and decompression.

Figure 7-15. Flight surgeons practicing surgery in the weightlessness of parabolic flight using a tent to protect the operative site. (Credit NASA)

4.3.2   Surgery in Space

Spacecraft are closed ecosystems with everything recycled, including the air. In the absence of gravity, microscopic particulate matter is dispersed in the air, rather than settling to the "ground." Surgical procedures must therefore be protected from this increased level of air contamination, and the solution to date has been to create canopies or tents to protect the operative site (Figure 7-15). Indeed, during open surgery in space, surgical debris would disperse throughout the spacecraft rather than being contained by gravity into the peritoneal cavity.

For those missions where surgery may be possible, procedures have to be developed to allow the surgeon to operate in microgravity. For example, both the patients and the surgeon must be restrained to prevent floating away from the operating table (Figure 7-16).

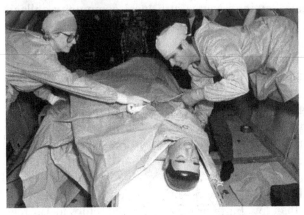

Figure 7-16. Lesson learned: when performing surgery in microgravity, both the patient and the surgeons must be restrained! (Credit Douglas Hamilton)

New training techniques must also be developed for instrument deployment and fixation, and to ensure a sterile environment. Cleanliness is particularly important, since, as mentioned earlier, there is an increased population of antibiotic resistant bacteria in space, and a decrease in the immune function has been documented. Other factors which may impact surgical procedures in space are the level of lighting (surgery rooms in hospital are equipped with very bright lighting ensuring the best exposure in all directions) and the possible decreased in proprioceptive sensitivity from the muscles, skin, and joints.

For long-duration space travels, "just-in-time" training, preprogrammed robotic surgical procedures, mentored-surgery performed by non-surgical personnel, or other techniques may be the only alternatives. In many instances, minimally invasive techniques can provide protection. These minimally invasive procedures can be conducted and viewed through a video monitor, which permits the opportunity now for telementoring, and, in the future, the potential for remote robotic telesurgery (Figure 7-17). There is, however, the issue of time delay in communications for planetary missions (40 min round-trip between Mars and Earth).

It is expected that the Mars mission will benefit from current general trends in medicine. Advances in microelectronics have enabled smaller, lighter, and less power-intensive components such as cardiac monitors, ultrasound imaging systems, and pulso-oximetry. There is also a trend toward less intensive therapies. Many once-major surgical procedures are being replaced by fiber-endoscopy approaches, vastly simplifying problems of sterile field maintenance and blood handling, which are magnified in microgravity (Campbell et al. 2001). Ultrasonic pulses applied externally are being employed in the process known as "lithotripsy" to treat kidney and gall bladder stones. Another more quiet revolution is in the area of advanced pharmaceuticals. In recent years new class of broad-spectrum antibiotics has been developed, which can be taken in pill form, replacing more complicated therapies that previously required intravenous administration. For example, the number of surgical procedures for peptic ulcer disease has been drastically reduced by the use of several classes of highly effective anti-ulcer medications (Barratt 1992).

These innovative solutions to operate in remote sites could lead to discoveries for new surgical applications on Earth for remote or small villages or less experienced medical personnel. By having such a critical need to provide medical and surgical support to space missions, space agencies will continue to push the envelope in leading edge surgical technologies and training techniques.

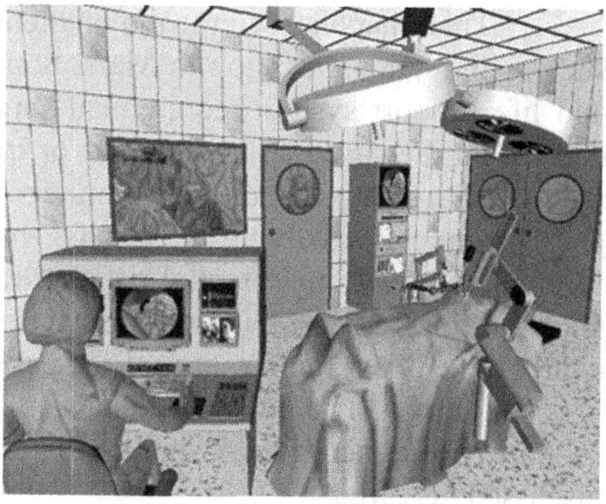

Figure 7-17. Artist concept of remote robotic surgery.

4.3.3 Evacuation

Recent studies at NASA based on nine categories of medical conditions and on 60 medical doctors review surveys, including such factors as on-orbit environment and medical facilities, have ranked the likelihood for medical conditions that would require an evacuation from the ISS (Billica et al. 1996). The results of these studies suggest that an evacuation of a crew of seven could take place every five years of a permanent utilization of the ISS. The most likely causes of evacuation besides medical emergency are a radiation dose event, collision with a micrometeorite, orbital debris, or major system failures.

In the past, several missions were aborted and required medical evacuations from space. In 1976, the crew on board Salyut-5 experienced chronic headaches secondary to life support problems, and the station was abandoned after 49 days. In 1985, the crewmember of Salyut-7 became ill with prostatitis and developed urosepsis, which required return to Earth after 56 days of a 216-day mission. In 1987, a crewmember of Mir-2 developed persistent dysrhythmia, and the mission duration was shortened from 11 to 6 months with a safe return in Soyuz. Also, in the 15 years of the Mir existence, three events could have prompted an evacuation: an O_2 candle fire, a collision with the Progress vehicle and the depressurization that followed, and an attitude control and power loss. The experience in Antarctica indicates that there are about 70 events prompting a medical evacuation per 2000 person-years.

A medical evacuation of the ISS is therefore likelihood. Soyuz is currently the ISS evacuation vehicle should such problems occur. However, flight surgeons are concerned because the Soyuz capsule has obvious limitations as a medical evaluation vehicle: it is small, not equipped with medical assistance equipment, and the decelerations during landing are considerable.[*]

5   FUTURE CHALLENGES

This chapter reviews the challenges posed by planetary exploration, such as the human mission to Mars. Some options of medical care programs and life support system requirements are unique to such a mission. Indeed, there is an increased risk of disease or injury with long-term habitation for a relatively large number of crewmembers. The amount of EVA work in a 0.38-g environment involving high-mass hardware will dramatically increase further compounding potential risks to trauma. If and when a serious medical situation does arise, rapid emergency transportation may not be available or appropriate. Also, rapid return of the patient to Earth for life-saving treatment is not an option. The potential impacts of the medical care options and the life support requirements on design and operations of a human Mars mission are discussed hereafter.

---

[*] Due to a software error, the Soyuz that returned the Expedition-6 crew from the ISS in May 2003 followed an even steeper descent path. As a result, the crew was subjected to about 8-10 g during reentry, and they missed their targeted landing site by nearly 400 km. When they opened the hatch, the crewmembers were still strapped into what had become the ceiling. Bowersox said to the journalist from the Washington Post. "As soon as you unbuckle your seat straps, you sort of fall down on the cosmonaut panel." Having come from almost six months in weightlessness, it took the three men more than an hour to drag themselves out of the hatch under the oppression of Earth's gravity and erect a folded communications antenna to help the searching planes and helicopters find them. "You'd crawl out... You'd move a little bit. Rest. Move a little bit more. Rest," Bowersox said. "It was pretty amazing." And he added, "I feel really lucky."

5.1 Human Needs for Long-Duration Missions

Yes, robots like the Mars Pathfinder have shown that a lot of scientific information can be learned about a planet's surface by sending robots instead of men and it can be done significantly cheaper.

But imagine you wanted to go to Paris. Would you be satisfied with a robot... taking very good pictures of the Eiffel tower and chemically sampling the French food?

We, humans, are really going into space as part of the human drive to explore and to understand. It's an intrinsically human thing. It is part of the human experience in the broadest sense and science is just one piece of that experience. Learning about the universe is only part of the equation; learning how the astronauts who participate in this exploration change and evolve is of equal importance.

However, before humans can embark in such a voyage, it is necessary to re-visit the current concepts of life support systems on board space vehicles.

5.1.1 Environment and Hygiene

The Earth's atmosphere is made up of 78% nitrogen, 21% O_2, 0.5% water vapor, along with very small amounts of argon, CO_2, neon, helium, krypton, xenon, hydrogen, methane, and other trace gases. We depend on the correct mixture of gases in the atmosphere to sustain our lives. We also depend of the pressure of our atmosphere to be able to breathe (at sea level, atmospheric pressure is 1 atm = 760 mmHg = 101.1 kPa = 14.7 psi). Therefore, space travelers must carry their own pressurized atmosphere with the correct mixture of gas.

Because spacecrafts are completely closed environments, CO_2 must be actively removed from this atmosphere. CO_2 level should be lower than 0.01%. If not, high CO_2 levels increase heart rate and respiration rate and cause problems with the acid-base balance of the body. In today spacecraft, the air moves through canisters of white, granular lithium hydroxide (LiOH) to remove the exhaled CO_2. The canisters also contain a layer of charcoal to trap odors in the air. Fans disseminated in the habitat pull the clean air constantly through screens that catch debris, such as lint, hair, and crumbs.

Although the Mars spacecraft will be assembled in a "clean room" where dust particles have been filtered out, bacteria nonetheless pervade all equipment and undoubtedly a colony of them will parasite our crew on their voyage. In all likelihood no virulent strains will find their way on board, but it is even possible that in the radiation and microgravity environment of space, some genetic mutations might develop and produce new forms of bacteria that humans have never encountered before. Just as Christopher Columbus had to fight scurvy and syphilis, so may the first Mars crews find that disease awaits them far from their homeport (Collins 1990)?

High humidity can promote the rapid growth of microbes or fungus. Low humidity can cause drying of the eyes, skin, and the mucous membranes of the nose and throat, thus providing less protection against respiratory infections. Temperature is also an important aspect of the body heat balance, and should ideally range from 18-27°C.

Skylab had a shower: while standing in a collapsible, cylindrical cloth bag, the astronauts squirted warm water at themselves using a water gun and scrubbed with liquid soap. In practice, the shower was a failure, since the two other crewmembers had to spend valuable time vacuuming water that escaped into the air and installations. There is no shower on the Shuttle or the ISS, and probably not in the spacecraft en route to Mars. Instead, crewmembers use sponge baths. According to the training procedure, astronauts draw a curtain from the toilet door to the side of the galley for privacy. A wash basin on the side of the galley provides warm water and a soap dispenser. Above the basin are a mirror and a light, and on the wall are strips of tape to attach towels, wash cloths and personal hygiene items. One cloth is used to wash, and another to rinse. At the rear of the basin is a fan that pulls the excess water toward a drain that leads to the wastewater tank under the floor. The wash cloths and towels go into the bag hanging on the bathroom door.

Since there is no washing machine on board, trousers (changed weekly), socks, shirts, and underwear (changed every two days) are sealed in airtight plastic bags after being worn. Garbage and trash are also sealed in plastic bags.

The toilets commode (or Waste Collection System) is in a private room. To remain seated, the user must insert his boots into foot restraints and snap together the seatbelt waist restraint. There are also handholds. Instead of water to flush away solid wastes, this toilet relies on a fan that draws away the "wastes" from the user and sends "them" to a compartment below. There, it is dried and disinfected. Urine is drawn into a contoured cup and flexible hose by airflow and the fluid is pumped into the wastewater tank under the floor.[*]

Then, there is the issue of contaminants. Placing humans on Mars might lead to a contamination of the Mars environment, complicating the search for extra-terrestrial life that might exist in ecological niches. Conversely, returning humans or soil and rock samples from Mars might contaminate species on Earth, although scientists regard the possibility as

---

[*] The Russian supplied toilet paper is not like what one normally thinks of as toilet paper. It consists of two layers of coarsely woven gauze, 10 by 15 cm in dimension sewn together at the edges with a layer of brown tissue sandwiched in-between. According to the astronauts, "it works very well for its intended purpose" (Pettit 2003).

extremely remote (Figure 7-18). Because of these possibilities, several nations have signed the *Outer Space Treaty* and agreed that "State Parties to the Treaty shall pursue studies of outer space, including the Moon and other celestial bodies, and conduct exploration of them so as to avoid their harmful contamination and also adverse changes in the environment of the Earth resulting from the introduction of extraterrestrial matter and, where necessary, shall adopt appropriate measures for this purpose".

It is generally agreed that the initial use of robotic devices would make much less impact on the planet than humans and their associated life-support infrastructure, and could provide advance information to lessen potential human impacts. However, the living bacteria found on one camera of the Surveyor-3 probe on the Moon (see Chapter 1, Section 1.1) is the evidence that contamination can also occur with unmanned vehicles.

Figure 7-18. The quarantine for the Apollo astronauts returning from the Moon was only used after the first two lunar landing missions. (Credit NASA)

5.1.2 Human Needs

Human basic requirements include atmosphere, food, and water. In one year, a 75-kg individual requires 4 times his/her weight in oxygen, 3 times his/her weight in food, and 17 times his/her weight in potable water. A much larger quantity of water is needed for hygiene, sanitation, than for nutritional requirements. Table 7-15 shows a perfect balance between the inputs and outputs needed for sustaining humans. However, simple things like food dislikes can dramatically disturb the equation.

	One day (per person)	One year (per person)	% of total mass
Inputs			
Oxygen	0.83 kg	303 kg	2.7 %
Food	0.62 kg	226 kg	2.0 %
Potable Water	3.56 kg	1300 kg	11.4 %
Hygiene Water	26.0 kg	9490 kg	83.9 %
Total	31.0 kg	≈11400 kg	100 %
Outputs			
Carbon dioxide	1.0 kg	363 kg	3.2 %
Metabolic solids	0.1 kg	36 kg	0.3 %
Water	30.0 kg	10950 kg	96.5%
including:		metabolic / urine	12.3%
		hygiene / flush	24.7%
		laundry / dish	55.7%
		latent	3.6%
Total	31.0 kg	≈11400 kg	100 %

Table 7-15. Average values for most human inputs and outputs. Potable water is used for drink and food preparation; hygiene water is used for hygiene, flush, laundry, and washing dishes.

5.1.3   Nutrition Requirements

According to the values in Table 7-15, an astronaut needs about 0.62 kilograms (dry weight) of food per day. However, this amount may vary depending on the activity level at which he/she is operating. The caloric requirements are determined by the following formula for basal energy expenditure (BEE):

Women BEE = 655 + (9.6 x weight) + (1.7 x height) – (4.7 x age)
Men BEE = 66 + (13.7 x weight) + (5 x height) – (6.8 x age).

Based on the activity levels of the astronauts, the estimated required energy ranges from 2300 to 3200 kcal/day. An additional 500 kcal/day is needed on EVA days. Diets for space missions are generally planned at caloric levels close to those needed for normal activity on Earth (Table 7-16). For missions lasting from 30 days to one year, the energy provided by each food group (nutrient breakdown) is the following: protein: 12-15%; carbohydrate: 50-55%; fat: 30-35%; fiber: 10-25 g/day; and fluid: 1.5 mL/kcal (>2 L/day).

Crewmembers assigned on a mission can select food from 150 items, ranging from commercially available items (e.g., crunch bars and cookies) to rehydratable, thermostabilized (canned food), irradiated (treated with radiation to kill bacteria), freeze dried, or intermediate moisture food (see Table 1-03 in Chapter 1). A tasting session takes place about 9 months prior to a flight, and each crewmember chooses his/her own menus approximately 5 months preflight. Menus consist of three meals per day, plus snacks. Each menu is analyzed by a dietician to ensure that if follows the recommendations to meet the Recommended Daily Allowances (RDA) of vitamins and minerals. For Shuttle mission, each menu weights about 1.7 kg (out of which 0.5 kg is packaging).[*]

Nutrient	Quantity
Protein	0.8 g per day per kg (minimum recommended)
Carbohydrate	350 g per day
Lipid	77-103 g per day (less than 30% of calories)
Kilocalories	2300-3100

Table 7-16. Nutritional requirements for a typical Shuttle mission. Note that these numbers represent the nutrients provided, not consumed calories. (Source Philipps 1997)

Despite this, observations during long-duration missions suggest that individuals do not crave continual variety in foods but rather tend to select foods in the same small range or limited number over months, stretching out to a lifetime. In absence of in-flight diet logs (who has eaten what and when), caloric consumption is generally derived by assessing food that has disappeared, and assuming equivalent intake by each crewmembers. Where data is available, it appears that crewmembers consume fewer calories than provided and recommended. Perhaps the dietary intake is inadequate. Some nutritionists claim that it represents only 60-70% of the recommended energy requirements (Lane and Schoeller 1999). This could explain in part the weight reduction in astronauts (see Figure 5-11). On the ISS, the food packaging includes a barcode, and crewmembers are required to scan this code on their personal computer for a more precise evaluation of daily caloric consumption.

In the past, many athletes and astronauts have been convinced that high protein intake builds muscle and strength. However, the physiological evidence indicates that protein is increased in muscle only when needed for the muscle hypertrophy required by continuing physical activity; excess calories of any kind are converted to and stored in the body as fat. In addition, numerous previous studies unrelated to space have indicated that increasing the protein intake increases the urinary excretion of calcium. Because this would add to the bone demineralization and the potential for kidney stones formation, the level of protein in the diets of astronauts, therefore, needs to be monitored. Some degree of uncertainty exists as to whether the high

---

[*] The history of space food and the methods of food preservation and preparation used by NASA are described in a .pdf document "Space Food and Nutrition. An Educator's Guide With Activities in Science and Mathematics", EG-1999-02-115-HQ, which can be downloaded from the NASA web site at http://spacelink.nasa.gov/products/.

phosphate content of meat is partially protective against the effect of high protein intake to increase urinary calcium. At the same time, there is concern not to accentuate the negative nitrogen balance associated with muscle atrophy in weightlessness by encouraging too low a protein intake. Since negative nitrogen balance in space has occurred at daily protein intakes of 85 to 95 g the recommended intake should not fall below this level (Philipps 1997).

It is well known that under time pressure, Shuttle astronauts often prefer to use snacks, i.e., food rich in carbohydrate, on their workplace rather than a full meal in the middeck galley. However, carbohydrates are of special concern because any dietary carbohydrate that elicits the secretion of insulin can, unless consumed with adequate amounts of protein, increase the synthesis and release of the brain neurotransmitter serotonin. This substance makes people drowsy and interferes with optimal performance. Menus and the time of consumption of particular items, especially snacks, might not be appropriate to the tasks required, particularly if they are complex and prolonged. It is possible that other food constituents also affect behavior, mood, and cognition. As carbohydrates are the likely products of future chemical synthetic systems, it is important to determine the type and maximum amount of carbohydrate that should be reasonably contained in a human diet.

Since no studies have yet been made on the effects of spaceflight on the metabolism of any of the trace elements, no comment can be made other than that care should be taken that space diets contain trace elements in the amounts recommended in the nutritional standards.

The important vitamin in long spaceflights is vitamin D, the "sunshine vitamin". Enclosure in a space vehicle prevents the normal conversion in the skin of the vitamin D precursor to vitamin D. This is normally accomplished by exposure of the face and arms to as little as 20 to 30 minutes of sunlight a day. Since vitamin D is essential for facilitating calcium absorption from the intestine, as well as other calcium-related effects in kidney and bone, a surplus of this vitamin needs to be supplied to astronauts. The space recommended dose is 10 microg per day, whereas the Earth recommended dose is 5 microg per day.

Other vitamins are not so critical since it is expected that adequate amount is taken in the diet, provided it is "balanced" and the vitamins are not degraded by the methods of food preservation in use. It has become customary, however, to provide astronauts with daily vitamin supplements.

In the early days of planning for human spaceflight, scientists believed that diets should be low-residue in character so that bowel movements would be small and infrequent. It was observed especially in longer flights that bowel function in microgravity is essentially normal. Hence diet is normal in residue, and adequate bulk is available to afford relatively easy passage of stools once or twice a day.

The ISS experience will help to make sure that there will be adequate and satisfactory food selection and storage for the 3-year flight of a human mission to Mars. The lessons learned form the ISS will also provide guidelines for the Controlled Ecological Life Support System (CELSS) and in particular for quantities of food to be produced by this system.

5.2   Controlled Ecological Life Support System

A controlled ecological life support system (CELSS) is attempting to create an integrated self-sustaining system capable of providing food, potable water, and a breathable atmosphere for space crews during missions of long-term duration. The control of these parameters is obviously an engineering question. The failure of any one component, however, immediately involves the medical support group. CELSS have been extensively described in Peter Eckart's book (1996). We will focus here on the issues related to their malfunction during a space mission.

5.2.1   Life Support System Fundamentals

The systems that provide life support functions are divided into two general categories: *regenerative* or *non-regenerative*. Both regenerative and non-regenerative systems provide and maintain key life support resources. Non-regenerative systems generally only effect a change in state or quantity of a resource, while regenerative systems involve resources that are physically or chemically altered but have the potential for being regenerated (hence the name). Non-regenerative systems provide temperature and humidity control, total and partial pressure control, atmosphere composition monitoring, and airborne contaminant removal. Water and gas resources are involved but not altered. Technologies used to provide these functions are generally well developed and will not be discussed further. Oxygen, food, water, and waste management are provided by regenerative systems. The remainder of this section will focus on how regenerative life support functions are provided.

There are two fundamental approaches for designing regenerative systems: open loop and closed loop. The first approach brings all life support resources from Earth and discards them after they are converted to a non-useful form. Systems which utilize this method are called "open loop" to signify the continuous flow of material into and out of the system. In this scenario, all food, water and oxygen are from stored sources (Figure 7-19). Oxygen can be transported as a cryogenic fluid or a high-pressure gas. High-pressure storage is ready-to-use, but introduces risk of tank rupture and has decreasing delivery pressure. Open loop technologies tend to be simple and highly reliable and have been extensively used in human spaceflight to date.

The big disadvantage to open loop systems in general is that resource requirements continue to increase linearly as mission duration and crew size increase. Using the numbers in Table 7-14 we can see that in three years a crew of four will use 2.7 tons of food, 3.6 tons of oxygen, and 129.5 tons of water! And that is without packaging.

Figure 7-19. Astronaut Susan Helms is photographed in front of the potable water storage on board the ISS. (Credit NASA)

The second approach for designing regenerative systems (closed loop) is to bring an initial supply of resources from Earth and then process the non-useful waste products to recover useful resources. These types of systems are called "closed loop" because once a resource enters the system it does not leave and the non-useful forms of material are recycled. The big advantage of closed loop systems is the one time transport of mass to orbit, of processing hardware and initial resource supply, and with minor subsequent re-supply of expendables. The disadvantages are lower technology maturity and increased power and thermal requirements. However, when mission duration becomes long enough, or re-supply is not possible (e.g., on a trip to Mars) there is a time when closed loop technologies provide the most cost-effective solution.

A *controlled ecological life support system* is the ultimate closed-loop regenerative system. An effective CELSS must have subsystems both for plant and animal (including human) growth, food processing and waste management. It must be much more than a "greenhouse in space", however. It must be a multi-specific ecosystem operating in a small closed environment. Duplicating the functions of the Earth without the benefit of its large buffers, that is the oceans, atmosphere, and landmasses, is challenging. Several experiments have been attempted, including the Biosphere-2 project, with limited success.* The main question is how small can the requisite buffers be

---

* The Biosphere-2 project supported a crew of eight inside a large glass building resembling a giant terrarium in the Arizona desert, including a rainforest, savanna, ocean, desert, human habitat, and intensive agriculture. The full spectrum of biological life support agents was used, with plants producing food, oxygen and clean

and yet maintain extremely high reliability over long periods of time in a hostile environment. Also, by necessity, space-based systems must be small, therefore a high degree of control must be exercised.

5.2.2   Engineering Solutions for LSS

In physical-chemical life support systems, human is the only biological component. Physical-chemical processes include use of standard engineering mechanical components such as fans, filters, etc., physical separation processes such as molecular sieves, reverse osmosis, or electrolysis, and chemical separation and concentration processes. These processes are well understood: engineers feel comfortable with them, they are relatively compact and low maintenance and have quick response times. Biological processes employ living organisms such as plants or microbes to produce or breakdown organic molecules. They are less well understood: they make engineers nervous, tend to be large volume, power and maintenance intensive, with slow response time (Doll 1999).

There are many Earth-based technologies which are capable of providing the five major life support functions; provide and maintain a comfortable and breathable atmosphere, provide oxygen, food and water, and manage waste. Traditional heating, ventilation and air-conditioning systems can control temperature and humidity and provide air distribution. Sophisticated contaminant removal systems are used commonly for cleanroom applications. Atmosphere regeneration techniques purify air and provide oxygen on submarines and municipal facilities routinely process water and waste.

One of the major challenges of adapting or improving these technologies to be used for space applications is related to the nature of the space environment itself. For example, because there is no gravity, there is no convection for mixing of gases or for natural convective cooling, and phase separation of gases and liquids requires special devices. There are severe power, weight, and mass constraints on hardware design as well as extremely limited local resources. Because life support is a critical system for survival of the human space travelers and it is operating in a totally isolated environment, safety and reliability requirements are also very strict. Ingenuity, an understanding of the space environment, and familiarity with multi-disciplinary tasks are key characteristics for successful life support system engineers.

---

water, people and animals as consumers, and microbes decomposing waste materials and metabolizing airborne contaminants. The living biomass was approximately 70 tons! This system, however, had severe problems maintaining the atmosphere levels and food required for the crew.

Atmosphere Management. The function of the atmospheric revitalization and control subsystem is to continuously control temperature and humidity, and regenerate atmosphere. It also monitors and removes the harmful trace contaminants that are generated by the crew and the equipment.

In the past, several contamination issues have occurred during missions on board various spacecrafts: a faulty fiberglass insulation (Apollo-10, 1969), CO_2 build-up (Apollo-13, 1970), propellants on reentry entered via vents (Apollo-18, 1975), acrid odor (Soyuz-21/Salyut-5, 1976), eye irritation from LiOH canisters and payload chemicals, and formaldehyde and ammonia from overheated refrigerator motor (Space Shuttle), as well as ethylene glycol and fumes from fire (Mir).

According to the Substance Maximum Allowable Concentrations (SMAC), toxic effects are classified into three groups: a) non-carcinogenic, acceptable endpoint (slight irritation, mild headache); b) non-carcinogenic, unacceptable endpoint (leading to anesthesia, blindness, disability); and c) carcinogenic (lifetime risk less than 0.01% per mission). The problem is that most accidental exposures involve a mixture of these effects.

Today space vehicles are monitored weekly by microbial air surface and water sampling. The Space Shuttle is monitored after each mission (using both gas chromatography and mass spectrometry) for the presence of contaminants. It is interesting to note, however, that bacteria and fungi (which proliferated on board Mir) are not considered as contaminants.

CO_2 Reduction. In a regenerative system, CO_2 is not to be discarded, but reduced to leave useful components inside the system. Thus, the output from the CO_2 concentration subsystem is used as the input for the CO_2 reduction subsystem. Currently the main competing regenerative subsystems for CO_2 reduction are the Bosch process and the Sabatier process (the ISS U.S. module uses the Sabatier process). Further technologies that are being considered for CO_2 reduction are the Advanced Carbon-Formation Reactor System, electrolysis, and superoxides.

Water Management. The water management system supplies the metabolic and wash water requirements of crew, and collects atmospheric condensate and wastewater. The basic processes fall into two categories: distillation and filtration processes. While the distillation method is mainly considered for urine recovery, the filtration method basically processes hygiene and potable water.

It is anticipated that very long-duration space missions will include two water recycling and storage subsystems: one subsystem will process concentrated feeds, such as urine and flush water, and the second a more dilute feed, such as laundry or shower water. Potable water will probably be recycled using a phase change process while lower quality water will be recovered by filtration.

In addition to the subsystem designed specifically for reclaiming space habitat waste water, by-product water is derived from other space habitat subsystems, such as H_2O_2 fuel cells, CO_2 reduction and the space habitat condensing heat exchanger used for cabin humidity control. Water from CO_2 reduction is of high quality since it is derived from a high-temperature process which destroys harmful bacteria. Also, the feed gases used for CO_2 reduction are clean. Water derived from fuel cells and the condensing heat exchanger may require post-treatment to remove chemical and biological impurities prior to being reused.

5.2.3 CELSS for Long-Duration and Exploratory Missions

For very long-duration missions in low Earth orbit or for exploratory missions, it becomes too expensive or impractical to rely on stored supplies. For a Mars mission, it is also impossible to depend on re-supply from Earth. The life support system must instead rely on regenerative system to regenerate food and oxygen, recycle waste, and purify air and water. Advanced life support systems for these missions will likely rely on plants to serve as sources for food, and to purify air and water. However, this solution has several drawbacks. For example, the introduction of plants and humans to a closed environment increases the chances of unwanted, potentially dangerous microbial contamination. Microbes are resilient organisms, capable of adapting to harsh environments, and able to colonize any surface that contains adequate nutrients and moisture. It is then far more desirable to monitor microorganism levels in a CELSS. Biochemical tests must be developed, which can be simply done by the crew or automated, not dangerous and low time-consuming. For example, computer algorithms could be used to identify a suspected organism by comparing its rRNA sequences with those of known organisms (the so-called "DNA chip" technique) (Sanz et al. 2001).

On Earth, animals breathe in O_2 from the air and breathe out CO2 as a waste. Plants absorb this CO_2 from the air, and using the energy of sunlight plus water and materials from the soil and air produce sugar, starch and other things, based on a process called photosynthesis. Plants emit O_2 as a waste. That completes the animal-plant cycle. In this cyclic manner, animals and plants are mutually dependent upon each other. Plants produce both food and O_2 for animals. In turn, animals produce CO_2 for plants. In addition, animals produce excrement wastes, which enrich the soil. Dead plants also enrich the soil and are not wasted.

Photosynthesis: $CO_2 + H_2O + light \rightarrow (CH_2O)n + O_2$
Respiration: $CH_2O + O_2 \rightarrow CO_2 + H_2O$

This natural cycle can be moved to space, in whole or in part. Early experiments in the 1950s and 1960s focussed on recycling air using algae, not food crops. Flat tanks of algae were put under artificial light in order to absorb CO_2 that humans had exhaled in closed chambers, and emitted the O_2 for the humans to breathe. It was found that each human required about 8 m^2 of algae for equilibrium. (The alga tanks were generally stacked as shelves so that they took much less than 8 m^2 of floor space.)

More recent research has expanded this to include production of edible food, and recycling of animal or human excrement wastes and dead plant wastes in the food cycle. Studies are required for defining the conditions required for optimum rates of dry matter production. Although most research has been done with open systems, experiments with closed systems have recently been initiated (e.g., the "Melissa" project at the University Autonomous of Barcelona, Spain; the "Aquatic Biosphere" developed by Paragon Space Development Corp. in Tucson, Arizona; the CEBAS Minimodule developed by OHB-System in Bremen, Germany).

For example, a "Biomass Production Chamber" has been developed at the NASA Kennedy Space Center, which consists in a sealed large steel chamber approximately 3.5 m in diameter and 7.5 m high, with two floors, each floor having multiple racks and lamp banks, duct work for air flow, and various equipment for controlling temperature and humidity. Total chamber plant growing area is 20 m^2. Wheat, soybeans, potatoes, sweet potatoes, strawberries, rice, peanuts, radishes, and other foods were grown in the facility. Instrumented test subjects were placed inside the chamber for limited duration. Subjects were exercising on electronic cycle ergometer in order to evaluate gas exchanges in various conditions. Adjacent laboratories were used for converting wastes into plant nutrients, plant fertilizer, carbon dioxide, and water. At every stage, careful and detailed measurements of many kinds were made in order to understand the processes in depth. Results indicated that a minimum surface of 40 m^2 of crop field was required per human for total recycling of oxygen and food (Doerr 2001).

Some factors that will need to be considered in establishing a CELSS for food production are using crops that provide a dependable yield, have a high edible biomass yield, are of small size, provide dietary variety, and can be combined to form a nutritionally complete diet. Based on consideration of primarily agricultural plant species, a small number have been selected for further investigation. These include wheat, potato, soybean, and tomato. It may well be that some of the plants will be genetically modified to increase levels of certain micronutrients, essential lipids or amino acids. An additional factor is that very intensive agriculture will be practiced to grow a maximum quantity of usable raw food in an area as small as possible. Although the development and growth of plants seem little affected by the space environment (both radiation and microgravity), at least during the first couple of generations (see Chapter 2, Section 4.2), no experiments have yet been performed in microgravity in real scale to determine if current systems can function in space. In short, a considerable increase in research efforts is required in order to reach the desired goals of a controlled ecological life support system.

Figure 7-20. Rice harvest in Pokhara, Nepal.

5.3 Terraforming

Planets are places, not vehicles. Surviving there will be different from camping in a spacecraft flying into space. An effort to build human communities either on other planets or on artificial new worlds might begin with altering the current climate and atmosphere to resemble more closely that of Earth. The process of transforming a planet to create a more Earth-like habitable living environment is called *terraforming*. Any terraforming process is likely to take the candidate world on a path from initial sterility through a continuum of improving habitable states. "Full" terraforming (the achievement of an entire environment suitable for Earth-like humans, animals, and plants) is likely to remain a distant, although not impossible, goal. Indeed, terraforming an entire planet into an Earth-like habitat would almost certainly have to be done over several centuries or even millennia. On the other hand, the initial stages of terraforming might take only several decades. For example, humans could begin by living in transparent domes. The domes would have radiation shielding that protects plants and animals. This would permit the construction of ambient pressure dwellings and the replacement of pressure suits with simple breathing gear. Such solutions could allow human habitation well before full habitability of a planet is attained.

The Earth was once terraformed. At the beginning, there was no oxygen in the atmosphere, only carbon dioxide and nitrogen, and the land was composed of barren rock. Using photosynthesis, organisms transformed the CO_2 in Earth's atmosphere into O_2, in the process completely changing the surface chemistry of our planet. The evolution of aerobic organisms then modified the Earth still more, colonizing the land, creating soil, and drastically modifying global climate. Once the biosphere had extended, humans accelerated its development rate by using irrigation, crop seeding, weeding, and domestication of animals (Zubrin 1999).

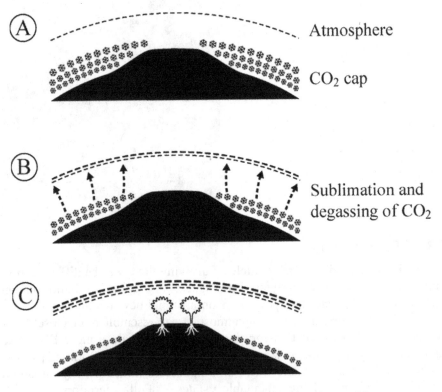

Figure 7-21. Schematic showing the principle of terraforming Mars as proposed by Zubrin (1996).

Terraforming Mars would use the same principle, but using the greenhouse effect would accelerate the process. The first step would be to set up factories to produce artificial greenhouse gases (e.g., perfluoromethane, CF_4) for release into the atmosphere (Figure 7-21A). Zubrin (1996) predicts that if these gases were produced at the same rate as chlorofluocarbon (CFC) gases are currently produced on Earth (about 1000 tons per hour), the average global temperature of Mars would be increased by 10°C within a few decades. This temperature would cause CO_2 to outgas from the regolith, which would warm the planet further due to the greenhouse effect (Figure 7-21B). This effect could even be amplified by adding bacteria releasing methane and ammonia, two very strong greenhouse gases. The net result of such a program would be a planet with acceptable atmospheric pressure and temperature, and liquid water on its surface (Figure 7-21C). Even though the atmosphere would not be breathable by humans, space suits would no longer be required when outdoors (just breathing gear). Crop could grow in the fields and aquatic life flourish in lakes oxygenated by algae. Humans would live in closed habitats until plants release enough oxygen in the outside atmosphere. A Japanese company, Obayashi Corporation, recently proposed a design concept for such a Mars habitat, including a farm and terrarium in an area of about 400 m x 500 m. It is envisioned that by the year 2057, that is for the 100th anniversary of Sputnik, this habitat could host 150 pioneers.

At the same time, the environment of space will also influence Earth-like biological and physiological processes. Organisms might begin evolving into forms suitable for the local environment. It is not certain that the gravity of Mars will be sufficient to prevent the bone demineralization and muscle atrophy observed in microgravity. It might be much healthier for crews to provide artificial gravity for long-duration habitation. This could be achieved by exposing the crew to intermittent artificial gravity produced by the centrifugal force generated by a centrifuge or a slow rotating room. How much artificial gravity and for how long are questions that need to be answered by investigations during long-duration space missions. For small habitats, rotating them to produce artificial gravity results in some very noticeable differences with real gravity due to the Coriolis effect. Ground-based experiments have led to guidelines for a "comfort zone" in artificial gravity, bounded by values of the radius of the rotating structure, head-to-foot acceleration gradient, rotation rate, and tangential velocity. However, this comfort zone is essentially terrestrial, since very little is known about artificial gravity in space. On the other hand, it might be better to facilitate biological adaptation to reduced gravity instead.

These thoughts also bring interesting questions, which go far beyond the area of space life sciences. For example, what are the costs and benefits of encouraging some or all organisms to adapt to the local environment of space rather than trying to make that environment Earth-like? What are the ethical concerns of whatever we do and to whom/what we have ethical obligations? For example, do intelligent beings only deserve ethical concern? Or all "life" forms (whatever "life" is)? And do nonliving environments (e.g., fossils, rocks) have rights or deserve ethical concern? And finally, what are the possible sequences in extraterrestrial settlement and expansion?

5.4   Conclusion

This, then, is space medicine in a nutshell. As a new field of medicine, it has developed quickly in a highly technical arena. It has caused us to redefine much of what we know about human physiology and performance for a new and challenging environment, and it will keep in step with the continued projection of humanity off the home planet. Like the ship's surgeon of a 16th century vessel of exploration, the crew medical officer of the first human Mars expedition will be equipped according to the best information possible and at the same time constrained by space and resources. He will, like his earlier counterpart, also have other duties, but will be responsible for the health of the crew. He can be expected to observe many effects for the first time and must be prepared to adequately describe new medical findings and adequately react if they are hazardous. And as always through history, he and all the crew will be anxiously awaited back at the homeport to share the discovery (Barratt 1995).

6   REFERENCES

Bagiana F, Clément G, Monchoux F, Barer A (1993) Human physiological model. Preparing for the Future. *ESA's Technology Programme Quarterly* 3: 12-13

Barratt M (1992) *Extravehicular Activity for Orbital Construction. Medical and Human Factors Considerations.* Lecture notes. Summer Session of the International Space University, Kitakyushu

Barratt M (1992) *The Onsite Medical Facility for Orbital Platforms.* Lecture notes. Summer Session of the International Space University, Kitakyushu

Badhwar GD (1997) Deep space radiation sources, models, and environmental uncertainty. In: *Shielding Strategies for Human Space Exploration.* Wilson WJ, Miller J, Konradi A, Cucinotta FA (eds) Washington, DC: National Aeronautics and Space Administration, NASA CP 3360, pp 17-28

Barratt M (1995) A Space Medicine Manifesto. Notes of the Lecture on *Current and Future Issues in Space Medicine* presented at the Summer Session of the International Space University, Stockholm

Bhardwaj R (1997) *Radiation and its Implications on a Human Mars Mission.* Essay prepared during the International Space University summer session in Houston

Billica RD, Simmons SC, Mather KL, et al. (1996) Perception of the medical risk of spaceflight. *Aviation, Space and Environmental Medicine* 67: 467-473

Campbell MR, Kirkpatrick AW, Billica RD, Johnston SL, Jennings R, Short D, Hamilton D, Dulchavsky SA (2001) Endoscopic surgery in weightlessness: the investigation of basic principles for surgery in space. *Surgical Endoscopy* 15:1413-1418

Collins M (1990) *Mission to Mars.* New York, NY: Grove Weidenfled

Comet B (2001) Study on the survivability and adaptation of humans to long-duration interplanetary and planetary environments (HUMEX). *European Initiatives in Advanced Life Support Developments for Humans in Interplanetary and Planetary Environments.* ESA TN-003, Nordwijk, NL: European Space Agency

Doer DF (2001) *Bioregenerative Life Support System for Long Duration Spaceflight.* Lecture notes. Master of Space Studies of the International Space University, Strasbourg

Doll SC (1993) *Life Support Systems during Spaceflight.* Lecture notes. Summer Session of the International Space University, Stockholm

Doll SC (1999) Environmental control and life support. In: *Key to Space. An Interdisciplinary Approach to Space Studies.* Houston A, Rycroft M (eds) Boston, MA: McGraw Hill, Chapter 8, pp 38-48

Durante M (2002) Biological effects of cosmic radiation in low-Earth orbit. *International Journal of Modern Physics* 125-132

Eckart P (1996) *Spaceflight Life Support and Biospherics.* Space Technology Libraries. Dordrecht: Kluwer Academic Publishers

Goldman M (1996) Cancer risk of low-level exposure. *Science* 272: 1821-1822

Hamilton D (1997-2002) *Operational Space Medicine.* Lecture notes. Summer Sessions of the International Space University, Vienna Austria 1996, Houston USA 1997, Cleveland USA 1998, Ratchasima Thailand 1999, Valparaiso Chile 2000, Bremen Germany 2001, Pomona USA 2002

Heimbach R, Sheffield P (1985) Decompression sickness and pulmonary overpressure accidents. In: *Fundamentals of Aerospace Medicine.* DeHart R (ed) Philadelphia, PA: Lea and Febiger, pp 132-161

Hills B (1985) Compatible atmospheres for a space suit, space station, and shuttle based on physiological principles. *Aviation, Space and Environmental Medicine* 56: 1052-1058

Holland AH, Marsh RW (1994) Psychologic and psychiatric Considerations. In: *Space Physiology and Medicine*, 3rd edition. Nicogossian AE,

Huntoon CL, Pool SL (eds) Philadelphia, PA: Lea and Febiger, pp 422-433

Institute of Medicine (2001) *Safe Passage: Astronaut Care for Exploration Missions*. Ball HR, Evans CH, Ballard JR (eds) Washington, DC: National Academy Press

Kilic F, Bhardwaj R, Trevithick R (1996) Modeling cortical cataractogenesis. In vitro diabetic cataract reduction by venoruton. *Acta Ophthalmologica Scandinavia* 74: 372-378

Lane HW, Schoeller D (eds) (1999) *Nutrition in Spaceflight and Weightless Models*. Boca Raton, FL: CRC Press

Lugg DJ (2000) *Antarctic Medicine*. JAMA 283:2082-2084

McCormick NJ (1981) *Reliability and Risk Analysis; Methods and Nuclear Power Applications*. New York, NY: Academic Press

Monk TH, Kennedy KS, Rose LR, Linenger JM (2001) Decreased human circadian pacemaker influence after 100 days in space: A case study. *Psychosomatic Medicine* 63: 881-885

National Research Council (1996) *Radiation Hazards to Crews of Interplanetary Missions*. Task Group on the Biological Effects of Space Radiation. Washington, DC: National Academy Press

Nelson BD, Gardner RM, Ostler DV, Schultz JM, Logan JS (1990) Medical impact analysis for the space station. *Aviation, Space and Environmental Medicine* 61: 169-175

Newman D, Barratt M (1997) Life support and performance issues for extravehicular activity. In: *Fundamentals of Space Life Sciences*. Churchill S (ed) Malabar, FL: Krieger Publishing Company, Volume 2, Chapter 22, pp 337-364

Osborn MJ (Committee Chairperson) (1998) *A Strategy for Research in Space Biology and Medicine in the New Century*. National Academy of Science. National Research Council Committee on Space Biology and Medicine. Washington, D.C: National Academy Press. Also available at NASA Web address: http://www.nas.edu/ssb/csbm1.html

Phillips RW (1997) Food and nutrition during spaceflight. In: *Fundamentals of Space Life Sciences*. Churchill SE (ed) Malabar, FL: Krieger Publishing Company, Chapter 10, pp 135-148

Pross HD, Kost M, Kiefer J (1994) Repair of radiation induced genetic damage under microgravity. *Fifth European Symposium in Life Sciences Research in Space Proceedings*. Nordwijk, NL: ESA Publications Division

Reifsnyder R (2001) Radiation Hazards on a Mars Mission. MIT Mars Society Youth Chapter. *The Martian Chronicles* 8. Available at: http://chapters.marssociety.org/youth/mc/issue8/index.php3

Sanz ML, Warmflash DM, Willson RC, Fox GE (2001) *Monitoring Microorganisms during Spaceflight.* Department of Biology and Biochemistry, University of Houston

Stolwijk J, Nadel E, Wenger C (1973) Development and application of a mathematical model of human thermoregulation. *Archives of Sciences and Physiology* 27: 303-310

Wilson JW, Miller J, Konradi A, Cucinotta FA (eds) (1997) *Shielding Strategies for Human Space Exploration.* NASA CP-3360. Also available from the NASA Langley Technical Reports Web address: http://techreports.larc.nasa.gov/ltrs/ltrs.html

Wright KP, Czeisler CA (2002) Absence of circadian phase resetting in response to bright light behind the knees. *Science* 297: 571

Zubrin R (1996) *The Case for Mars.* New York, NY: The Free Press

Zubrin R (1999) *Entering Space: Creating a Spacefaring Civilization.* New York, NY: Tarcher Putnam

Additional Documentation:

Acceptability of Risk From Radiation. Application to Human Spaceflight. April 30, 1997. Symposium Proceedings No. 3. Bethesda, MD: National Council on Radiation Protection and Measurements

Astronaut Medical Evaluations Requirement (1998) Houston, TX: NASA Johnson Space Center: NASA Reference Document JSC 24834, Revision A

International Workshop on Human Factors in Space (2000) *Aviation, Space and Environmental Medicine* 71, Number 9, Section II, Supplement

Medical Guidelines for Space Passenger (2001) Aerospace Medical Association Task Force on Space Travel. *Aviation, Space and Environmental Medicine* 72: 948-950

Modeling Human Risk: Cell & Molecular Biology in Context (1997) Berkeley, CA: Ernest Orlando Lawrence Berkeley National Laboratory Report LBNL-40278

Principles Regarding Processes and Criteria for Selection, Assignment, Training and Certification of ISS (Expedition and Visiting) Crewmembers (2001) Multilateral Crew Operation Panel. Washington DC: National Aeronautics and Space Administration. Available at: http://www.nasa.gov/hqpao/isscrewcriteria.pdf

Radiation Hazards to Crews of Interplanetary Missions: Biological Issues and Research Strategies (1996) Task Group on the Biological Effects of Space Radiation. Space Studies Board Commission on Physical Sciences, Mathematics and Applications, National Research Council. Washington, DC: National Academy Press

Safe Passage: Astronaut Care for Exploration Missions (2001) Board on
    Health Sciences Policy, Institute of Medicine. Washington, DC:
    National Academy Press

Chapter 8

SPACE LIFE SCIENCES INVESTIGATOR'S GUIDE

In recent years, space life sciences investigations have been conducted on board the Space Shuttle (also called the *Space Transportation System*, STS) and in space laboratories such as Skylab, Spacelab, and Mir modules. New experiments are being prepared for flying on board the International Space Station (ISS). Special laboratory equipment and experimental procedures must be specifically designed for use in space. In addition, flight experiments must fit within physical limits of the spacecraft and its resources constraints. Yet, as many experiments as possible are to be conducted on each mission to achieve maximum scientific return. This chapter reviews the constraints of space life sciences missions and the step-by-step procedures "to fly" an experiment.

Figure 8-01. View of the Spacelab module, which flew in the cargo bay of the Space Shuttle Columbia during 15 space missions ranging from 7 to 16 days between 1983 and 1998. (Credit NASA)

1   RESOURCES AND CONSTRAINTS OF SPACE LIFE SCIENCES MISSIONS

For many reasons, progress in human physiological research in space has been limited. The dearth of flight research opportunities reflects the low priority given to life sciences research in general. In addition, there has been a long absence of flight opportunities between Skylab and Shuttle programs and between the last Spacelab mission (Neurolab in 1998) and the time where ISS will be fully operational.

The nature of the current space program is such that there is much to do and a few flight opportunities that must be shared. In addition, these flight experiment opportunities are constrained in a number of ways, such as mass, volume, power, re-supply of consumables, and crew time. Finally, the limited opportunities impose that a space experiment must be as much reliable as possible. Both the hardware and the protocol must be evaluated and tested in order to guarantee that they will function properly in orbit. Consequently, experiments that might take weeks on Earth take years to plan and execute in space.

1.1   Opportunities for Space Life Sciences Experiments

Three types of flight experiments are currently solicited: a) *on-orbit experiments* that can be implemented on the ISS, the SpaceHab, or the Space Shuttle; b) *pre- and post-mission studies* involving data collection and analysis of biological specimens prior to and on return from space missions on the ISS, Space Shuttle, Soyuz, or unmanned biosatellites; and c) *laboratory ground-based investigations*.

1.1.1   International Space Station (ISS)

The arrival of the first permanent crew of the International Space Station (ISS) took place on November 2nd 2000. It signified the beginning of continuous stay in space of at least 15 years, with crew rotations about every 120 days.

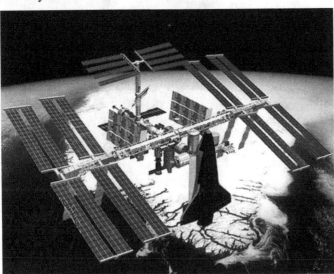

Figure 8-02. Artist view of the International Space Station in its final configuration, with the Space Shuttle docked to it. (Credit NASA)

Research opportunities are available on a limited basis during the construction phase of the ISS. Some research is accomplished during Space Shuttle missions when the Shuttle visits the ISS and during the time period between the Space Shuttle missions when the ISS crew acts as experiment operators and, if necessary, as subjects. It is also possible to conduct experiments on board the ISS during this period. However, ISS experiments are severely constrained by limitations on resources such as mass, volume, power, re-supply of consumables, and crew time.

Once assembly is completed (Figure 8-02), which was planned for 2006 before the *Columbia* tragedy, a permanent crew of seven astronauts should allow 160 hours/week of time dedicated to experiments (Table 8-01).

Parameters	*Characteristics*
Truss length	*108 m*
Total module length	*74 m*
Mass	*about 420 tons*
Maximum power output	*110 kW (30 kW for payloads)*
Total pressurized volume	*1200 m³*
Atmospheric pressure	*1013 mbar (1 atmosphere)*
Orbital altitude	*350-460 km*
Orbital inclination	*51.6 deg*
Orbital velocity	*about 8 km/sec*
Maximum crew	*7*
Crew time for science activities	*160 hours/week*
Data rate uplink	*72 kbits/sec*
Data rate downlink	*43 Mbits/sec (150 Mbits/sec from 2006)*
Ku-band coverage	*68% of time*
S-band coverage	*50% of time*
Anticipated lifetime	*>10 years*

Table 8-01: ISS characteristics at completion of assembly. The variation in altitude is a direct consequence of the 11-year solar activity cycle that causes the Earth's atmospheric density profile to vary (expanding the atmosphere at solar maximum). Due to the overall low altitude of the ISS, atmospheric drag causes a decrease in altitude of approximately 200 m per day. To counteract this height reduction, a periodic re-boost (that occurs approximately every 10 to 45 days) of the ISS is required, which increases the altitude temporarily. (Credit ESA)

1.1.2   SpaceHab

The current SpaceHab is modeled on the Spacelab orbiting laboratory, which was built by the European Space Agency for use in the Shuttle cargo bay. Spacelab was a pressurized module, 4 m in diameter and 7 m in length (Figure 8-01) equipped with standard experimental racks (0.48 m) which flew during dedicated life and material science missions of the Space Shuttle *Columbia* between 1983 and 1998.

SpaceHab was particularly useful to conduct life and material sciences experiments during dedicated missions while waiting for the completion of the ISS.* SpaceHab modules are also added to Space Shuttles missions visiting the ISS to carry supplies requiring a pressurized environment

SpaceHab is designed for housing a four-person crew in a pressurized laboratory within the Space Shuttle cargo bay. This laboratory includes temperature and moisture control, and power supply with AC and DC current supplied to all experiment locations, and high-data rate communications. The Research Double Module (RDM) is proposed on a commercial basis to microgravity experiments: it includes 6 standard double-rack locations, and storage lockers (up to 27 kg and 0.36 m³), for a total payload capacity of approximately 4000 kg. The crew has access to the RDM through a pressurized tunnel connected to the Shuttle middeck airlock.

Like its older brother (Spacelab), SpaceHab relies on the high bandwidth Ku-band signal processing of the Space Shuttle. However, during periods of communication blackout (loss of signal or LOS) data can also be stored on board and downlinked later.

A SpaceHab mission starts several years before the flight, when compatible experiments are selected for a mission payload. Experiment designers, called *Principal Investigators* (PIs), form an Investigator Working Group under the direction of a Mission Scientist. Together the group develops a complex timeline of science operations, squeezing the maximum experiment time into the SpaceHab flight. Missions generally last between 10 and 16 days.

In addition to the Commander, Pilot, and three Mission Specialists that composed a Shuttle and SpaceHab crew, scientists designate 3 or 4 Payload Specialist candidates, based upon their special skills or qualifications for the science payload. The candidates participate in crew training, lasting from 18 months to 2 years. Approximately one year before the flight, NASA selects one or two of the Payload Specialists as additional crewmembers. The remaining candidates, called *Alternate Payload Specialists*, work in the ground control center during the mission. They serve as a key link between the researchers on the ground and their colleagues in orbit.

---

* The Space Shuttle *Columbia* accident of 1 February 2003 occurred at the end of a dedicated SpaceHab scientific mission (STS-107). The downstream effects of this tragic event in terms of logistics and utilization scheduling of Shuttle, SpaceHab, and ISS are still being worked out. Assuming a successful recovery of the Space Shuttle program, regular Shuttle flights will be re-established to complete, maintain, and utilize the ISS. Some of these flights will presumably use the SpaceHab pressurized storage capabilities. The frequency of these Shuttle flights to ISS could even increase to compensate in part for the time delays occasioned by the loss of *Columbia*. However, this scheduling eventually compromises the opportunities for dedicated SpaceHab science missions.

1.1.3 Shuttle *Small Payload* Flight Experiments

Opportunities for flight experiments continue to be extremely limited at this time. For the period until complete construction of the ISS, a majority of the capacity of the Space Shuttle fleet is dedicated to assembly and operation of the ISS. Therefore, opportunities for Shuttle-based experiments are limited.

Nevertheless, the Space Shuttle can accommodate flight experiments with typical flight duration of 8 to 12 days. Equipment can be stored in the storage lockers (up to 27 kg and 0.36 m^3) on the forward bulkhead of the middeck (Figure 8-03). Each drawer has foam-rubber spacers to hold its content in place. The experiments themselves must require only limited crew training and involvement to execute. Experiment hardware which occupies or requires a large volume to operate will not likely be accommodated. Experiments that do not require Shuttle power (i.e., battery-operated) are more easily accommodated, since in general there is no power available in the middeck lockers during ascent and reentry.

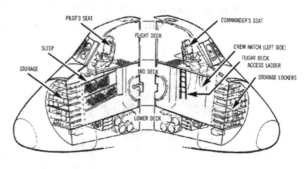

Figure 8-03. The living quarters in the Space Shuttle are the flight deck on top, which contains the flight controls and crew stations for launch, orbit and landing, the middeck, below, with the accommodations for eating, sleeping, hygiene, and waste disposal, and the lower deck, which contains the life support systems and a stowage area.

Opportunities are also available to perform experiments, collect samples, and take physiological measurements of the astronaut crew both prior to their space mission and following their return to Earth. Comparison between postflight and preflight measurements of a given parameter (for example: the maximal exercise capacity, or the locomotion behavior, etc.) is then used to evaluate the effects of spaceflight. Such proposals are also considered flight experiments.

Accessible to anyone with a good idea (and considerable money), the Small Self-Containers (SCC) or "Getaway Special" payloads, can also be used to conduct space life science experiments. The SCC are small (90 kg, 1.4 m^3) cylindrical containers attached to the inside wall of the Space Shuttle cargo bay. There are placed on board the Shuttle when allowed by space and weight restrictions. These containers must contain their own systems for power, handling data, and environmental control. Some of the systems may be turned on or off from the flight deck, but otherwise they are completely automatic. They must, however, adhere to flight safety guidelines. Many experiments proposed by students have flown in these containers.

1.1.4   Biosatellites (Bion/Foton)

As mentioned in Chapter 1, unmanned biosatellites were once used by the U.S. and Soviet Union to test the effects of spaceflight on animals before the first humans were sent into orbit. These biosatellites are now being used for research. For example, unmanned Bion/Foton capsules are regularly launched from Russia by a Soyuz launcher on an orbit of about 280 km for duration of approximately 15 days. Unlike Bion, Foton capsules are pressurized and temperature-controlled and can host a payload of 700 kg in a volume of 4.3 m^3. 800 W of electrical power are provided for the entire duration of the mission. After the flight, the capsule lands between the border of Russia and Kazakhstan (Figure 8-04). The biological specimen are removed from the capsule by a ground team and placed in refrigerated containers. The Foton capsule is then transported, first by helicopter, then by aircraft, back to the Soyuz plant in Samaria (Russia). The samples are then dispatched to the participating science teams via Moscow.

Figure 8-04. Russian Foton capsule after landing in Kazakhstan desert. (Credit InterKosmos)

The Foton capsule provides unique opportunity for flying biological specimen (animals, cells, and plants) when no crew activity is needed. Telemetry can be used to activate some procedures (e.g., fixation of cells, turn on or off the light, etc.) during the flight. Small onboard centrifuge generating centripetal accelerations of up to 1-g can also be utilized, to provide comparison with ground controls and make sure that the observed effects of the flight on the specimen are not due to the stress of launch and landing or to atmosphere changes. The samples are loaded in the capsule up to a few hours prior to launch.

For example, past missions have included studies of the gravity-sensing organs of plants roots (e.g., *Brassica*), the expression of genes that is modified in microgravity in plant cells (e.g., *Arabidopsis*), the influence of gravity on differentiation and tumor formation in cancerous cells (e.g., breast epithelial cells), and on calcium balance in muscular or bone (osteoblasts) cells (Gasset 2001).

Method	Microgravity	Duration
Bed rest	Simulated	3-12 months
Clinostat	Simulated	Unlimited
Centrifuge	$> 1\,g$	2-month (animals)
Drop tower	$< 10^{-4}$	seconds
Parabolic flight	10^{-1}-10^{-2}	20 seconds
Sounding rocket	10^{-5}	6-15 minutes
Biosatellites	10^{-5}-10^{-6}	15 days
Space Shuttle	10^{-4}	10 days
Spacelab/SpaceHab	10^{-4}	< 16 days
ISS	10^{-4}	Unlimited

Table 8-02. The various methods used to access to actual or simulated microgravity condition. Drop towers are usually evacuated tubes in which an experiment capsule is released and allowed to free-fall. Biosatellites include the Russian retrievable capsules Bion and Foton.

1.1.5   Ground-Based Investigations

Flight investigations must represent mature studies strongly anchored in previous ground-based research or previous flight research. Ground-based research may, and usually must, represent one component of a flight experiment proposal. Ground-based studies can be performed in reduced gravity, hyper-gravity, or normal gravity environments.

Facilities which provide a reduced gravity for limited duration on Earth include drop towers, parabolic flight, and sounding rockets (Table 8-02). Other ground-based laboratory facilities where the effects of gravity, or hypergravity, can be evaluated consist of clinostat (for cells), animal- or human-rated centrifuge (short- or long radius) (Figure 8-05), slow rotating rooms, and bed rest clinics.

However, to be supported by a space agency, it is usually recommended that this ground-based research be limited to activities that are essential to the final development of an experiment for flight and for the completion and publication of the scientific results of the experiment. Preparatory ground research designed to define a mature spaceflight experiment can also be proposed separately and in its own right as part of the ground-based program.

Figure 8-05. The NASA Ames Research Center centrifuge. This 8.5-m radius centrifuge is used to evaluate flight hardware as well as to test the effect of hypergravity on human and non-human (rodents, cells, plants) subjects. Human studies range from 1 to 12.5 g. (Credit NASA)

1.2   Constraints

Life sciences research typically flows in a progression, beginning with challenging problems and questions crucial to our advancement into understanding the relationship between gravity and life; the search for solutions and answers through research in the biological and medical areas; validation and demonstration of crucial concepts in-flight which eventually lead to the knowledge, experiments, and progress which will enable us to meet the challenges of the future (Figure 8-06).

The challenge in developing a flight experiment is to package the science objective and the experiment apparatus in ways that satisfy the spacecraft requirements and safety, but also the ethical considerations and the unique constraints of human space missions.

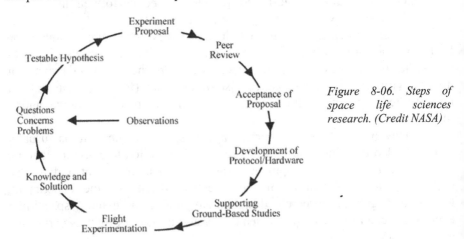

Figure 8-06. Steps of space life sciences research. (Credit NASA)

1.2.1 Ethical Considerations

All use of human subjects for research in conjunction with experiments on Shuttle, ISS, or pre- and post-mission studies must comply with NASA Policy Directive NPD 7100.8C, *Protection of Human Research Subjects*. In order to get approved, all experiments must receive approval from their local ethical committee, from their respective space agency ethical committee, and from NASA ethical committee (also called the *Institutional Review Board*).

Informed consent of human subjects must be obtained prior to carrying out any study in space, and potential applicants should be aware that obtaining such informed consent involves a uniform process regardless of the country of origin of the applicants. The availability of consenting subjects may impact the probability of achieving experiment objectives within the expected timeframe. Human life science experiments generally require at least two crewmembers (experimenter and subject). Although operating most experiments will not pose a problem to astronauts, there is no assurance that all crewmembers will agree to participate as subjects in experiments.

All proposals for the use of vertebrate animals must be accompanied by a certification of approval from the Investigator's institutional Animal Care and Use Committee and a Public Health Service Animal Welfare Assurance number. The use of animal subjects in all NASA ground and flight research is governed by NASA Policy Directive NPD 8910.1, *Care and Use of Animals*, and NASA Procedures and Guidelines NPG 8910.1, *Care and Use of Animals*.

1.2.2 Other Considerations

The major difference between a flight experiment and the same experiment in a standard laboratory on Earth is the smaller number of subjects or specimens and observations in the flight experiment. Experimental sample size has been and will continue to be small. Due to limited space and power, a finite number of animals (including a limited number of species) or specimens is available for in-flight research. This limited number often requires the development of elaborate and detailed sharing plans to maximize their use. For example, sharing plans for human blood samples in orbit are a prerequisite since the volume of blood draw is limited.

Also, in Earth laboratories, it is common to repeat experiments. This is even the basis for a scientifically sound investigation. Every published scientific manuscript contains (or should contain) a Method section detailed enough to allow other scientists to repeat the experiments, to verify and confirm the proposed hypothesis or interpretation of results. For space experiments, it is very rare when the exact same experiment is repeated on a second or a third mission, because of the financial and time constraints. The investigators are doomed to success in the first trial.

In order to limit the risk of failure in scientific return from a mission, and to give more opportunities to investigators, it has become common practice to "integrate" several experiments. In this process, multiple investigators must share the same equipment. The inconvenient is that some common procedures or conditions must be "negotiated" between the investigators (for example, a given centrifuge velocity, or a given temperature for animal habitats), and that a malfunction of the equipment might alter several experiments.

For a life sciences mission, crew time is the most precious resource. It is also known that activities require more time to be performed in 0-g: For example set-up of complex experimental equipment takes approximately 40% longer in space than on the ground. Other activities may require extra operators (e.g., dissections or hazardous operations, such as rotating chair). As a result, 4 hours of crew time in 0-g correspond to only 2.4 hours on the ground.

The availability of the crew for training prior to the mission is also limited. Training requirements depend on the complexity of both the individual instruments and the integrated payload.

Also, during the flight, the investigators have limited access to real-time data. In an Earth laboratory, a flaw in one experimental protocol is immediately detected and corrected before the experiment continues. During a spaceflight experiment, it is difficult to assess the exact situation remotely, and to suggest changes in an experimental protocol that has been designed over several years. Also, the suggested changes could have an impact on other experiments. Perhaps for these reasons, the results of flight experiments are mainly unexpected results. In some cases, results of space investigations have confirmed classical or generally held hypotheses. However, most results have been startling and unexpected, requiring researchers to reexamine their assumptions about the intricate relationship between gravity and life.

1.2.3   Space Shuttle Constraints

On Shuttle missions, the crew typically consists of seven crewmembers. The Commander and Pilot are responsible for spacecraft operations and are available as subjects or operators on a limited basis. Two Missions Specialists are trained for EVA and are available after the EVA has been completed. Three Missions Specialists are dedicated to payload activities as operators, and as subjects on a voluntary basis.

A typical Space Shuttle on-orbit day is as follows (per crewmember): 8 hours for sleep; 6 hours for pre- and post-sleep activities (e.g., debriefing, revising schedule, e-mail, leisure time); 2 hours for lunch, hygiene, and exercise; and 8 hours for payload activities.

In addition, two half-day off are required on 14-day missions. The first day and the last two days of the mission have limited time available for payload activities (Mission day 1: approximately 2 hours per crewmember; last day minus one: approximately 3 hours per crewmember; last day (entry day): approximately 1 hour per crewmember).

The number of crew subjects available to perform short-duration human studies is restricted due to the limited amount of crew time available for such experiments, and there is no assurance that all crewmembers will agree to participate as subjects in experiments.

Equipment and supplies that do not have a long shelf life may be loaded onto the Shuttle days or weeks before launch. It is possible to arrange for late preflight installation (approximately launch minus 20 hours) (Figure 8-07) and early postflight recovery (landing plus 3 hours) of equipment, supplies, and data which have time- or temperature-critical sensitivities. Note that there are periods of time before the flight and after landing when no access to the experiment is possible and maintenance of the equipment integrity must be assured. The availability of Shuttle resources for experiments that require animal as subjects is also extremely limited for short-duration experiments.

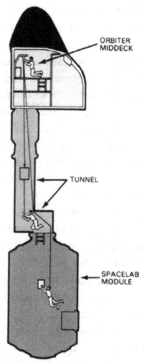

Figure 8-07. Only limited science activity occurs at the launch pad. A final test is run to verify the integrity of the entire set of pad/Shuttle/module system connections. Biological specimens and perishable items may be loaded in the Shuttle middeck or module only a few hours before launch. When it is necessary to install scientific samples and specimens in the module, it can be entered through a vertical access kit. (Credit ESA)

1.2.4 International Space Station (ISS) Constraints

During its initial phase of operation, each ISS increment includes three crewmembers (also called the *Expedition* crew) for duration of 3-4 months. All crewmembers participate in life sciences experiments as operators, and as subjects on a voluntary basis.

Transport frequency, power during transportation, and mass of transported items are all severely constrained throughout the ISS assembly period. The primary opportunities to transport scientific equipment, supplies, and samples are on the periodic logistic flights of the Shuttle to ISS specifically dedicated to this purpose. In addition, modest capabilities for research-related deliveries and sample returns are available on the Shuttle flights dedicated to assembly of the ISS, and on the Soyuz "Taxi" flights. Refrigerated and frozen transport of samples on the Shuttle and Soyuz is very limited, and during certain timeframes refrigerated and frozen storage may not be available on ISS. Power outages may also be experienced during the assembly of ISS.

Experiments with few or simple crew-supported in-flight activities have the greatest potential for selection during this time frame due to limitations on crew time and crew training. The availability of the crew for science operations and as subjects of research is also extremely constrained during ISS assembly. On average, a total of 2 to 5 hours of crew time per week are allocated to the entire field of life sciences experiments, including experiment operations and equipment maintenance. Estimates of crew time required to complete the experiment must include the time needed for crewmembers to both operate an experiment and serve as subjects.

Based on the Mir experience, after assembly completion, a typical on-orbit day on the ISS averaged over a 3-month stay including one EVA (per crewmember) should look as follows: 8 hours for sleep; 4 hours for pre- and post-sleep activities; 5 hours for lunch, hygiene, and exercise; 1.5 hour for preparing EVA; 1.5 hour for public relations; and 4 hours for payload activities.

In addition, two days off will be required every week. The first two days of the mission have no time available for payload activities. The first measurement on-board ISS are not performed before Mission Day 4. EVA preparation by the ISS crewmember prevents payload activities for one week. Moreover, crew time for data collection before and after flight is extremely limited and consideration of current exercise countermeasure protocols is strongly recommended

When the ISS will be fully operational, these resources will considerably increase, with crew time of 120 hours per week for science activities. Shuttle flights to ISS will occur approximately every 80-120 days. This allows returning periodically to Earth samples or specimens from experiments. Depending upon the duration of the active phase of the experiment, storage of samples up to 120 days are possible. During this phase electronic data may also be transmitted to the ground or stored on the ISS.

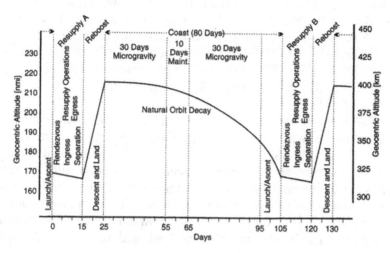

Figure 8-08. ISS experimental cycle. A visiting vehicle (e.g., Progress or Shuttle) rendez-vous with the ISS when the orbit is relatively low, and during the next 15 days, the crew performs the necessary re-supply operations. During the following 10 days the re-boost operations are performed to raise the altitude of the ISS, after which the ISS will "coast" for a period of 80 days. This ensures two 30-day periods of continuous, good quality (less than 10^{-6} g) microgravity. (Credit ESA)

Time-critical supplies or specimens can be loaded in the Space Shuttle between 40 and 20 hours before launch. There is a minimum storage period of 5-6 days before starting an ISS experiment, since the Shuttle has to travel to (2 days) and dock on the ISS and the experiment has to be transferred to its ISS facility. The experiment can then stay on board the ISS for the duration of one or several increments (Figure 8-08). After the last increment, the samples will be transferred back into the Space Shuttle for a minimum of 5-6 days, and then returned to Earth, where they will be made available to the scientists approximately 3 to 5 hours after landing.

1.2.5 Soyuz "Taxi" Flights Constraints

Samples and specimens can also be transferred to and from the ISS using the Soyuz capsule. The Soyuz "Taxi" flights are dedicated to exchange the Soyuz emergency return vehicle on ISS and are therefore planned exactly every 6 months. Two to three cosmonauts (with one seat being commercially available for about 20 million dollars!) participate in such flights to the ISS, offering approximately five effective onboard ISS experiment days.

The total length of flight is around 10 days with 2 days in orbit before docking with ISS. The time between de-orbit and recovery is in the order of a few hours. The Soyuz capsule has a pressurized volume of about 4 m³. The mass of payload that can be carried to the ISS is about 250 kg, and the mass that can be brought back to Earth is about 150 kg. However, most of this mass is used to carry supplies to the ISS cosmonauts (water, food, and personal items). Therefore, only a limited number and simple experiments can be accommodated. Based on the European experience, a passenger of a Soyuz "Taxi" mission is allowed to carry 12 kg of equipment or samples up (volume 0.4 x 0.4 x 0.4 m), and to return 4 kg of equipment (e.g., tapes, films) or samples down to Earth.

The experiments are usually performed in the Russian section of the ISS. Crew time can also be used to perform experiments using hardware already on board the ISS, in the Russian module, or brought up by the *Progress* cargo re-supply ship. Sample return immediately after termination of the mission is extremely limited and without temperature control capability (room temperature only). These flights are best suited for monitoring human adaptation process on orbit, or the activation of (semi-) automatic experiments. "Taxi" flights are also well adapted for pre- and postflight experiments.

1.2.6   Constraints of Pre- and Post-Mission Studies

Opportunities are available to perform experiments, collect samples, and take physiological measurements of the astronauts both prior to their space mission and following their return to Earth. However, access to all crewmembers immediately before a space mission is extremely limited due to their very busy training schedule. Preflight baseline data collection generally takes place several weeks prior to launch. Preflight measurements can be repeated several times, for example at a one-month interval, in order to evaluate the test-retest repeatability and the variance of the responses studied.

On a typical Space Shuttle entry day, the crew has to be awake 8-14 hours before landing (3 hours for post-sleep activity, 4 hours for de-orbit preparation, 1 hour from de-orbit to landing). A mission rule also imposes that the maximum wake time for a crewmember on landing day is 18 hours. The duration of crew transfer (Figure 8-09) from the runway to the flight clinic takes about 1.5-2.5 hours (depending whether the crew decides for a "walk-around" on the runway). The first activities there include a medical exam (0.5 hour), a visit with the family, a meal and a shower (1 hour). Another mission rule states that if data collection during experiments on landing day exceeds 4 hours, a 1-hour break must be observed. Consequently, the Space Shuttle crewmembers' availability on landing day rarely begins before 2 hours after wheel-stop and does not exceed 4 hours for all scientific investigations. Access to long-duration ISS crews for a pre- and post-mission study is even more limited. Availability of ISS astronauts for research tests on the day of return to Earth, or the day after, may be as little as one hour per day total.

On a typical Soyuz entry day the crew spends up to 10 hours in the Soyuz up to the landing. A medical exam is performed on the landing site. The crew is then transferred first by helicopter and then by plane to the Gagarin Cosmonaut Training Center at Star City near Moscow. During the transfer, which can last 6-8 hours, the crew takes some rest and has a meal. The first postflight data collection does not generally take place before 20 hours after landing, its duration does not to exceed 2 hours, with the crew remaining in the supine or sitting position.

Figure 8-09. Crew Transport Vehicle (CTV) docked with the Space Shuttle after landing. Once access to the Shuttle is possible, physicians board the Shuttle and conduct a brief preliminary examination of the astronauts. They assist the crew in leaving the vehicle and removing their launch and reentry suits. (Credit NASA)

2 HOW TO "FLY" AN EXPERIMENT

2.1 Flight Experiment Selection

Proposals are solicited through Announcements of Opportunity (AO) issued at intervals of about 18 months. As of May 2003, AO could be found at the following URL address: http://research.hq.nasa.gov/research.cfm. The AO are internationally coordinated between the space agencies for research in specific areas of life and physical sciences. The International Space Life Sciences Working Group (ISLSWG) coordinates space life sciences research. The international AO allow for a worldwide scientific competition and cooperation, by forming the best scientific teams, as well as offering access to an instrument pool from all ISS partners.

Participation in these AO is usually open to all categories of organizations, industry, educational institutions, other non-profit organizations, research laboratories, and government agencies. Proposals from entities within countries from the ISS partners are made to the solicitation from their corresponding space agency. Present or prior support by any space agency of research or training in any institution or for any investigator is not a prerequisite to submission of a proposal or a competing factor in the selection process. Selections through AO can be for periods of many years, involve budgets of many millions of dollars for the largest programs, and usually are awarded through contracts, even for non-profit organizations, although occasionally grants are also used.

Independent internationally recognized experts, so called *peers*, first evaluate the proposals based on their space relevance and *scientific merit*. As a result of the peer review, all proposals are ranked according to an absolute scale (from 0 to 100) or a designation of "not recommended for further consideration" based upon the intrinsic scientific or technical merit of the proposal. This score reflects the consensus of the review panel.

A second review is an evaluation of the *feasibility* of implementing the proposed work using available facilities on a space platform. The flight feasibility review is conducted for each flight experiment proposal that receives a scientific merit score greater than a threshold score (usually 75) agreed upon by the International Life Sciences Working Group Steering Committee. This study includes a safety review, an assessment of accommodation possibilities on board ISS, as well as a verification of the availability of the required utilization resources and flight opportunities. An international team of engineers and scientists experienced in the development of spaceflight experiments conducts this review.

Once the proposal is recommended by the peer group and its technical assessment confirms feasibility, the next decision layer is based on programmatic aspects, such as funding confirmation for the project realization, confirmation on relevance to the program priorities by the sponsoring space agency, and availability of the utilization resources requested.

The next phase, the *implementation* of the flight experiment implementation, is actually a multi-step process (Figure 8-10). Following the complete review of flight proposals, successful investigators will receive a letter informing them that their experiment has been selected for entry into a definition phase. During the definition phase, the agency with management responsibility for the experiment will interact with the investigator to determine specific hardware and operational requirements needed to achieve the proposed objectives. Identification of issues that will affect implementation of the spaceflight experiment and refinement of the funding requirements are key components of the definition phase. After successful completion of this phase, the experiment will be selected for flight and enters into a development phase, leading eventually to implementation on a space mission. Proposals are usually funded in one-year increments until the experiment is completed. Detailed budgets are refined or negotiated for each flight experiment during each phase. The flight experiments selected can be reviewed periodically for technical progress, availability of flight opportunities, implementation feasibility, and to ensure that the science continues to be relevant. This review may result in a decision to deselect a flight experiment prior to its implementation or completion.

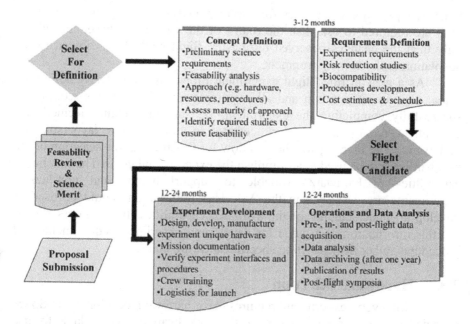

Figure 8-10. Steps of the experiment proposal selection procedure. The time from proposal submission to acceptance takes typically 6 months. This period can be shorter for smaller projects or longer for larger projects. The time from acceptance to the actual launch of a payload on the ISS is payload-specific. The time period can typically take 1 to 5 years depending on the first instance on the time needed for payload development. (Credit NASA)

2.2 Experiment Design

The experiment objectives can include applied aspect of space research (e.g., the assessment of operational issues in support of NASA's or other space agency's programs) or fundamental research. In the first option, the so-called "Detailed Supplementary Objectives" (DSO), the experiments are usually performed in the Space Shuttle. The number of subject per flight is very limited (1-2), but there are many flights available. So, a DSO experiment is typically performed on ten or more subjects. However, in these experiments, the hardware and protocol must be extremely simple, at least for the flight portions. In fact, most DSOs are performed pre- and postflight.

When the experimental objectives fall into fundamental research, the experiments are performed on a dedicated science mission using SpaceHab, or during one (or more) increment aboard the ISS. The number of subjects will obviously be smaller (2-8). Therefore, the flight experiments require control studies for small N data set. In particular, the dependent variable needs to be well defined (i.e., valid, reliable, relevant, practical); the astronauts' preflight data must be compared with those of control groups on Earth; preflight tests must be repeated several times for variance analysis; and multiple postflight tests are required to establish the return of the variable to baseline. Indeed, only then the changes observed during the flight could be attributed to the adaptation to the new environment.

As a general rule, flight experiment proposals must clearly define the actual experiment duration and all requirements and conditions required to successfully completing the experiment. Be sure to explain succinctly all experiment requirements and procedures in terms that a layperson can understand. The investigator should allow for flexibility in the selection of the best hardware to be used to accomplish the experimental goals. The functional capabilities of hardware available to support human and non-human experiments are described in the AO. This information should be used to develop an understanding of the available capabilities. Investigators should use this information as a guide for developing experiment requirements and procedures rather than selecting specific hardware items.

2.3   Hardware Selection

Not every experiment requires new hardware development. Space agencies have a growing inventory of flight hardware that is available for use by investigators (Table 8-03).

Some investigators may wish to develop their own special experiment hardware to work in conjunction with the facilities and functional capabilities of existing hardware. Design, construction, and flight of major experiment-unique equipment hardware items or facilities usually require the commitment of large quantities of resources (power, crew time, volume, and budget). Below are some tips for the design or selection of flight experimental hardware:

- **Weight and volume** must be as small and compact as possible. It must be simple and intuitive to use because crew training will be limited.

- **Power and data** management needs add immensely to complexity. It is preferable to use non-powered or battery-operated equipment if possible, and to store the data in the instrument. Although it is possible to store the data on board the Shuttle or ISS, or to use the downlink capability, providing for data recording within the instrument reduces the interface count, chance for failure, and competition for spacecraft capability. This option does not provide a data quick-look capability, but quick-look data are of benefit only if one can act on the information.

- **Data management** should use identified standard and meet laboratory data analysis capability (e.g., provides useful information). Since experiments can take years to fly, non-standard ground-support equipment might not be available by the time the experiment is performed.

- **A long shelf life**, "bullet-proof" technology desired. The utilization of the ISS should cover a period of more than 10 years[*], and some of the equipment (e.g., racks, freezers, treadmill) will stay on board for all that period.

- **Hardware must be modular** and should be build such as it is easy to replace and upgrade components. At forty-thousand Euros per kilogram, the cost of returning failed equipment to the ground for repair and returning it again to space becomes prohibitive. In addition, the disruption of the associated research program, use of limited crewtime, and impact on the restricted upload/download transportation capability would be significant. A well-balanced reliability and maintainability approach is thus necessary.

- **Last but not least, think zero-g!** Things people hardly notice can be big problems in space. A broken test tube, a spilled liquid, or a dropped screw can float through the cabin instead of falling harmlessly on the floor, creating a potential hazard for the crew. As we seldom are exposed to gravity levels other than 1-g for any length of time, we have developed a "1-g mentality". We use gravity in our daily life without even thinking about it and have difficulty comprehending the appropriate design of space hardware or human-machine interfaces in reduced gravity. To design space hardware, engineers must develop a microgravity mentality, rather than solely a 1-g mentality.[†]

In addition, equipment that is carried within the Shuttle, Soyuz, or ISS must meet the requirements for structural integrity, safety, flammability, odor, and toxicity. Usage of materials that produce toxic off gassing is avoided in habitable areas, except in controlled enclosures, such as the "glove box" for the manipulation of biological samples.

---

[*] The Russians have a saying: "if it is not broken, don't fix it!". The Mir space station was originally planned to stay eight years in orbit, and was actually utilized for more than fifteen years.

[†] According to NASA, approximately 40% of equipment flown in space for the first time does not work, often due to heat build-up from lack of convection, lack of dissipation of air bubbles, or designs more appropriate to normal gravity than microgravity.

Experiment Hardware	Shuttle	ISS	Agency
Physiological Monitoring			
Manual Blood Pressure Device	X	X	NASA
Automatic Blood Pressure System	X		NASA
Continuous Blood Pressure Device		X	NASA
Combined Blood Pressure Monitoring		X	NASA
Cutaneous Electrical Muscle Stimulator	X	X	NASA/ESA
Pulmonary Function System		X	NASA/ESA
Gas Analyzer Mass Spectrometer		X	NASA
ECG / EMG / EEG	X	X	NASA
Holter Monitor	X	X	NASA
Pulse Oximeter	X	X	NASA
Respiratory Impedance Plethysmograph	X	X	NASA
Ultrasound Doppler		X	NASA
Venous Occlusion Cuff and Controller		X	NASA
Sample Collection and Stowage			
Human Sample Collection Kits	X	X	NASA
Exercise			
Cycle Ergometer	X	X	NASA
Treadmill	X	X	NASA
Interim Resistive Exercise Device		X	NASA
Muscle Strength, Torque, and Joint Angle			
Muscle Atrophy Research and Exercise System		X	NASA/ESA
Resistive Exercise Device		X	NASA
Hand Grip/ Pinch Force Dynamometer	X	X	NASA/ESA
Cardio-Vascular Loading			
Lower Body Negative Pressure	X	X	DLR
Activity Monitoring			
Activity Monitor		X	NASA
Medical Procedures			
Injection and Infusion System	X	X	NASA
Eye Movements			
3 D Eye Tracking Device	X	X	DLR
European Physiology Modules			
Multi Electrode EEG Mapping Module		X	ESA
Bone Analysis Module		X	ESA
Body Movement Analysis Instrument		X	ESA
CARDIOLAB		X	ESA
Physiological Pressure Measurement Instrument		X	ESA
Xenon Skin Blood Flow Measurement Instrument		X	ESA

Table 8-03. Summary of available hardware to support human subject research on board the Space Shuttle and the ISS. (Source NASA)

2.4 Feasibility

Of particular concern regarding the evaluation of the feasibility of a proposal is the identification of risk factors that could impact the implementation of an otherwise meritorious proposal. For example, the feasibility of implementing a scientific experiment and associated risks will be evaluated using the following technical criteria:

- **Functional requirements:** Will the planned flight and ground hardware meet the requirements of the experiment? What experiment-unique hardware will be required, and can it be developed in time for projected flight opportunities? Are the numbers of subjects or specimens required attainable within a reasonable period of time (1-2 years) considering projected flight opportunities and other competition for those flight opportunities?

- **Operational feasibility:** How complex are the experiment procedures? Will the crew have sufficient time to be trained to perform the experiment? Will they have sufficient time in space to perform the experiment? Are the requirements for launch vehicle loading and unloading of the experiment specimens compatible with the capabilities of these vehicles? Can requirements for data collection on human subjects be accommodated in the preflight and postflight schedules for the astronauts? Has the experiment protocol taken into account the unavoidable period of time between the launch of an experiment and the actual initiation of the experiment? Will the experiment requirements for crew time, experiment volume, mass, power, or other features of on-orbit operations (such as temperature-controlled storage) affect the completion of this or other experiments? What other impacts will the experiment have on activities or experiments planned for the same mission?

- **Environmental health and safety:** Are there elements of the proposed ground or flight activities that pose concerns for the health and safety of personnel or the environment? For experiments that utilize the crew as research subjects, could the implementation of these experiments, even if considered safe, lead to an impact on the performance of the human subjects with respect to their other crew duties? Is it possible that specific restrictions on the human subjects (such as diet, exercise, etc.) will interfere with their other activities?

Using the risk factors identified in the evaluation of the feasibility of a proposal, a score is assigned to indicate this level of uncertainty. The proposal are scored "low risk", "medium risk", or "high risk" when the risk to the successful achievement of objectives is considered minimal, moderate, or extreme, respectively.

Figure 8-11. Artist view of the European Columbus module for the ISS, and the mock-up used for crew training prior to the flight. (Credit ESA)

2.5   Experiment Integration

The primary goal of the experiment integration process is to assemble a group of experiments on a space mission in a way that maximizes the scientific return of the mission while effectively utilizing the resources of the space vehicle (Figure 8-11).

During this process, the Principal Investigator is responsible for providing the Mission Manager with a complete description of the experiment, its equipment, and the interface with the space vehicles. Before the final decision to build the equipment, and during its construction, the Principal Investigator participates in technical reviews. When the equipment is ready and "integrated" in the space vehicle, the Principal Investigator is again asked to participate in the functional evaluation of the equipment and its utilization procedures.

2.5.1   Key Documentation

To facilitate the interface between the experiment and the mission management, the Principal Investigator is requested to submit an Experiment Requirements Document (ERD) that describes the science objectives (summary), the characteristics of the equipment to be used (mass, size, power, thermal control, interfaces), when (pre-, in-, postflight) and how (command, data management, software, man-machine interfaces, data analysis) it will be used. Besides formally specifying the requirements for performing the proposed experiments, the ERD provides hardware developers and mission management with a convenient source of information for the implementation of life sciences flight experiment projects. Inputs from this document are placed into a controlled database which will be used to generate other documents.

Figure 8-12. Astronaut Rick Husband reads over the Payload Flight Data File in the cabin of Space Shuttle Discovery. (Credit NASA)

2.5.2   Reviews

For most research projects, a series of formal program reviews play an important role in coordinating the experiment integration process. Each review occurs at a natural transition point in the integration activities and has a specific purpose associated with it:

- **The Experiment Requirement Review (ERR)**. The experiment requirements by several Principal Investigators are integrated. As a result of this review, the available space resources (i.e., Shuttle or ISS) will be allocated to the instruments and subsystems elements.

- **The Experiment Preliminary Design Review (PDR)**. Its purpose is to finalize the mission requirements (resources, crew time), baseline the equipment design interfaces, finalize the safety verification methods and begin their implementation, begin planning for physical integration and flight support, and finalize the planning for crew science training.

- **The Experiment Critical Design Review (CDR)**. Its purpose is to review the final design and compare with the requirements to verify compatibility with the space vehicle and the other experiment equipment, and to verify overall system safety.

After the equipment has been delivered, the Principal Investigators (PIs) are expected to support the integration of their hardware into the space vehicle and its preparation for flight. Documents such as procedures for interface tests, calibration, special tests, servicing, and maintenance should be provided. The PI is also expected to participate in science verification testing of their equipment, both before (in a high fidelity mock-up) and after integration in the space vehicle.

Integration in the Space Shuttle typically occurs three months before flight, but the equipment should be delivered to NASA KSC not less than one year prior to launch. Experimenter activity in the SpaceHab or the Shuttle is very limited during integration, but there may be provisions for hand-on work if needed (i.e., loading samples or specimens) and if the requirement is identified early. Small experiments and equipment can be installed in the Shuttle middeck lockers up to 12 hours before launch (see Figure 8-07).

Figure 8-13. Astronaut Pedro Duke prepares an experiment with the help of the Payload Flight Data File. (Credit NASA)

2.6   Crew Science Training

While the flight equipment is being integrated in the space vehicle, exact copies of this equipment (so-called "training models") are used for training. Training is aimed at ensuring ground personnel and flight crew involved to perform an experiment safely and effectively under both nominal and off-nominal operations situations.

The first step is to *inform* the crewmembers selected for the given mission about the rationale of the research and the associated hazards. Once the crewmembers have accepted to participate in an experiment and signed the Informed Consent Form for this experiment, they are briefed in more depth about the scientific rationale for the experiment. This phase is important should the astronauts discover new research approaches during the flight, which were not foreseen by the investigators. With a complete understanding of the scientific background and rationale for the experiment, the astronauts could then suggest, in agreement with the investigators, appropriate changes in the flight protocol to maximize scientific return.

The second step is to *train* on a copy of the flight equipment, in order to refine the experiment operations flow. The activities related to an experiment are described in sequences of operations (or Functional Objectives, FO) that satisfy a specific engineering or scientific goal. A FO typically consists of a number of functional steps, such as activation, calibration, various steps of the experiment operations, standby, and deactivation. The experiment operations flow is a basis for developing step-by-step procedures that will be used by the crew, and followed from the ground, during the actual mission. Procedures for both nominal and contingencies (e.g., trouble-shooting of the equipment) operations are developed. These procedures and the reference material for each experiment are assembled in a Payload Flight Data File. This document is stowed on board for use by the science crew during the mission (Figures 8-12 and 8-13).

The third step is to *evaluate* the time required to perform the required operations (proficiency) on the ground. This time is multiplied by a factor of 1.4 to take into account the microgravity factor. Once each experiment has done the same evaluation, the mission management can decide if and when several experiments can be executed in parallel. These evaluations allow putting together the operations flows during the mission, which will lead to the Crew Timeline (Figure 8-14). At this point the crew is trained as a team to use the experiment procedures according to the timeline. Occasionally, integrated simulations are conducted with the support of ground controllers in the same configuration as during the actual mission.

Science briefing and familiarization courses are conducted approximately one year before launch. Hands-on familiarization using a copy of the flight model occurs 3-6 months before launch. Simulations of on-orbit operations are conducted in the final months before launch.

2.7 In-Flight Science Operations

2.7.1 Organization

As with all space missions, the crew in orbit gets instructions during flight from a Mission Control on the ground. The ISS flight operations are performed in a decentralized manner, under the overall responsibility of the Mission Control Center in Houston (MCC-H) for system operations, and the Payload Operations and Integration Center in Huntsville (POCC) for payload operations. Other control centers have responsibility for operations of certain modules (such as the Columbus Control Center) and its associated payloads, interfacing with the MCC-H and POIC, respectively. Scientists can control their experiments and payload operations from a Facility Responsible Center (FRC) or from their own User Home Base (USB). More commonly, the scientists will be staying in the Science Control Room at the MCC-H during the periods where their experiments are performed. Some science teams can remotely control their instruments, sending uplink commands directly from the MCC, or from remote centers via the POIC, to their equipment in orbit.

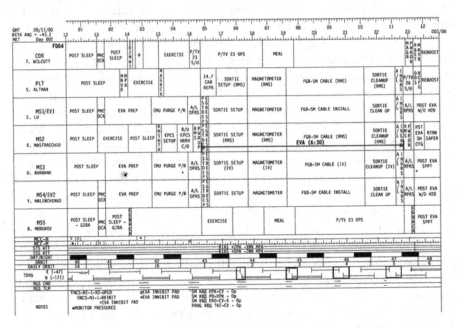

Figure 8-14. Timeline for 12 hours of crew activities during the Space Shuttle STS-106 mission. Crewmembers are identified on the left column (CDR: Commander; PLT: Pilot; MS: Mission Specialists). Flight attitude of the Space Shuttle and the visibility of the communication satellites are provided on the lower rows. (Credit NASA)

For SpaceHab missions, a few hours after launch, the crew begins to activate the laboratory and check out equipment. Then, the science begins. Mission and Payload Specialists work through their minute-by-minute schedule of experiment activities, often staffing the SpaceHab around the clock with 12-hour shifts. Control team members make sure that activities are on schedule and everything is operating properly, and determine the impact of mission events on the science timeline. The Mission Scientist and the Principal Investigators meet once or twice every day during the mission to review progress and make schedule adjustments for the upcoming day (Figure 8-15).

SpaceHab missions are short-term space missions with a large crew size. The SpaceHab crew can concentrate on science activities, while the orbiter crew takes care of the routine Shuttle maintenance and flight control. Therefore, the schedule for scientific activities is densely packed. The schedule for experiments on board the ISS is much lighter. This is because the ISS crew is now composed of three persons only, whose activities are shared between maintenance, repair, and research. The ultimate promise of space research on the ISS will be realized only when a permanent crew of seven will be available. Similar experiments will then be performed for months on board the ISS, instead of just days or weeks at a time on the Shuttle and SpaceHab. The lessons learned over more than 15 years of Spacelab flights lay the foundation for the international partners of the ISS as they design laboratory equipment and make plans for science research on board their full-time laboratory in orbit.

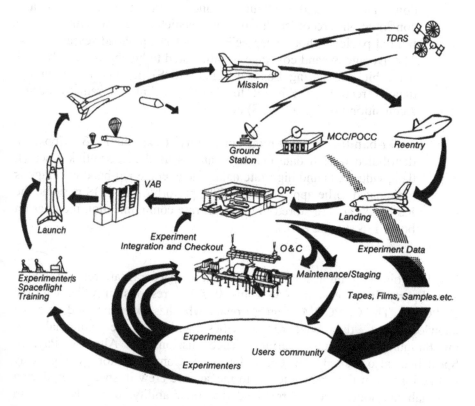

Figure 8-15. Shuttle-Spacelab operations cycle. Operations comprise integration, training, real-time monitoring, and pre-and postflight baseline data collection. Similar tasks are required during an ISS science mission. (Credit NASA)

2.7.2   Communications

The NASA Tracking and Data Relay Satellite (TDRS) system is the primary communications link between the Space Shuttle and ISS with the ground (to the ground station at White Sands in New Mexico). The data is then routed to various locations around the world, including Europe, Russia, and Japan.

Communication between Space Shuttle and ISS and ground is performed via S-band (bi-directional, i.e., uplink and downlink, or to and from the ISS, respectively) and Ku-band (downlink data only):

> **- S-band:** The uplink capability (maximum 72 kbps) is used for uplink of payload and system command, files and audio data to the ISS. The downlink capability (maximum 192 kbps) is used for downlink of audio, caution and warning data, command confirmations, recorder dumps and telemetry, i.e., system data and limited payload "housekeeping" data but not payload research data. The average S-band coverage is anticipated to be approximately 50% per orbit, and during Loss-Of-Signal (LOS) periods, the downlink data is recorded on board for later downlink on request during Acquisition-Of-Signal (AOS) periods.

> **- Ku-band:** The downlink capability (maximum 43 Mbps) is distributed over 8 data channels, and is used to downlink research data, video data and high-rate data. The average Ku-band coverage is anticipated to be approximately 68% per orbit. For LOS periods, the Ku-band data are stored in one high rate communication recorder on board ISS (Figure 8-16).

Usually, voice communications from ground to spacecraft are relayed though a third party. In the case of NASA missions, this responsibility lies with the CAPCOM. The CAPCOM is an astronaut who has extensive and personal knowledge of the onboard activities and of the potential for miscommunication issues. During science missions, the Alternate Payload Specialist, also called the Crew Interface Communicator (CIC), relay instructions from the science teams to the science crew in space. Yet, during Spacelab missions, researchers found that their ability to collaborate with crewmembers-operators was significantly impaired by this two-step process. Alternately, if many individuals were to have access to the communication system, the aggregated messages could become overly burdensome and interfere with effective communication.

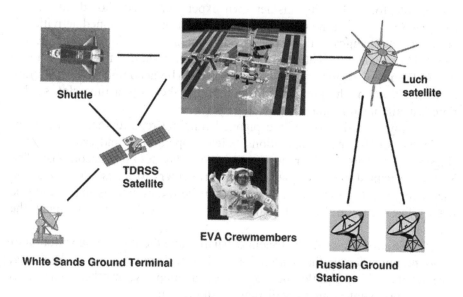

Figure 8-16. Communication pattern between the ISS, Shuttle, and both the U.S. and Russian ground segments.

When useful, a Principal Investigator may talk personally to the crewmember experimenters in orbit, for congratulating them for the good work, discussing an unexpected result, or for determining together the best methods for modifying a failed experiment.

By analogy with the amateur radio communication "etiquette", there is a code for communication to and from orbit, with the first word pronounced being the organization or the person to be called. For example, the CAPCOM in Houston will use "Alpha" (the nickname for the ISS, by analogy with the Star Trek series) or "MS1" (for Mission Specialist 1) as a call sign from the ground. Similarly, the crew in orbit will use "Houston" as the call sign from orbit.[*] Other codes are used for acknowledging good reception ("Roger") or for answering positively ("Affirmative") or negatively ("Negative") to a question.

2.7.3 Re-Planning

Prior to flight, an operations timeline will have been developed and practiced during the flight simulations. During the flight, each investigator is

---

[*] A popular poster in Houston claims "Houston, first word from the Moon". Indeed, just after the LEM had landed on the Moon in July 1969, the first communication between Astronaut Neil Armstrong and Earth began with those words: "Houston...Tranquility Base here... The Eagle has landed."

responsible for monitoring his/her own experiment, and should inform the Mission Scientist and the CIC about deviations from the planned activities. It is likely that changes will be made to the planned operation timeline during flight. Major changes that must be made within a short time (such as those due to failure of major equipment) will usually be accomplished by shifting an entire block of the preplanned timeline into the current time slot, so that time is available to re-plan the affected operations.

Any modification of the planned timeline requires consideration of the impact on spacecraft operations, science operations, and crew activities (Figure 8-17). This usually requires at least one day or two to accomplish. The Mission Scientist will meet with the experimenters to assess the impact on each experiment and reach a consensus on the modifications that should be made. The Mission Control management team will then implement the changes.

Major changes to the timeline usually take the form of substitution of one experiment for another within the timeline resource allocations. However, changes that affect Shuttle operations, crew availability, or other interactions among experimenters are very difficult to implement.

Experimenters can request changes in the operation of their own equipment in response to results from the flight. Such changes should not impact crew time or other experiments, though.

Figure 8-17. A Space Shuttle pilot consults the flight documentation in preparation for reentry and landing. (Credit NASA.

2.8 Data Analysis

Following completion of on-orbit and postflight operations, the investigators process the data for later publication in peer-reviewed scientific journals. The data remain the property of the Principal Investigator for one year following the end of a mission. After this period, the data belong to the public domain. After one year following the mission, the Principal Investigator must deliver a Final Report and file-in the information required for the mission databases.

Databases exist at space agencies, which provide access to information regarding space life sciences experiments carried out since the 1960s. Generally, an experiment description in the database (submitted by the investigators themselves) provides information about the Principal Investigator, the flight or mission on which the experiment was performed, the experiment equipment, and an extended abstract (objectives, procedures, and results of the experiment).

For example, the Space Life Sciences Data Archive (LSDA) at the NASA Johnson Space Center is an online database containing descriptions and results of completed NASA-sponsored flight experiments. This database includes descriptions of the experimental objectives and protocols, hardware, biospecimens or data collected, personnel, and documents. A limited number of experiments contain final reports and spreadsheet data suitable for downloading at the URL address http://lsda.jsc.nasa.gov. Data from human subjects are unavailable online for reasons of privacy.

The ESA Microgravity Database (MGDB) is a unique repository of the ESA-sponsored microgravity experiments results and reference documentation. As of may 2003, the URL address: of the MGDB is: http://www.esrin.esa.it/export/esaHS/ESA6CMOC_research_0.html.

This database is soon to be updated and enriched where possible with multimedia material. It will be called the Erasmus Experiment Archive (EEA).

Further information on the Russian and Japanese central repository for space life sciences mission data can be obtained at http://www.rssi.ru/ and http://www.nasda.go.jp/guide/researcher_e.html, respectively.

The ESA, NASA, and NASDA archives are all part of the International Distributed Experiment Archives (IDEA), and can be searched as if they were a local archive. A well-established Interoperability Document rules the exchanges of records.

It's now your turn to propose space experiments… Good luck!

3 REFERENCES AND DOCUMENTATION

Carey W (2001) *The International Space Station European Users Guide.* Noordwijk, NL: European Space Agency, UIC-ESA-UM-001

ESA (1979) *Spacelab Users Manual.* Paris, France: European Space Agency, DP/ST(79)3

ESA (2003) *The European Utilization Plan for the International Space Station.* Paris, France: European Space Agency Scientific and Technical Publications Branch, ESA SP-1270

Gasset G (2001) *Foton M1 IBIS. Microgravity Research to Further Biological Knowledge.* Toulouse, France: Centre National d'Etudes Spatiales

International Space Life Sciences Working Group (2001) *Space Life Sciences and Space Sciences Experiments Information Package.* A Companion Document to Agency Solicitations in Space Life Sciences and Space Sciences

Joels KM, Kennedy GP (1982) *The Space Shuttle Operator's Manual.* New York, NY: Ballantine Books

Jones TD, Benson M (2002) *The Complete Idiot's Guide to NASA.* Indianapolis, IN: Alpha Books

Longdon N (1983) *Spacelab Data Book.* Paris, France: European Space Agency Scientific and Technical Publications Branch, ESA BR-14

NASA (1986) *STS Investigators' Guide.* Huntsville, AL: NASA Marshall Spaceflight Center, PMS-021

NASA (1994) Spacelab. *Information Summaries.* Huntsville, AL: NASA Marshall Spaceflight Center, PMS-021

NASA (2000) *Experiment Document. Format and Instructions for Human Flight Research Experiments.* Houston, TX: NASA Johnson Space Center, Biomedical Systems Test and Project Management/SM3

NASA (2001) *Research Announcement. Research Opportunities for Flight Experiments in Space Life Sciences and Space Sciences.* Washington, DC: NASA Office of Biological and Physical Research, NRA 01-OBPR-03

Space Life Sciences and Space Sciences Flight Experiments Information Package. Available at the following URL address (as of May 2003): http://peer1.idi.usra.edu/peer_review/nra/01_OBPR_03.html

Space Life Sciences Ground Facilities Information Package. URL address: http://peer1.idi.usra.edu/peer_review/nra/ILSRA_2001

Wilson A (ed) (2003) *European Utilisation Plan for the International Space Station.* Nordwijk, NL: ESA Publications Division, ESA SP-1270

INDEX

A

Absorbed Dose 269, 272, 274
Acceleration
    Angular 41, 92, 95, 121-122, 132, 134
    Gravitational 4-6, 109, 116, 150-151
    Linear 21, 76, 92, 95-96, 99, 111, 120-122, 132, 307, 318
    Threshold 120, 150-151, 158, 169, 217
Adaptation 25, 37-38, 72, 91, 94, 97, 99, 101, 105, 109, 112-113, 117, 121, 130-131, 133-135, 156, 161, 169, 191, 226, 248, 307, 326, 330
Aerospace Medicine 13-14, 246, 252
Aging 4, 10, 43, 66, 161, 183, 200-201, 262-263
Alternate Payload Specialist 316, 341
Ames Research Center (NASA) 42, 134, 320
Amino Acids 48, 51-52, 192, 198, 201, 304
Amphibians 1, 71, 86
    Avians 85, 87
    Development 64-65
    Frog 64-66, 71, 74, 100, 123
    Newt 71, 73
    Tadpole 64, 66, 71-72, 86
    Xenopus laevis 64, 86
Amyloplast (*see Statolith*)
Analogs 208-209, 221, 225, 227, 233, 237, 240, 259, 279
Animal 1, 3, 55, 58, 160, 174, 271, 278, 300, 303

    Care and Use 84-86, 107, 306, 321, 323
    Development 43, 47-49, 61, 64, 66, 68, 70, 73-74, 123, 187, 190, 192, 199
    Flights with Animals 13-17, 42, 100, 318
AO (Announcement of Opportunity) 327, 330
Anemia 61, 63
Antarctica 52, 206, 209, 214, 227, 225, 259, 262, 287, 292
Antibiotics 57, 248, 291
Antibody 63, 263
Antigravity
    Bones 191
    Muscles 187, 192, 201
    Suit 148, 167
Anxiety 27, 205-206, 208, 223, 260
Aorta 147
Apical Dominance 78
Apoptosis 66, 271
Appendicitis 248
Applications (*see Spin–Offs*)
Arabidopsis (*see Plants*)
Arrhythmia (*see Dysrhythmias*)
Arteries 143-144
Artificial Biosphere 300, 304, 306
Artificial Gravity 19, 36, 38, 40, 42, 139, 307
    Radius (Short-Arm) 41
    Coriolis force 41
    Intermittent 41, 307
Artificial Light 7, 80, 304
Asymmetry 80, 123, 133
Astrobiology 52
Astronaut
    Candidate 30, 214, 216, 218-221, 232, 249, 252-253, 255, 275, 279, 316

Mission Specialist 30, 249-252, 255-256, 316, 338, 341
Office 238, 252
Participant 30, 250-252
Payload Specialist 30, 250-252, 316, 338, 341
Pilot 14, 17, 21, 30, 33, 93, 140-143, 148, 158, 228-231, 249, 251-252, 255, 258, 316, 322
Qualification Requirements 217, 247, 249-252, 287, 316
Spouses Group 238
Training 30-32, 38, 130, 165, 237, 249, 252, 255-259, 281, 287, 289, 294, 316, 322, 324, 327, 330, 335-336
Women 24, 209, 224, 230, 249, 274, 296
Atmosphere 5, 8, 19, 27, 50, 68, 80, 102,
Composition 28, 49, 264, 293, 315
Monitoring and Control 162, 273, 299, 302
Regeneration 3, 162, 300, 305, 307
Requirements 266, 295, 299, 301
ATP (Adenosin Triphosphate) 48-49, 177, 179
Autogenic Feedback 134

B

Bacteria 1, 28, 82, 293, 302, 307
Extremophile 49-52
Growth 28, 48, 57
Mutation 48, 60, 290, 294
Balance 10, 19, 27, 74, 91, 95, 111-113, 116, 164
Balloon Flight 13-15, 268
Baroreflex 155

Baseline Data Collection 326, 339
Bed Rest 20, 139, 163, 174, 184, 192-194, 198-201, 319
Behavior 7, 56, 68-70, 73, 87, 101, 112, 116, 118, 123, 206, 208, 213, 218, 221, 225, 228, 232-240; 248, 298, 318
Bends (the) (*see Decompression Sickness*) 259, 265
Benefits from Spaceflight 2, 4, 70, 307
Biomass 52, 301, 304
Biorhythms 6, 38
Biosatellites 314
Cosmos 16, 100, 190, 194
Photon 318
Biosphere 2, 28, 49, 51, 306
Biotechnology 2, 56, 61
Bladder 82, 169, 273, 291
Blood 19, 32
Components 20, 60-61, 143, 149, 263
Pressure 16, 139, 143-151, 153, 160, 167-168, 251, 280, 282, 286, 332
Forming Organs 82, 269-270, 273
Body
Height 26, 251-252, 296
Restraint System 165, 196, 282-284, 286, 294
Weight 177, 183-184, 192
Water 146, 154
Bone
Demineralization 3, 173, 175, 193, 246, 297, 307
Density 28, 68, 183, 188, 190, 194
Formation 182, 190, 194, 198-199
Fracture 163, 194
Growth 190, 198

Marrow 82, 175, 263, 270, 275, 278
Minerals 30, 34, 182
Modeling 180-182, 198
Resorption 162, 182, 194, 200
Boundary Person 228
Bowels 25
Bradycardia 160
Bubbles 172, 265, 331
Bungee Cords 19, 111, 165, 198

C

Calcium 34, 61, 76, 100, 162, 173, 181, 188-189, 193, 199-200, 298
Calcium Carbonate 68, 95
Calories 34, 199, 296-297
Cancer Risk 30, 36-37, 82, 267, 270-277, 379
CAPCOM 341
Capillaries 143-145, 180
Carbohydrate 186, 296-298
Carbon Dioxide (CO_2) 28, 49, 296, 304, 306
Cardiac
Defibrillator 280, 282, 286
Function 16, 267
Muscle 142
Output 146, 149-150, 155, 159, 282
Cartilage 2, 182
Catastrophe 194, 207, 249
Cataract 82, 270-271, 274
Catheter 153-154
Cell
Components 55, 59, 143, 148, 182, 270,
Culture 2, 56-60, 82, 84, 193, 272
Division (Mitosis) 66, 77, 86, 271

Membrane 47, 56, 58-59, 75, 177
Suspended 58
Transduction 75-76
CELSS (Controlled Ecological Life Support System) 28, 299-300, 303-304
Central Nervous System 81, 91, 96, 100, 116, 120, 125, 132, 187, 267, 271-274
Central Venous Pressure (CVP) 150, 153
Centrifuge 8, 10, 40, 69, 76, 151, 254, 319
Accommodation Module 22, 84-85
Principle 41
Short-Arm Centrifuge 41, 307, 318
Neurolab Centrifuge 41-42, 99, 121-122
Centripetal Acceleration 121, 318
Chemical Compound 7, 48, 50, 59, 82, 194, 280, 285, 299, 302
Chicken 70
Chicken Legs 26, 163, 184
Chloroplasts 78
Chromosomes 49
Aberrations 82, 272
Breakage 82, 271-274, 277
CIC (Crew Interface Communicator) 341
Circadian Rhythm 6-7, 38-39, 70, 262
Circumnutation 76
Clarity 27
Cleanliness 29, 290
Clinic (Onsite) 31, 33
Clinostat 10, 58-59, 319
CMO (Crew Medical Officer) 280, 288-289
Cognition 125, 298

COF (Columbus Orbital Facility) 22-23
Collagen 181-182
Commander 19, 211, 230-231, 249, 286, 316, 322, 338
Communication 8, 27, 32, 106, 191, 209-213, 227, 233, 236-239, 289, 316, 339-341
Compatibility 209, 228, 236, 335
Concanavalin 60
Conference 17, 27, 238
Confinement 6, 9, 185, 205, 208-211, 221, 229-230, 262-263
Conflicts 35, 93, 128, 131-133, 206, 209, 211-212, 214, 225, 227, 229, 231, 233
Congestion 26, 106, 15&, 153, 166, 260,
Contamination 33, 51, 162, 285, 289, 294-295, 302-303
Coordination of Movements 113, 115, 119-120, 123, 148, 173, 174, 233
Coriolis Force 41, 134, 307
Cosmic Rays (*see Radiation*)
Cosmonaut (*see Astronaut*)
Countermeasures 30-32, 36-40, 42, 125, 133-134, 163-169, 173, 176, 189, 195-200, 206, 213-214, 238- 239, 245-247, 274, 278-281, 324
Counter-Rotation (*see Eye*)
CPR (Cardio-Pulmonary Resuscitation) 284
CRV (Crew Return Vehicle) 23
Crew
    CHeCS (Crew Health Care System) 281, 286
    Interface Communicator (*see CIC*)
    Medical Officer (*see CMO*)
    Return Vehicle (*see CRV*)

Crewtime Requirements 43, 314-315, 322-324, 326, 330, 333, 342
Crisis Management 211
Critical Path Research Plan 194
Critical Period 66, 169
Crop Field 2-3, 77, 304, 306-307
Cross-Sectional Area 179, 186-187, 201
Crowding 224
Crystal Growth 4, 56, 62-63
CT (Computerized Tomography) 162, 189, 190, 193
Cube 129-130
Cuffs 160, 165, 196, 280, 332
Culture (Social) 130, 214, 221-223, 225-227, 230, 235, 237
Cumulative 11-12, 58, 193
Crystallography 56, 62-64
Cycle Ergometer 27, 39, 156, 165, 185, 196, 198, 265, 281, 304, 332
Cysteine 82
Cytoplasm 49, 54, 64-65, 76, 86
Cytoskeleton 58-59, 76

D

Database 33, 86-87, 112, 272, 279, 287, 335, 342, 347
Data Management 287, 330-331, 334
Deconditioning 31, 36, 38, 42, 139, 158, 165-166, 169, 246
Degeneration 100, 170
Denervation Atrophy 186
Dental Care 31, 246, 253, 261
Depression 7, 205-206, 208, 213-215, 239, 264
Depressurization 126, 265, 292
Design 2, 8, 34-35, 38, 41, 103, 111, 130, 166, 196, 234, 239,

245, 258, 264, 266, 279, 292, 301, 307, 329-331, 335

DSO (Detailed Supplementary Objective) 329

Determination 66, 82

Diagnostic 195, 247, 281-283, 287, 289

Diapause 69

Diaries 209, 239, 262

Diastole 144

Diet 198-200, 274, 286, 296-299, 304, 333

Differentiation 58-59, 66, 78, 319

Digestion 146

Disease 2, 7, 25, 30, 32, 38, 87, 112, 141, 183, 200-201, 214, 247-248, 252, 254, 260, 262, 279, 287, 291-292, 294

Disorientation 26-27, 30, 32, 91, 97, 123, 126, 135, 206

Disuse Atrophy 185-186

Documentation 334, 342-343

Dorso-Anterior Axis 64-66

Dose Equivalent 269, 272-273

Dosimeter 83, 85, 275-276, 284

DNA (Desoxyribo-Nucleic Acid) 48-49, 51, 82, 193, 270-272, 274-275, 277, 303

Drawing Test 129

Drop Tower 4, 319

Drugs 2, 72, 94, 134-136, 153, 163, 176, 198, 236, 275, 282

DEXA (Dual-Energy X-Ray Absorptiometry) 189-190

Dust 293

Dysrhythmias 32, 157, 259, 261, 267, 292

E

Earth
Atmosphere 5, 49, 80, 102, 264, 266, 293, 306-307

Evolution 1, 4
Gravity 4-6
Magnetic Field 4, 80, 267-268

ECG (Electrocardiogram) 157, 168, 253, 286-287, 332

Echocardiography 162, 282

EEG (Electroencephalogram) 130, 168, 253, 332

Eggs 1, 65-66, 68, 70-72, 82, 85-87

Egress 21, 33, 115, 140-141, 151, 169, 257

Electrolysis 301-302

Ellipse 117, 129

Embryo 1, 64, 66, 70-72, 82, 86-87, 273

Embryogenesis 66, 74

Emergency 20, 23, 27, 30, 33, 115, 126, 140, 146, 151, 235, 257, 264, 280, 283-284, 287-289, 291-292, 335

EMG (Electromyogram) 168, 332

EMI (Electromagnetic Interference) 283

Endocrine System 7

Endolymph 95

Endoscopy 253, 291

Energy 5, 8, 28-29, 35-36, 47-52, 69, 79-82, 84, 146, 177, 186-187, 189, 206, 267-269, 272, 275-278, 296-297, 303

Environment (Space) 1, 4, 6, 8, 21, 30, 37, 44, 47, 59, 64, 73, 76, 205, 208, 248, 301, 304

Environmental Health System 281, 284, 333

Equilibrium (*see Balance*)

Equipment 3, 10, 22, 28, 105, 167-168, 196, 226, 234, 250, 257-258, 279-282, 284, 286, 292-293, 302, 304, 313, 317, 322-326, 330-331, 334-343

ERD (Experiment Requirements Document) 334
Ergometer (*see Cycle Ergometer*)
Escher Staircase 126
Eukaryote 49
EVA (Extra-Vehicular Activity) 9, 27, 31-32, 94, 123, 157, 196, 256, 257-259, 264-267, 292, 296, 322, 324
Evacuation 27, 126, 274, 287, 291-292
Evolution 1-4, 48, 51-53, 65, 67, 142, 278, 306
Exercise 20, 27, 31, 38-39, 42, 139, 146-147, 165, 176-179, 192-193, 195, 198, 207, 221, 247, 253, 265, 286, 332-333
    Capacity (Capability) 139-141, 155-157, 196, 266, 317
    Regimen 30, 39, 166-167, 185, 188, 199, 322, 324
    Interim Resistive Exercise 165, 196-197, 281, 332
Exobiology (*see Astrobiology*)
Expanders (*see Bungee Cords*)
Experiment
    Design 329
    Integration 334-335
    Selection 327
Exposure Limits 272-273
Extremophiles 49
Eye
    Counter-Rotation 121
    Eyeball 80, 118, 261
    Eye Contact 223
    Eye-Hand (or Eye-Head) Coordination 119-120
    Movement 10-11, 70, 91, 96, 102, 118-122, 133, 332

F

Facial Expression 166, 236, 260

FRC (Facility Responsible Center) 338
Fainting 140, 158
Family 27, 30-31, 212-214, 222, 226, 236-238, 247, 326
Fatigue 14, 34, 131, 163, 174, 176, 178, 187, 238, 259, 260, 262, 271
Feasibility 287, 328, 332-333
Fertilization 2, 64, 66, 68, 71, 73, 86-87
Fire 126, 213, 261, 289, 292, 302
Fish (*see Zebrafish*)
Fitness 30, 34, 38-39, 42, 156, 179, 286
Flight
    Opportunities 10, 313, 328, 333
    Status 30, 33
Fluid
    Dynamics 57, 61
    Loading 38, 164, 169
    Shift 26, 44, 104, 124, 148, 150, 152-154, 163,, 166, 175, 184, 191, 195, 223, 280, 282
    Shift Theory 131-132
Fetus 101, 273, 277
Food 27, 48-49, 52, 146, 277
    Processing 187, 297-298, 300, 303
    Production 3, 28, 75, 267, 299-300, 304
    Requirements 34, 295-296
    Selection 35, 106, 206, 226, 230, 234, 237-238, 297-298
Free Radical 81, 270-271, 277
Freezer 85, 331
Frog 64-66, 71, 74, 100-101, 123
Functional
    Hypothesis Theory 44
    Objective 337
Fungi 48, 60, 3022

G

Gait 113-114

Gamete 61, 64, 66, 71

Gas Analyzer System 168, 282, 332

GCR (Galactic Cosmic Radiation) 268-270, 273-274

Gender 110, 112, 179, 195, 211, 221, 224-225, 230, 273

Gene 48, 51

  Mutation 60, 69, 82-83, 86, 272, 275, 277, 293

  Genetic Alteration 60-61, 82, 87, 274

  Genetic Factors, 68, 183, 193

  Genetical Engineering 56, 64, 304

GEO (Geosynchronous Earth Orbit) 268

Glove Box 85, 331

Glucose 58, 166, 177, 200

Glycogen 177, 186

Graviception 54, 75,

Gravitropism 7, 53, 75

Gravity

  Artificial (*see Artificial Gravity*)

  Gradient 307

  Gravito-Inertial Force 40, 99-100

  Perception 54, 76

  Threshold 68, 85

  Vector 40-41, 54, 59, 64, 94, 132

Greenhouse 300, 307

Ground Control 32, 57-58, 68-69, 71, 73, 77, 83, 101, 191, 206, 318

Group

  Dynamics 213, 230, 234, 240

  Fission 230-231

  Fusion 231

Goals 52, 227-228, 234, 305, 330

Growth

  Factor 57, 63, 181

  Hormone 63, 166, 189

  Plate 181-182

Gypsy Moth 69

H

Habitability 33-34, 237, 257, 306

Habitat 9, 37, 72, 74, 84-85, 87, 107, 130, 205-208, 224, 234, 288, 303, 305-307

Hair Cells 95-97, 101-102

Hardware (*see Equipment*)

Harvest 305

Hazards of Spaceflight 1, 14, 17, 176, 233, 259, 264, 267, 272, 279-280, 282, 336

Headache 153, 206-207, 259-261, 266, 291, 302

Health

  Maintenance 30-33, 163, 245, 247-248, 280-282, 289

  Stabilization Program 30, 38

Hearing 101-102, 105-106, 251

Heart

  Muscle 142

  Rate 14, 139-140, 146-147, 149, 154-160, 165-169, 260, 264, 266, 286, 287, 293

  Size 154-155

Heat 10-11, 34, 56, 61, 143, 216-217, 254, 266, 294, 301-303, 331

Height

  Body Height (*see Body Height*)

  Perceived Height 129

Hematocrit 61

Hemorrhage 248, 259-260

Hoffman Reflex (or H-Reflex) 110-111

Homeostasis 47, 139, 182

Hormone
  Abscisic Acid 55
  Adrenaline 146
  Anti-Diuretic Hormone 148-149
  Erytropoetin 61, 63
  Gibberellin 55
  Indoleacetic Acid 55
  Melatonin 7, 262-263
Housekeeping 224, 340
HRF (Human Research Facility) 168, 281-282
Human Factors 34, 109,
Humidity 294, 299, 301-304
Hydrazine 280, 285
Hydrogen (H₂) 14, 35, 63, 80-81, 268, 293
Hydrostatic Gradient (or Pressure) 56-57, 143-144, 149, 160, 163, 195
Hygiene 27, 34, 206, 224, 234, 247, 293-296, 302, 317, 322, 324
Hygiene water 296
Hypoxia 14, 254, 265
HZE (High Atomic Number and High Energy) Particle 268-269, 272, 273

I

Ideotropic Orientation Vector 124
Illness 9, 20, 30, 32, 208, 212, 245, 247-248, 262, 280, 283, 287
Illusion
  Inversion 124-125
  Proprioceptive 108,
  Self-Motion 114, 121-123, 135
  Visual 115, 126, 132
Immune System 28, 31-32, 60, 63, 82, 195, 263, 271, 290,

Infection 32, 63, 82, 163, 259, 260-263, 294
Injury 38, 67, 112, 176, 225, 247-248, 261-262, 265, 267, 270, 280, 283, 287, 289, 292
Inner Ear 1, 3, 10, 91-92, 95, 112
Insects
  Beetle 69
  Fruit Fly 69
Insomnia (*see Sleep*)
Institute of Bio-Medical Problems 252
Institutional Review Board 321
Internal Estimate of Gravity 99, 116
Interpersonal Relation 32, 206-209, 212-217, 227-229, 234, 236, 238
Interplanetary 36-37, 209, 224, 229, 267-268, 274-276, 278,
Interstitial Space 146, 154
Interview 9, 17, 33, 209, 215, 217, 223, 233, 236, 252-253
Invertebrate 68-69
Ionizing Radiation 6, 34, 79-81, 267, 269, 272-273, 278
IRED (Interim Resistive Exercise Device) (*see Exercise*)
ISEMSI (Isolation Study for European Manned Space Infrastructure) 210-211, 227
ISLSWG (International Space Life Sciences Working Group) 327
Isolation 6, 56, 184, 205-206, 208-211, 217, 227, 230, 239, 262-263
ISPR (International Standard Payload Racks) 23
ISS (International Space Station) xix, 10, 12, 22-24

ISU (International Space University) xviii, 222-223

J

Jellyfish 67-68
JEM (Japanese Experiment Module) 22
Johnson Space Center (NASA) 33, 252, 255, 343
Joints 26, 92, 95, 102, 107-108, 112, 177, 179, 181, 183, 265, 290, 332
Journey 15, 25, 37, 275

K

Kennedy Space Center (NASA) 21, 304
Kibo (*see JEM*)
Kidney 61, 83, 139, 143, 147-149, 154, 161-163, 175-176, 273, 285, 298

L

Laboratories 85, 304, 313, 321, 327
Lamp 253, 304
Landing 13, 17, 19-21, 30, 32-33, 35-37, 60, 94, 98, 100, 113, 115, 119, 140, 151, 158, 237, 258, 292, 318, 323-327
Launch (or Lift Off) 5, 7, 15-18, 25, 31, 35, 68, 140, 150-154, 212, 257-259, 318, 323, 325, 333, 336-338
Laundry 296, 302
LBNP (Lower Body Negative Pressure) 164, 167, 254
LCG (Liquid Cooling Garment) 266
Leadership 206, 210, 221, 225, 227, 230-231, 233

Leg Volume 153, 167, 184-185
LEO (Low Earth Orbit) 268
LET (Linear Energy Transfer) 272, 276-277
Lizard 74
Life
  Definition 47
  Expectancy 33
  Fossil Life 48, 52-53, 308
  Life on Mars 51
  Search for Life (*see Astrobiology*)
Life Support System
  Closed or Open Loop 299-300
  Physical-Chemical 28, 44, 301
  Ecological (*see CELSS*)
  Regenerative 299, 300, 302-303
Lifetime 11, 273-274, 288, 297, 302, 315
Light 2, 6-7, 22, 27, 65, 70, 75, 78, 80, 102, 104, 118, 207, 256, 262, 294, 303-304
Liquid Cooling Garment 266
Liver 83, 253, 273, 285
Localization 86, 106, 125
Locomotion 70, 73, 113-115, 175, 192, 317
Longitudinal Studies 33
Lungs 72, 82, 86, 139, 143-145, 154-156, 162, 168, 259, 282, 284-285
Lymphocytes (*see White Blood Cell*)

M

Magnesium 194, 199-200
Magnetic Field 4, 75, 80, 162, 267-268, 276, 278, 283
Mal de Débarquement 94
Mammals 2, 43, 66, 72-73, 86-87, 101

Mars 4, 8, 10, 35,
    Gravity 36
    Direct 35
Mass Perception 107, 109
Materials 28, 43, 62-64, 80-83,
    107, 236, 301, 303, 331
Matrix (Extracellular) 30, 55-56,
    59, 95, 181-182
Maturation (Cell) 59-60
MEDES (Institute of Space
    Medicine) 266
Medical
    Debriefing 260-261
    Monitoring 31, 287
    Requirements 247, 251-252
    Screening 30, 208, 214, 217,
      245, 249, 254
Medication 38, 134-135, 280,
    282, 284, 287, 291
Membrane 47, 51, 56-59, 75, 177,
    294
Menopause 182
Mental
    Representation 128-130
    Rotation 126-127
Mentality 331
Metabolism 49-50, 58, 66, 100,
    163, 174, 199, 279, 298
Metabolic Expenditures 266
Mice 15-16, 84, 190
Microbes 52, 294, 301, 303
Microgravity 1-4
Microlesion Concept 272
Micronutrients 304
Microorganisms 49-51, 60, 303
Microtubule 65-66
Middeck 125, 298, 316-317, 336
Mission
    Control 9, 150, 212-213, 228,
      231-232, 236, 256, 337, 342
    Manager 334
    Scientist 316, 338, 341-342
    Specialist (*see Astronaut*)

Mitochondria 49, 177
Mitosis 66, 82, 270, 272
Mixing 43, 56-57, 301
MMPI (Minnesota Multiphasic
    Personality Inventory) 215,
    218-220, 232
Mockups (*see Simulations*)
Model 51, 56, 64, 68, 70, 85, 87,
    109-110, 117, 139, 162-163,
    190-193, 264, 266, 268-269,
    271-272, 274, 336,
Monitoring 31, 33, 110, 147, 167-
    168, 214, 234, 236, 238, 247,
    253, 275, 284, 286-287, 299,
    326, 332, 339, 341
Monkey 15, 16, 18, 20, 72, 74,
    101, 278
Moon 4-5, 8, 17, 20, 42, 51, 94,
    104-105, 157, 188, 267, 269,
    295, 341
Mood 7, 135, 206, 208, 238, 260,
    298
Morale 207, 230, 236
Motivation 16, 23, 34, 92, 131,
    206, 207, 216-217, 230-231
Moving Platform 111-112
Muscle
    Atrophy 58, 173-175, 183-184,
      192-193, 247, 298, 307, 332
    Breakdown 184
    Contraction 116, 174, 177, 179
    Extensors 116, 167, 174, 185
    Flexors 116, 174, 185
    Fiber 58, 173-174, 177-179,
      186, 187, 190, 192, 201
    Myofibril 177, 190
Mutation (*see Gene*)
Myocardial Infarction 158, 169

N

Nausea 21, 26, 32, 41, 92, 131,
    134-135, 270-271

Navigation 116, 125-126, 128, 130, 255, 258
Nitrogen (N₂) 34, 51, 80, 173-174, 184, 192-193, 265-266, 282, 293, 298, 306
Nobel Prize 10
Noise 8, 17, 28, 31, 33, 105-106
Neural Map 86, 125
Neutral Buoyancy 68, 125, 148, 256
Norepinephrine 60, 160-161
Nucleotides 52
Nutrition 35, 39, 50, 176, 187, 193, 199, 201, 247, 279, 295-298, 304
Nutrients 2, 29, 57-58, 139, 142-143, 177, 199, 297, 303-304
Nystagmus 10, 118
    Caloric 10-11
    Optokinetic 122, 253
    Vestibular (*see Vestibulo-Ocular Reflex*)

O

Observatory Satellites 276
Ocular Saccades 120, 153
Oogenesis 61
Organization 9, 47, 59, 126, 143, 209, 213, 225, 227, 238-239, 327-329, 341
Organogenesis 64, 66
Orthostatic
    Hypotension 20, 30, 139-142, 160, 164, 275
    Intolerance 33, 38, 139-141, 146, 158-159, 162, 166-169, 247
Oscillopsia 114-115
Osteoblast 182, 194, 319
Osteoclast 182
Osteocyte 182

Osteoporosis 2, 68, 176, 182-183, 188, 193, 195, 200-201
Otoconia 3, 95, 100, 133
Otolith 73, 95-101, 109, 110, 116, 121-124, 132-135,
OTTR (Otolith Tilt-Translation Reinterpretation) 97, 99, 122, 133
Outer Space Treaty 295
Outgassing 279
OVAR (Off-Vertical Axis Rotation) 121
Oxygen (O₂) 15, 18, 28-29, 35, 48-50, 57, 61, 78, 80, 140, 143-145, 155, 162, 168, 177, 186, 266, 270, 295-296, 299-307

P

Parabolic Flight 4, 74, 254-255, 284, 289, 319
Parachute 18, 25, 151, 234-235
Parasympathetic Nervous System 146
Participants (Space) 30, 250-252
Particles 36, 53, 55-56, 61, 75, 79-84, 104, 261, 268-278, 280, 293
PAT (Preflight Adaptation Training) 134
Payload 23, 31, 187, 250, 256, 302, 315-318, 322-324, 326, 329, 335, 338, 340
    Specialist (*see Astronaut*)
    Flight Data File 335-337
Peer Evaluation 13, 233, 328, 342
Penguin Suit 167, 197
Perception
    Distance Perception 104-105, 115, 129,
    Vertical Subjective 121, 124

Performance 7, 16-17, 31, 39, 103-106, 111, 130-131, 135, 139, 173, 194, 206-209, 213, 216-217, 221, 225-234, 239-240, 298, 308, 333

Personal Hygiene (*see Hygiene*)

Personal Space (or Distance) 34, 221-225

Personality Tests 215-216, 218, 232, 253

Phantom Torso 83, 269, 281

Phosphorus 194, 199-200

Photosynthesis 49, 78, 303, 306

Phototropism 6-7, 55

Phylogeny 67

Physical Examination 251, 253

Pilot (*see Astronaut*)

Pleurodele (*see Newt*)

Plant
Arabidopsis 77, 319
Development 76
Flowering 77-78
Germination 76-77
Growth 6, 78
Wheat 75, 77, 304
Yeast 274
Zea Maize 53

Plasma 57, 60-61, 75, 145-149, 154, 156, 169, 263

Plasticity 72, 101, 113, 187

Platelets 148

Postnatal Development 66-67, 73, 101

Potassium 34, 157, 194, 199

Power (Muscle) 177-179, 187, 196

Pre-Breathing 265

Predictive Tests 132, 135

Pregnancy 73, 254, 273

Prevention 30, 32-33, 133, 247, 259, 275, 277

Principal Investigator 41, 316, 321, 334-336, 338, 341-343

Privacy 105, 206, 224, 234, 294, 343

Prokaryotes 48-49

Products 4, 28, 57-58, 61, 63, 261, 270, 279, 285, 298, 300

Proposal
Selection 329
Submission 327, 329

Proprioception 92, 95, 102, 107-108, 111, 126, 290, 297

Propulsion Gun 18

Protein 2, 48, 51-52, 56, 58, 61-65, 81, 100, 174, 184, 186-187, 192-193, 198-201, 274, 296-298

Protists 48

Prostatitis 259, 292

Protoplasm 55-56

Protozoa 49

Psychiatric Disorders 205, 208, 214-218, 239

Psychological
Services Group 238
Test 211, 217-218, 236, 253

Puffy Face (*see Congestion*)

Q

Quail 70-71, 73, 87

Quality Factor 269, 272, 274

Quarantine 30, 295

R

Radiation (Ionizing)
Gamma Rays 269, 272, 276
Particles (*see Particles*)
Protons 79-81, 268-272, 274
Neutrons 79, 81, 269, 272, 281
X-Rays 17, 64, 79, 188-189, 193, 195, 253, 268-269, 272

Radiation Sickness
Syndrome 270-271, 274
Radioprotectants 82

Shielding 28, 36-37, 82-83, 268-270, 274, 277-278, 283, 306

Source 80

Teratogenesis 277

Rat 15-16, 73-74, 84, 100-101, 107, 126, 175, 185-186, 190-192, 194

Re-Adaptation 97, 99, 112-113, 156, 169

Receptors

Baroreceptors (Pressure) 147-149, 152, 154, 160, 164

Neck 109, 133

Proprioceptive 92, 95, 108, 110-111

Tactile 92, 95, 109

Vestibular 100, 102, 133

Visual 66

Recovery (*see Re-Adaptation*)

Recycling 3, 28, 302, 304

Recreation 34, 210, 221, 236-238, 247

Red Blood Cell 20, 61, 143, 148, 263

Reentry 8, 16-17, 25, 27, 68, 94, 115, 140, 158-159, 164, 169, 292, 302, 317, 323, 326-328, 342

Reference Frame 110, 117, 124, 126, 128-130

Rehabilitation 27, 30, 33, 38, 113, 141, 163-164, 247

Requirements 35, 49, 106, 155-156, 199, 207, 230, 247, 249-252, 264, 266, 287, 292, 295-297, 300-302, 320, 322, 328-335

Rescue 30, 205, 235, 280, 283-284, 287-288

Resistive Exercise (*see Exercise*)

Respiration 16, 49, 168, 293, 303

Rest 34, 146-147, 158, 168, 237-238, 262-263, 282, 286, 292, 327

Posture 109, 185

Resuscitation (*see CPR*)

Retina 7, 80-82, 95, 101, 104, 118, 125, 128, 253

Review 321, 328, 334-335

Rewards 207

Right Stuff 117, 206, 215

Righting Reflex (or Response) 73, 101

Risk 30-31, 33, 37-38, 57, 82, 106, 133, 142, 169, 175-176, 183, 193-195, 214-215, 245, 248-249, 270, 272-275, 277-280, 292, 299, 302, 322, 332-333

Roadmap 194

Robot 8-10, 23, 249, 256-257, 267, 298, 290-293, 295

Rock 49, 52-53, 294, 306, 308

Roots 1, 7, 48, 53-54, 75-79, 84, 319

Rorschach Ink Blot Test 216

Rotating

Chair 120, 131, 322

Slow Rotating Room, 307, 319

S

Saccule 95-97, 102, 108

Safe Haven 289

Safety 38, 44, 194, 215, 226, 246, 249, 288, 301, 318, 320, 328, 331, 333, 335

Saline (Fluid) Loading 38, 139, 164, 169

Samples 3, 19, 60, 62, 201, 275, 282, 285-286, 318, 321, 323-326, 331-332, 336

Sarcopenia 201

Schedule 23, 31, 34, 156, 165, 168, 187, 198, 206, 212-213, 225, 234, 237, 262, 322, 326, 333, 338-339

Sea Urchin 68, 71

Seats
Reclining 148
Recumbent 169-170

Sedimentation of Particles 54-57, 61-62, 75

Seeds 3, 44, 54, 76-79, 82, 306

Selection Criteria 30, 214, 216-218, 221, 231-233

Semicircular Canals 41, 95-96, 101, 116, 118, 132-133

Senses 102, 153, 107

Sensory
Conflict 131-133
Organs (*see Receptors*)

Separation Techniques 56, 301

Sheep 13, 199

Shielding 28, 36-37, 82-83, 268-270, 274, 277-278, 283, 306

Show Stopper 278

Shower 166, 206-207, 294, 302, 326

Side Effects 135, 275

Silence 106

Simulations 21, 208, 234, 266, 337, 341

Sled 120, 151

Sleep 19, 25-26, 34, 80, 135, 205-209, 212-213, 237-238, 247, 253, 260-263, 317, 322, 324, 326

SMAC (Substance Maximum Allowable Concentration) 285, 302

Small Self-Containers 317-318

Smell 27, 35, 102, 106-107, 153, 236

Smooth Pursuit 120

SMS (*see Space Motion Sickness*)

Snakes 74, 148

Solar
Flare 36-37, 80, 84, 268, 270, 274, 276-278
Particle Events (*see SPE*)

Sopite Syndrome 131

Sounding Rocket 15, 76, 79

South Atlantic Anomaly 27, 80

Spacecrafts
Agena 19
Apollo 19-20, 25, 51, 80, 93, 123, 135, 140, 157, 188, 217-218, 222, 229, 295, 302
Gemini 15, 18-20, 25, 93-94, 102-103, 140, 188, 217-218, 222, 233
Mercury 15-20, 25, 93-94, 140, 216-218, 222, 229
Progress 23-24, 104, 236, 238, 292, 325-326
Soyuz 15, 18, 22-24, 32, 115, 123, 140, 158, 188, 222, 235-236, 240, 251, 258, 288, 292, 302, 314, 318, 324-327, 331,
Space Shuttle (*see STS*)
SpaceHab 314-316, 319, 329, 336, 338-339
Spacelab 10, 43, 76, 100-101, 107, 109, 120, 153-155, 186, 191, 313, 315-316, 319, 323, 339, 341
Surveyor 51, 104, 295
Vostok 11, 15, 188, 235

Space Life Sciences Data Archive 343

Space Motion Sickness 21, 38, 92, 107, 125, 127, 130, 131, 135, 153, 259, 261

Space Station
International Space Station (*see ISS*)

Expedition 24, 324, 236, 238, 292

Increment 12, 22, 25, 230, 324-325, 329

Mir 12, 22, 39, 44, 70, 77-78, 100, 105-106, 141, 157-158, 166, 174-175, 184, 189, 196, 199, 205, 209, 211, 213, 222, 226, 230, 232, 235-237, 248, 261, 269, 281, 292, 302, 313, 324

Salyut 21, 77, 116, 141, 188, 212, 222, 239, 291, 302

Skylab 1, 21, 57, 69, 80-81, 93, 103, 109-110, 123, 131, 135, 156-157, 185, 188, 193, 196, 198-199, 207 212, 222, 246, 246, 269, 294, 313

Space Suit 9, 27-28, 151, 157, 252, 255, 264-267, 286, 307

Space Tourism 24

Spatial Orientation 91, 94, 107, 123-125, 128, 130, 133

SPE (Solar Particle Event) 268-269, 274

Specimens (*see Samples*)

Spermatogenesis 61

Spider 69-70

Spin-Offs (Terrestrial Applications) 2-4, 47, 70, 228

Spine 163, 175, 189, 198, 200, 253

Sponge Bath 27, 206, 294

Spores 51, 82

Skeleton 176-177, 181, 183, 190, 193, 198-199

Skill 209, 213, 217, 230, 233, 287, 316

Standing 25, 33, 37, 68, 111, 121, 124, 139-140, 143-145, 151, 158-159, 164-165, 167, 174, 176-177, 198-199, 294

Statolith 54, 75-76

Sterility 271, 305

Stones (Kidney or Renal) 32, 162-163, 175, 194, 248, 254, 286, 291, 297

Storm Shelter (Safe Haven) 36-37, 277-278

Strain 32, 55, 163, 197-198, 259

Strand Break 82

Strength (Muscle) 32, 69, 164, 173-174, 176-177, 179, 184-185, 187, 192-193, 196-197, 200-201, 297, 332

Stress 9, 16, 25, 30, 32, 59-60, 66, 146, 157, 174-175, 180, 182-183, 193-194, 198, 205-209, 216-217, 224, 226, 230-239, 245, 247, 253, 260, 262, 267, 279, 318

Stroke Volume 146-147, 150, 159

STS (Space Transportation System) 21

Submarines 49, 105, 208-209, 248, 259, 301

Surface Tension 56-57, 72

Surgery 2, 101, 248, 289-291

Surveyor-3 Probe (*see Spacecrafts*)

Survival 16, 51, 229, 231, 234-235, 254-255, 258, 282-283, 301

Suspended Rats 191-192

Symmetry 66, 80, 123, 127-128, 133

Sympathetic Nervous System 60, 146-147, 149, 159

Synapse 67, 101-102, 110

Syncope 140, 151, 158-160

Systemic Pressure System 16, 143, 153

Systole 144

T

Tadpole 64, 66, 71-72, 86
Taste 102, 106-107, 153, 206
Taxi Flights 24, 288, 324-326
TDRS (Tracking and Data Relay
　Satellite) 339
Team Player 213, 216, 228
Telemedicine 286-287
Telemetry 15-16, 265, 318, 340
Telesurgery 290
Tendons 102, 108, 177, 194, 282
Teratogenesis 277
Terraforming 305-307
Terrestrial Applications (*see*
　Spin-Offs)
Thermal Convection 55, 266
Third Quarter Phenomenon 208,
　210
Thirst 9, 28, 150, 153-154
Time Perception 225
Timeline 256, 259, 316, 337-338,
　341-342
Toilets 294
Touch 91, 102, 107, 113, 116, 132
Tourism (Space) 24
Toxicology 33, 284, 287
Training
　Medical 31, 258, 289
　Parachute 234
　Psychological 32, 38, 233-234,
　　239
　Survival 229, 234-235, 254-
　　255, 258
　Science 335-336
　Water 255
Transport (for Cell) 47, 56, 61,
　143
Treadmill 27, 39, 42, 114-115,
　140, 185, 196, 198, 281, 331-
　332

Treatment 20, 94, 113, 135, 199,
　201, 247-248, 263-264, 272,
　279-281, 288, 292, 303
Turnover 56, 192, 201, 270
Twitch 178, 186-187, 201

U

Ulcer 248, 291
Undergarments 150
Urine 19, 147, 153, 162, 184, 188,
　194, 201, 254, 260, 286, 294,
　296, 302
"Us vs. Them" Syndrome 212-
　213
User Home Base 338
Utricle 95-97, 102, 108

V

Vacuum 1, 4, 28, 51, 253, 266,
　281
Van Allen Belts 80, 268
Vascular Resistance 160-161
Veins 26, 143-147, 153-154
Velocity 6, 8, 40-41, 116-118,
　121-122, 179, 307, 315, 322
Venous
　Compliance 154
　Pressure 150, 153
　Return 145, 147, 161, 167
Venules 143
Vertebrates 44, 64, 70, 72, 74, 86,
　125, 321
Vertebrate Disk 26
Vessels 32, 143, 146-149, 152-
　155, 160-161, 167, 253, 282
Vestibular Apparatus 66, 118
Vestibulo-Ocular Reflex 118-120
Vestibulo-Spinal Reflex 110
Vibration 8, 17, 31, 33, 85, 105,
　108, 166, 198-199, 217, 237
Virtual Reality 21, 105, 125, 130,
　240, 257

Virus 60, 263
Vision 2, 19, 91, 101-102, 104, 107, 115-118, 123, 151, 251, 253, 265
Visual Orientation 123
Vitamin 7, 189263, 274, 277, 297-298
Voice Analysis 236, 341
Vomiting 21, 41, 92, 132-134, 271, 275

W

Walking 27, 33, 94, 98, 113-116, 121-122, 164, 174, 177, 179-180, 196, 198
Washing 285, 294, 296
Waste Collection System (*see Toilets*)
Water 28-29, 33-34, 36, 38, 43-44, 51-53, 70, 73, 81, 102, 146, 148-149, 154, 169, 184, 206, 208, 255-256, 266, 270, 277, 281, 284-286, 293-296, 299-303, 307, 326
Watertropism 6
Weight (Body) 177, 183-184, 192
Weightlessness (*see Microgravity*)
Wheat (*see Plants*)
White Blood Cell 60, 143, 148, 271
Women 18, 24, 160-160, 182-183, 209, 211, 220, 224-226, 230, 249, 273-275, 296
Work 4, 9, 24-25, 29, 34, 131, 140, 155, 173, 175, 192, 196, 206, 208-209, 212-213, 221, 225, 227-228, 230-231, 234, 237-239, 256, 258, 262, 267, 292,
Work-Efficiency Ratio 233

Workload 131, 157, 196, 210, 212,, 266-267
Worms (*C. elegans*) 86
Writing Test 117, 129

X

X-Ray (*see Radiation*)
Xenopus laevis (*see Amphibians*)

Y

Yeast (*see Plants*)
Yuri Gagarin 11, 17, 252, 258, 235

Z

Zea Maize (*see Plants*)
Zebrafish 84, 87, 96
Zeolites 63
Zubrin Robert 35, 229, 274, 307